# Soft Matter

Roberto Piazza

# Soft Matter

## The stuff that dreams are made of

COPERNICUS BOOKS

An Imprint of Springer Science+Business Media

Roberto Piazza
Dipartimento di Chimica, Materiali ed Ingegneria Chimica
Politecnico di Milano
Sede Ponzio (edificio CESNEF), via Ponzio 34/3
20133 Milano
Italy
roberto.piazza@polimi.it

Translation from the Italian language edition:
*La materia dei sogni* by Roberto Piazza
Copyright © Springer-Verlag Italia 2010
Springer-Verlag Italia is part of Springer Science+Business Media
All Rights Reserved

Published in The Netherlands by Copernicus Books,
An imprint of Springer Science+Business Media, LLC

ISBN 978-94-007-0584-5          e-ISBN 978-94-007-0585-2
DOI 10.1007/978-94-007-0585-2
Springer Dordrecht Heidelberg London New York

Library of Congress Control Number: 2011921124

*Publishing Editor*: Maria Bellantone
*Copy-editing*: Lindsay Nightingale
*Cover illustration*: "Bubbles" by sir John Everett Millais, 1886, oil on canvas, Lady Lever Art Gallery,
Port Sunlight Village, UK. With kind permission from Unilever Nederland Holdings BV, Rotterdam, the
Netherlands
*Cover design*: Paola Parise

Printed on acid-free paper

Springer is part of Springer Science+Business Media (www.springer.com)

To Silvia, Andrea, and Giulia
the stuff of my dreams

# Foreword

Some 25 years ago, Pierre-Gilles de Gennes and Philip Pincus coined the word Soft Matter to address the behavior of colloidal suspensions, surfactant solutions, and polymeric systems. These encompass materials ranging from plastics and cement to biological matter. De Gennes and Pincus showed that a number of simple concepts could describe these materials in a unified and quantitative manner. This has lead to great enthusiasm among not only physicists but also physical chemists, chemical engineers and biophysicists. The field is now mature with its own journals (such as *EPJ E*, *Soft Matter*, and *Soft Materials*, to mention a few) and textbooks (for instance *Soft Matter Physics*, by Maurice Kleman and Oleg Lavrentovich).

The book by Roberto Piazza adds to this a fresh and original treatment which should appeal from high school students (our future!) to professional scientists. First of all, this book shows in a most attractive way that soft matter accompanies our daily lives from early in the morning to late at night. The list of polymers and surfactant that matter are presented with a light touch but clearly, and with relevant details showing why these materials make the difference. He also demonstrates in a convincing way that an understanding of these materials can be reached starting from elementary considerations.

Roberto Piazza trained as a physicist *pur et dur* in the school of Vittorio Degiorgio, and is now professor in Politecnico di Milano. He gave highly significant contributions to the field, and his enthusiasm for the material springs from the pages. It is also clear that he is not only a gifted teacher and physical scientist, but equally well at home in history and literature. I particularly enjoyed his account of the triumph of Roman concrete based on the use of pozzolana, with lasting results such as the Pantheon. The book is beautifully illustrated and the last chapter, a dictionary of soft matter from A to Z, is a gem. Enjoy and profit.

Henk Lekkerkerker
Royal Academy Professor
Utrecht University

# Preface to the English edition

I have always thought that writing on popular science is what scientists do when they grow old. Now that I am actually in the closing stages of a popular science book, I have fathomed that even if you are not so advanced in years at the start, you will certainly look older by the end. In fact, I realized it well before finishing the original Italian edition of this book, "La materia dei sogni", published last year by Springer Italia. So, when my Editor asked me whether I was "enthusiastic" about preparing an English translation, I thought that, according to the original Greek meaning of this word, I would have to be "possessed by a god" – not necessarily a benevolent one – to accept.

In addition, I am also rather skeptical about "popular" science in general, in particular when I bump into those books pretending to address in "popular" language formidable mathematical conjectures, or esoteric topics such as black holes, superstrings, and dark matter. Quite often, skimming through their first chapters, the non-professional reader gets the impression that everything is as clear as day, to realize well before the end that it is in fact quite a foggy day. Worse than that, books of this kind may convey an idea of science as "magic", rather than as the fruit of the humble but strenuous labor of many people. And this is, to my mind, a major sin for a popular science writer.

Why then write a popular book on "soft matter", whatever this means (wait and see)? In short, because it is challenging and fun. Challenging, because explaining simple things in a simple way is much harder than presenting hard stuff to a scholarly audience – you are generally so accustomed to the matter that you hardly realize how much technical jargon has impregnated your day-to-day language. Fun, because writing this book has possibly made me grayer, but definitely amused me a lot. Anthony Zee, author of one of the most popular books in theoretical physics (popular because physicists love it, not because of its content – although Zee is a master of popular science too), claims that "beauty is fun". Giving Keats his due, I think this is deep truth too.

So, is this an easy book? If by "easy" you mean that it is self-contained, requiring little background knowledge to be appreciated, yes, it is an easy book. Definitely. Leafing through the pages, you will find hardly any hideous expressions like "It is well known" or "It can be easily shown" (if you do, just let me know), and very little math – although I hope you will gather that reading physics without math is like listening to music with no idea of what notes and chords are. But if you mistake "easy" for "requiring no effort", you are wide of the mark. Americans say, "There is no free lunch", which is often assumed to be what Fiorello La Guardia meant, on becoming mayor of New York, by saying "È finita la cuccagna". In fact, this Italian idiom rather translates as "The party is over" and, as far as this book is conerned, I think it is quite the other way around: the party is just starting, and I hope you'll have fun, but this requires a little commitment on your side. For there is no free fun in science (and life in general): real, long-lasting pleasure necessarily stems from personal effort. Einstein proclaimed, "Physics should be made as simple as possible, but not any simpler." Anthony Zee recasts this as "Physics should be made as fun as possible, but not any funnier." Let me dare to mix these two insightful remarks by stating that physics should be made simple enough to be amusing, but not so trivial as to spoil real fun. Nevertheless, I have tried to approach this matter very gently – a goal that, eventually, turned out to be less challenging than I expected, and not by virtue of my modest skill, but because the subject we are going to tackle is so deeply rooted in our daily experience that my only job has been to turn it into plain words.

This work is only the tip of an iceberg arising from the joint efforts and heated debates I had with many colleagues, and from the little I managed to remember from them. In fact, many years in this field led me to look upon those who share with me the fascinating enterprise of doing science much more as friends than colleagues. Truly, together with the wonder of discovering so many brilliant minds among my students, this is the most precious gift I have gained from my job. I am sure that, knowing my amnesiac tendencies, these friends will not expect me to remember them all. Let me, however, mention explicitly some fellow scientists who made a direct contribution to this book by providing original material and ideas, or simply pointing out stupid mistakes. So, many thanks to Piero Baglioni, Giuseppe Caglioti, Marina Carpineti, Roberto Cerbino, Paul Chaikin, Luca Cipelletti, Mario Corti, Daan Frenkel, Marzio Giglio, Albert Philipse, Peter Pusey, Francesco Sciortino, Gaetano Senatore, Joel Stavans, Bill van Megen, and David Weitz. I also thank Claudio Beccari for his precious help as a great expert in LaTeX, Pierluigi ("Piero") Durat for presenting me with the nice picture that closes this book, and finally my coworkers Stefano Buzzaccaro and Daniele Vigolo, begging their pardon for the amount of time I dedicated to this *divertissement*, leaving aside hard science (not to mention my duties as husband and father). No thanks would be adequate for Vittorio Degiorgio and Walter Goldburg, to whom I owe most of the little I know about science. Last but surely not least, a big hug to Henk Lekkerkerker, and not only because of the nice foreword he prepared for this book. Everyone who has ever listened to one of his talks agrees that Henk is crystal clear in his explanations. What is probably less known is how painstakingly he "distils" each word he says, even if he has perfectly mastered the subject: just try to invite him out for dinner the night before his talk, and you'll realize this in full. I regard this as a basic lesson in what it means to be professional.

Let me also acknowledge the precious help given by Springer in setting up a perfect framework allowing me to address, with this English translation, a far wider audience. Compared with the Italian edition, this book has been rather enlarged, by adding a somewhat unconventional glossary, a special index of all common stuff mentioned in the text, and many new color figures. If you like them, don't thank me, but rather the strenuous insistence of Maria Bellantone. Maria is originally from Sicily, which is farther from Milano than London. Yet our endless arguments about what should versus what *could* reasonably be done to improve this book convinced me that Italians, despite growing up so far apart, share at least one virtuous bad habit: stubbornness. Nevertheless, all this would have remained mere wishful thinking without the invaluable contribution by Lindsay Nightingale who, besides turning my rather uninspired English into a sparkling language, was crucial in clarifying many scientific aspects (Lindsay insists that she is just an "ex"-physicist, but I claim that physics is like malaria – once you are infected, you can hardly get rid of it). Finally, let me mention Mieke van der Fluit, who carried out all of the file-handling job, patiently bearing with this messy author. I am also indebted to her for the best appraisal of this book I have received so far. When asked if she had suffered too much in reading the book, she said that on the contrary, it was like eating a cookie. I hope you'll suffer as much.

Milano, October 2010

# Contents

1   Overture: a special day ................................... 1

2   A life in suspense ........................................ 7
    2.1   A big cast of little characters ........................... 10
    2.2   When it pays to be superficial ........................... 13
    2.3   Colloidal Waterage, an award-winning firm ................ 15
    2.4   Rock and roll in suspense ................................ 17
    2.5   Osmosis, the breath of a dispersed world ................. 24
    2.6   Colloidal Lego, matter made to measure ................... 29
    2.7   Softness without limit: fractal aggregates ............... 33
    2.8   Concrete: united by charge ............................... 37
    2.9   Particles spreading waves: colloidal light and colors ......... 40
    2.10  A very particular particulate ink ........................ 46
    2.11  Flying colloids: deceptive beauty of the aerosols ........... 49

3   Freedom in chains ........................................ 55
    3.1   Long and disordered queues .............................. 55
    3.2   A tale of cross-links and double-crosses .................. 59
    3.3   Necklaces for all tastes ................................. 63
    3.4   Plastics: false solids with a biddable disposition ............ 68
    3.5   Snake dance ............................................. 74
    3.6   Entropy: disorder or freedom? ........................... 80
    3.7   Elastic by chance ....................................... 82
    3.8   The secret of Mr Fantastic .............................. 84
    3.9   *Panta rei* ............................................. 86
    3.10  Nightmares for Indiana Jones ............................ 93
    3.11  Charged polymers: polyelectrolytes ...................... 95

**4    Double-faced Janus molecules** ............................. 99
   4.1    Striding on water: the physics of Jesus bugs ................ 100
   4.2    Surfactants, a split personality ........................... 104
   4.3    Soap bubbles: a paradise for kids and math nerds ........... 108
   4.4    Micelles: when surfactants find peace ...................... 114
   4.5    As white as can be: the science of cleaning ................ 116
   4.6    A large and varied family .................................. 117
   4.7    Questions of shape ......................................... 124
   4.8    A mischievous break: watch the label! ...................... 130
   4.9    Small but mighty emulsions ................................. 131
   4.10   Black gold ................................................. 136

**5    Nanoarchitecture** ........................................ 139
   5.1    Kepler, Bernal, and your greengrocer ...................... 140
   5.2    Colloidal crystals: ordered by entropy .................... 144
   5.3    Glasses and gels: when hate and love yield similar results ..... 151
   5.4    The world is not (just) a ball ............................ 155
   5.5    Sandcastles and shifting sands ............................ 161

**6    Dreamtime** ............................................... 165
   6.1    *Concludo, ergo sum* ...................................... 166
   6.2    Proteins, a matter of molecular origami ................... 173
   6.3    Little chemists ........................................... 182
   6.4    Truck drivers and intelligence ............................ 187
   6.5    Freemen of Flatland ....................................... 192
   6.6    Yard workers .............................................. 197
   6.7    Body builders ............................................. 199
   6.8    A (protein-rich) lunch break .............................. 204
   6.9    Artificial respiration .................................... 206
   6.10   The Chieftain and his Shaman .............................. 208
      6.10.1 The secret of simplicity ........................... 208
      6.10.2 Message in a bottle ................................ 210
      6.10.3 Double helices and strategies to pull them apart ...... 211
      6.10.4 The great contortionist ............................ 212
      6.10.5 The queen bee and her workers ...................... 215
      6.10.6 From Gladstone Gander to Donald Duck ............... 216
      6.10.7 The factory of dreams .............................. 218
      6.10.8 Inner secrets of the Chieftain ..................... 220
      6.10.9 Time dust or stardust? ............................. 222
   6.11   Back to the future ........................................ 224

**7    Weird words: soft matter from A to Z** .................... 227

**Index of common things (or almost so)** ...................... 277

# 1

# Overture: a special day

*We are such stuff*
*as dreams are made on*

W. Shakespeare, The Tempest

Lend me your support, reader! Of course, you already know that a writer needs you in order to qualify himself as a writer and, more mundanely, to earn his living. Yet when you start reading a book, you may not expect the author to ask you to sign a kind of Faustian pact. Don't worry: although slightly unusual, this is not binding. I would simply like to take you by the arm and walk together through any ordinary day, just asking you, in exchange, to mull over a lot of the stuff you come across in your everyday life. Yourself included, perhaps. Obviously, *you* have the whip hand. Whenever you like, you can put this book in cold storage, give it as a gift to a sworn enemy, or use it for less noble purposes. I just hope we can delay this as long as possible. But if you manage to follow me, I promise you will be rewarded with a wealth of useful knowledge about an exciting research field in physics, without getting a splitting headache from thoughts of quarks, superstrings, black holes, and "theories of everything", whatever these weird terms may mean. If this does not work out ... well, considering its title, this book may not be out of place on your bedside table.

Is that a deal? Then let us start. For the sake of argument, I shall assume you are a gentle lady (otherwise, just take a look at the closing remarks), one of the many who have not one, but *two* jobs. Besides your demanding profession, you must take care of a home, and above all of a husband and children, the latter too young, the former too inept, to deal with those simple tasks that our Stone-Age ancestors accomplished with little difficulty. So you wake up early, and first prepare breakfast for the family. Something that must surely go on the table is milk, so rich in the sugars, proteins, vitamins, and fats which are essential for the balanced growth of your children (but perhaps not so good for the spare tyre your partner is beginning to show).

R. Piazza, *Soft Matter*, DOI 10.1007/978-94-007-0585-2_1,
© Springer Science+Business Media B.V. 2011

Hang on. You may have only a vague idea of what proteins really are (though this may improve by the end of this book), but you do know that fats are akin to oil, and that oil and water (which makes up 80% of milk) hate to mix. How is it that the white marvel that is milk keeps together, while the mayonnaise you try to make (itself a strange mixture of fats with stuff that prefers to mix with water) goes runny so depressingly fast? Moreover, why is milk so white? Better still, are you really sure it is *always* white? Let us do a simple experiment. Take a glass of water and shine a flashlight through it. Than add milk (skimmed, for preference) drop by drop, and mix carefully. If you do this properly, you should see that this dilute solution becomes bluish. Where on earth did this "milk for Smurfs" come from?

Before having a full breakfast, you may like to wake yourself up with a cup of strong coffee, maybe prepared with one of these new (and expensive) machines that make an espresso as tasty and creamy as those served in Italy. But if you think this is a simpler kind of liquid, you are wide of the mark. Just take a look at the curious rim that is growing around the coffee stain you have already made on your brand-new nightdress, while it dries out (if you really want to make the experiment, you are allowed to use an old rag). On the table, besides milk and coffee, there are other foodstuffs with a very distinct physical appearance, such as butter and jam. You know that, whereas you call milk a liquid, this stuff is solid: in other words, it is "able to stand". Yet butter and jam are rather strange solids since, using a knife, you can spread them with no effort on your toast, where they slide almost like liquids. Try doing the same with a piece of wood, or with the knife itself!

When breakfast is over, and before getting ready for your day, you had better have a look at the clothes that your beloved "little ones" (already high-school students, in fact) intend to wear today. As you do so, you realize that these funny clothes (to use a euphemism) they insist on wearing are made of strange synthetic fibers, often with remarkable properties. For instance, they are waterproof but let the skin breathe, and they save you the trouble of ironing. You have heard that their fibers are made of something called "polymers", but what ever makes them so versatile? Next, take a look at their shoes, which may go by the names of sneakers, trainers, pumps, or kickers, but aren't so different from what we used to call just "tennis shoes". This makes you think of rubber, and what it would be like in a world *without* rubber (the world where your great-grandparents grew up). Maybe it is worth inquiring into the nature of this wonderful stuff.

At last, your moment has come. The bathroom is free, and you can get yourself ready. Before doing this, however, pause for a moment to examine what you are going to use. First of all, there's soap, always a symbol of the fight for health and life (how many newborn babies should thank the midwife who carefully washed her hands before helping?), or at least since we convinced ourselves that getting clean is not a sin. How is it that a small amount of this stuff, dissolved in water, manages to carry away substances that otherwise would cling to your skin like glue? To tell the truth, today our homes are full of stuff that plays a role similar to soap, such as bubble bath, shampoo, toothpaste, and shaving foam (the latter often spread, for baffling reasons, all around your sink and mirror). Why so many versions of detergents? Try cleaning your hair or teeth with washing powder, and you will realize that this is not *just* a question of marketing. Toothpaste is moreover one of those things that behaves as a solid only when needed, namely when it must stay on the toothbrush, whereas it flows with no problem when you brush your teeth. Besides this, your cabinet is crammed with bottles filled with stuff rather like milk, such as cleansing milk (precisely!), and day, night, or twilight beauty creams, where water and oils quite happily remain mixed in spite of what elementary chemistry predicts.

You are running late for work. But, before getting out of the house, you must remember to print and sign the papers your son needs to hand over to school today (and of course he only mentioned this last night). And you are the only one who can do it, for your husband, clean-shaven, sweet-swelling, and well-fed, has already left home (let's hope he did not forget to give you a quick kiss on the way). Anyway, you know how to do it, since the very same husband who apparently can't cook a couple of scrambled eggs has "explained" (to use a slight exaggeration) how to use your new home printer, while asking himself condescendingly why on earth a modern woman cannot deal with such trivial things herself. As usual, the ink cartridges have run out, despite being so expensive! But why are they so dear? Well, think about this for a moment. That wretched thing in your ink-jet printer spits out ink non-stop, generating accurate type at an impressive speed. As we shall see, ink is not a simple fluid either, but rather a suspension of particles. Preventing the nozzles that spray this liquid from clogging is not easy (quite the opposite, if you think of all the times the printer notifies you that it is "self-cleaning"). However, it's our ability since ancient times to suspend colored particles – the pigments – in water or in other liquids that enables us today to enjoy the sight of Monet's water lilies or of frescos in the Sistine Chapel, or to read Dante's Comedy and Whitman's poetry. And, in my own small way, it is what allows me to write this book.

Eventually, you manage to leave home, hastening to reach the office, the factory, the school or the shop where you work, finding your way through a foggy and wistful morning; fog, another white mystery, white as milk, made only of air and water that, in themselves, are not white at all. If it were a nice sunny day, your walk would be more pleasant. But is this really true? You may have heard that, in low-lying lands, the air can sometimes be worse for your health *just* when it is, apparently, so clear, because of those mysterious particles named "PM10". In fact, your son claims that they are worse than hard rock at full blast, citing as his source the similarly obscure acronym "AC/DC". You don't know what they are, exactly, but you sense they have something to do with the fact that you'd better not hang out the washing if you want it to remain white. Or perhaps they are related to those unpleasant dark streaks on the walls, just over the radiators or behind the fridge, which are already spoiling the rough coat of paint that your husband, suddenly turning into a home improvement expert, slapped on last spring (that paint too is only supposed to flow when the brush forces it, although the results do not seem to confirm this). In fact, as we shall see, these particles are not too different from those we found in your printer's ink, but in air, not water. You may be still more surprised to discover that even that beautiful opal you are wearing on your finger, fruit of a moment of unexpected generosity by your spouse, owes its gorgeous colors to humble glass particles, arranged with care and infinite patience by Nature in a very peculiar manner.

Since you still have a long way to go, to escape the melancholy of this foggy morning you start thinking about those lovely sunny days last summer, with the sky so blue (by the way, why blue?), the sea so enchanting, the sand so white and fine. Right, sand. Quite a strange substance too, don't you think? It flows like water, so much so that we can make hourglasses out of it, yet it is not a liquid, but rather a shower of grains *behaving* (almost) as a liquid. If you think carefully, there is a lot of similar stuff you encounter in your everyday life: sugar, table salt, flour, washing powder, to name but a few. What do these "granular fluids" have in common? Many interesting features, as you might guess. What you probably do not expect is that they can also teach us something about that traffic jam your husband's car is trapped in right now.

To leave you some freedom rather than clinging leechlike to you all day, and also because I have not the slightest idea about your real job, I shall pass over those eight or more hours that make up your working day, and just wait for you at the exit so that we may stroll back home together. Before going home, however, you have to pop to the supermarket, since you have to cook something for those starving vultures waiting at home, although, dead beat as you are, you would rather put *them* in the frying pan sometimes. While for you this is just hurried shopping, to me it is a paradise of opportunities to point out several things we shall talk about. Apart from foods that share many aspects with those we have already met at breakfast (such as butter, chocolate, ice-creams, jam, whipped cream, mayonnaise, jelly, creamy yoghurts), ranged on the shelves you will find a huge amount of inedible stuff such as detergents, softeners, adhesives, wax, silicon rubber, lubricating oil, starch, diapers, sanitary napkins, conditioners, gel deodorizers, even non-stick pans, which have a lot to do with what we shall meet later.

I do not linger over your supper and evening time, unless to mention that the newspaper you eventually managed to skim through could soon be replaced by a screen with wonderful properties. Unlike your computer screen, this doodad will allow you to read in full sunlight; indeed, it will perform *better* in these conditions, with a ludicrously low battery consumption. You may be surprised to discover that the stuff it is made of, so-called "electronic paper", has a lot in common both with fine dusts and with your opal ring. I then assume that, once in bed, your husband rapidly sinks into the arms of Morpheus, immediately starting that concert for woodwind and brass that has long made your nights shorter. Once you have taken a bedtime tea to calm your irritation, for you know that tomorrow morning he will deny the whole thing (as, by the way, I do too...), you may perhaps make use of this forced insomnia to ask yourself some of those questions for which your daily life leaves so little room. Who are *you*, actually? How does it work, this wonderful machine that, apart from the occasional ache, has allowed you so far to enjoy a happy life?

Try, then, to think back until you are once more a teenager smiling at life or, better still, that young child who marveled at each new discovery. Now for some more effort. Try, in your imagination, to return to that little sea we all come from, a sea we call mom, till you revert to that single cell from which you grew. Even that tiny droplet, whatever its existential state, was able to do many things. Above all, it (you!) clearly grasped the difference between what was *outside* and what *inside*, and could decide with great skill what to take in and what to leave out. Think about it. Somehow, you had already marked very clearly the boundary between yourself and the world. Whatever makes such a "declaration of identity" possible? We shall discover in the following pages that this enclosure that surrounds our "self" is basically made of something not too different from soap, and shall find out how the machines – the proteins – that allow the cell to work exploit in a very elegant way the same principles that drive soap molecules to cluster together in solution. In fact, we shall discover much more, because that tiny, insignificant cell is a true and proper city, where microscopic citizens perform with incredible efficiency the same jobs we towering humans do. We shall meet chemical and mechanical engineers, truck drivers and garbage collectors, surgeons and nurses, aerial fitters and sentries, architects and masons, bank clerks and reporters, policemen and gangsters, who volunteer for all these roles under the creative leadership of a great director and of a wonderful set designer. The little secret of this remarkable city we call a cell is that it manages to take unpromising individual players – the particles, polymers, and soaps – and orchestrate them to perform the most magical of concerts. In the most literal sense, it brings them to life.

As you see, there are many things we can talk about. If you will follow me, I shall also try to show you how other materials of great technological interest are made and work – crude oil, concrete, pesticides, foams, adhesives, inks, thickeners, many drugs, and so on and so forth. You may find it rather strange, if not unbelievable, that things as disparate as fog, tennis shoes, chocolate mousse, and proteins have something in common. Actually, many real experts (chemists, material engineers, biologists) would share your perplexity. However, physicists are well known to be a breed apart, often prey to a Peter Pan syndrome that drives them to look for new toys to play with, new interlocking pieces of that complex jigsaw puzzle we call reality. So these children who never grew up have come up with a new discipline they call "soft matter physics", and are stubbornly persuaded that the same basic ideas may help us understand how all these things work, from Titian's colors to the organization of life – a life that, according to Shakespeare, is made of the same "stuff as dreams are made on." And what could be softer than a dream?

And now, at long last, it's your turn to indulge in those dreams. Your husband has ended his nightly concert, and you may enjoy your well deserved rest. Thanks for your collaboration, and enjoy your reading.

*CLOSING REMARKS: If, instead of a gentle lady, you are a* man, *slightly annoyed by this opening, or simply convinced that addressing a generic public would be more proper, note that the fairer sex has surely the right to claim more attention in the opening words of a book. But if it makes you happier, remember that what you have just read should persuade you that speaking to your spouse, partner, mother, daughter, or sister has been quite useful. For what we are going to talk about, their day is far more interesting than yours or mine. Yet perhaps I am wrong? Maybe you do the laundry, shop at the supermarket, change your babies' diapers too. If this is the case, good for you. In this book, you will discover how many chances these tasks offer to satisfy your scientific curiosity. If not, don't you think it is time to start?*

# 2

# A life in suspense

*Tipsy, clinging, and very superficial: the colloids – Strolling though
restless molecules: Brownian motion – Osmosis, the breath of a scattered world – Molding a microworld: how to keep colloids under control – Softness without limits: colloidal aggregates and fractal geometry
(with a yoghurt for dessert) – A love/hate relationship: electric ambivalence, volcanic landscapes, and concrete – From lowlands fog to
Titian's palette: scattering light and colours – E-paper: when colloids
become writers – Gorgeous sunsets and blue moons: the aerosols' de-
ceptive beauty.*

The time has come, my lady reader, to leave our *tête-a-tête* temporarily and
address a wider audience, hoping there is one. So, what would I like to tell
you about? Many, many things. Starting with the most palatable stuff, my
tentative wishlist would include:

tea and coffee, ricotta and whipped cream, mayonnaise and yoghurt,
rice and sand, soap and toothpaste, inks and paints, milk and airborne
particles, contact lenses and jelly, ice and shaving creams, tyres and
proteins, spiders and artificial fabrics, polystyrene and bath foams,
concrete and chocolate, glass and opals, crude oil and thermal waters,
camera films and clays, drugs and bacteria, cells and soap bubbles. . .

. . . and continue for quite a while. To avoid making my editor nervous, I shall
only let myself hop here and there among these and other topics, trying at
least to make you scent their fragrance. But what does all this stuff have in
common? To be honest, it looks like a hodgepodge of things, some obviously
related, others less so. No problem for tea and coffee, soap and toothpaste,
drugs and bacteria, you may say, but milk and dust particles, tyres and proteins, cells and soap bubbles seem to be totally unrelated to each other.

No impression could be more wrong. The main aim of this book, perhaps
the only message I wish to deliver, is that *all* these materials share a common

R. Piazza, *Soft Matter*, DOI 10.1007/978-94-007-0585-2_2,
© Springer Science+Business Media B.V. 2011

basic feature. I hope this will be much clearer by the end, but I will just touch on this point with a simple comparison. Consider a beautiful mountain forest. To the forest ranger, the woodland is basically made up of pines, firs, spruces, larches, and brushwood. As a first step, he does not need to know much more to judge whether a tree is diseased, plan where to plant new trees, sketch new trails, or guard the wood from pyromaniacs. However, understanding *why* some plants are diseased and deciding whether they can be healed or must sadly be cut down, is another story. There, the ranger must resort to the botanist's help.

For a botanist, a wood is a quite different thing, for each pine, spruce, or larch is made of roots, trunk, branches, and seed-bearing cones. To get to the bottom of the mattter, the botanist has in a sense to descend several floors and, instead of considering structures with a scale of tens of meters, look carefully at much smaller objects. But certainly not the smallest. For a biologist any tree, and in the end the forest itself, is an ensemble of cells and stuff – like wood – that these cells produce. And within each cell there are smaller organelles that we must peer at, if we really want to understand why a plant becomes diseased or how it reproduces. We have gone many steps further down, and now the wood is made of little building blocks the size of a thousand of a millimeter or less.

However, they taught us at school (or they should have) that everything, including a cell, is made of molecules, which are in turn little families of atoms. For an atomic physicist, therefore, a wood is basically made of atoms. You might now think we have reached the ground floor, since these little building blocks are ten or a hundred thousand times smaller than a cell. But in fact there is a basement too, and a very deep one. A nuclear physicist is interested in what we can find *inside* the atom, things such as the nucleus that, although the largest object inside, is 100,000 times smaller than an atom. Our descent to the underworld seems to have no end. To tell the whole truth, physicists believe today that, to grasp what the world is really made of, we have to look at things and distances that are dwarfed by the atomic nucleus as much as the latter is dwarfed by a fir tree.

A wood may therefore have many increasingly refined *description levels*, each of them perfectly legitimate and self-contained, although related to the others. For instance, the forest ranger does not need to know that the forest is made of atoms, and he could not care less about electrons, protons, or quarks. If we look carefully at our descent to the underworld, however, we may notice that at one point we suddenly made a big jump, for two of these levels are much farther apart than the others. And the level that we missed, the one in between, is the richest one of all. Let me explain. To describe the basic structures of a conifer and their functions, a botanist does not need too many concepts. The biologist's task is a bit more challenging, since the different kinds of cells and biological materials that make up these structures are quite numerous, but that is enough for her (or at least it was until a few decades ago). At the other end of the scale the list is not too long either, for there are only about a hundred different types of atom, and only a few of them are plentiful in living beings. For the particle physicist, finally, the task is even easier, because the basic constituents of nature are really very few, far fewer than we believed some decades ago. The missing link between cells and atoms, however, is radically different, since the different kinds of molecules that make up a cell are numbered in *billions*.

Sure, you may say that we cheated by overlooking the *chemists*, who are able to find, study, modify, or even design from scratch an endless number of molecules. Absolutely true. Chemistry is wonderful, and I am deeply respectful of chemists and of their terrific ability. But *this is not enough*. Agreed, molecules are the real building blocks of the whole Universe, or at least of that part of the Universe we usually care about (atoms and the particles they are made of are only useful for understanding how molecules are made or react, but then we can almost forget about them). But if you were shown a heap of bricks, tubes, and cables you would probably find it hard to picture the house that could be made out of them. It is far easier if you are shown a roof, bearing and partition walls, pre-assembled plumbing and electric lines. It is not just the endless number of different molecules which makes biological materials so complicated, but rather that they, in turn, join into larger structures, each of them with a precise identity and function.

Simple stuff does not usually share this property. In a glass of water, in a pencil lead, even in the chips of your mobile phone, there is little that lies in between atoms or molecules and the object you see in front of you. In contrast, for the materials we shall talk about these intermediate blocks do exist, even if (despite being much larger than molecules) they are usually too small to be seen. Sometimes they exist only *within* the material, and disintegrate into simple molecules if we try to take them out. These materials differ from simple stuff in rather the way that a prefab, where prepackaged elements are simply assembled, differs from a traditional house built brick by brick. Compared with this simple example, however, building new materials starting from these blocks, rather than from simple atoms and molecules, opens up many possibilities still not fully explored. We shall gradually meet these little blocks, and learn to tell one kind from another. All of them belong to a kind of "Middle-earth" between the tiny, *micro*-scopic world of molecules and the large-scale, *macro*-scopic stuff we meet in our everyday life, and we shall therefore call them *meso*-scopic objects. My purpose is to suggest that you embark upon a journey through this Middle Earth.

Before starting, let us give a name to these peculiar "pre-assembled" materials. To tell the truth, there isn't a single name to label them, though in most cases, for reasons we shall see, they are dubbed *soft matter*. Now, this term is surely suitable for ricotta, whipped cream, toothpaste, and to a lesser extent for tires, or even spiders. Yet coffee, milk, or crude oil are really *too* soft, to the point that they are not even solids, but liquids. Fog is not even strictly a liquid (though we get damp enough walking through it), but rather something suspended in air: more than just soft, it is insubstantial matter.

In these cases we would speak more properly of *complex fluids*, to distinguish them from simple fluids such as air or pure water. But things such as opals or, even worse, concrete, are surely neither soft nor fluids! Probably, to describe these systems effectively, we ought to recall that they are built from units that are much larger than molecules, and call them *supermolecular materials*. But although that is a more precise and general term, it's also a rather awkward and pedantic one, which has not caught on in the scientific community. In what follows I shall use any of these expressions without worrying too much one way or the other, for the real point is to grasp what they mean.

Conversely, a term I shall try to avoid as much as possible is "nanomaterials", today as fashionable as "nanotechnology" not only in scientific literature, but also in newspapers, TV shows, and obviously science fiction. This is not because it is entirely incorrect, but because it is often an inaccurate and hackneyed word. "Nano" has a well defined physical meaning (for the curious, a nanometer means a billionth of a meter) and quite often the little blocks we are talking about are anything but "nano". Even in science, luckily, fashions change or fade with time, and, in a sense, I hope this one will. Sometimes, indeed, I fear that desperately looking for "nanotechnology applications" is a dangerous attitude, which may possibly... dwarf (this is actually the original meaning of "nano" in Latin, and in my native tongue too!) a promising young scientist.

## 2.1 A big cast of little characters

As in a good drama piece, or an Agatha Christie crime story, we had better introduce right from the start the "microscopic characters" lying at the heart of soft materials, complex fluids, and all the weird things we shall talk about. Obviously, as in any decent thriller, I do not expect you to grasp right now the look, the temper, the pet manias of these characters (let alone who is the murderer). You will have to be patient, please, and wait for the plot to sort itself out. Take it rather as a shopping list, useful to tell us at least the trade names of what we are going to buy. Chapter by chapter, we shall get to know each of these personages, according to the following order of appearance.

### Colloids and aerosols

Take a little fine dust, pour it into a glass of water, mix with care, and you get a colloid. That's all. A colloid, in its simplest form, is just a suspension of solid particles in a liquid. Actually, you don't even need a liquid. Candle smoke coiling and spreading in air is a colloid too, but one where the particles are suspended in a gas (since a colloid is also known as a "sol", here we more properly call it an aerosol). This looks like rather tedious stuff, but we shall see that many pretty interesting materials are, at heart, nothing but colloids. Moreover, understanding how a simple particle suspension works will provide us with the basis to understand much more complex systems. All things considered, most of what this book is about could actually be called colloids.

## Polymers

This is a word you may already know, for instance because you have heard that plastics are made of the stuff. Polymers are in a way the exception to the basic rule that our building blocks are made of many molecules. To be precise, they are long chains made of many "basic units", which in the simplest case are all identical, but can also be of different kinds. When dissolved in a liquid, these chains wind into little balls, similar to suspended particles in colloids, but much softer. When there are large numbers of chains, however, they no longer form separate coiled balls; instead, they grow intertwined into a kind of mesh, which is the fore-runner of what we know as rubber.

## Micelles, vesicles, and emulsions

You are likely much less acquainted with these terms, but you are of course familiar with soap. Soaps, and in general what a chemist would call surfactants (don't worry, it will be explained later), are made of very peculiar molecules, displaying a kind of double nature. Part of the molecule loves water, the rest cannot stand it. As a result, dissolved surfactant molecules huddle together to form large structures with the water-hating portions hidden in their midst, and these clusters are dubbed micelles. Surfactants are a typical example of a large class of chemicals called *amphiphiles* (amphi-, meaning both, here hints that they happily mix with both oil and water), having in common a readiness to organize themselves spontaneously into structures that exist *only* in solution (there is absolutely no way to take a soap micelle out of water). Not all amphiphiles form micelles. Some of them prefer to combine into more complex aggregates such as vesicles, which are, roughly speaking, water droplets surrounded by more water, but separated from it by a double layer of these particular molecules. If we add some oil, besides water and surfactant, the structures that form are stranger still. They are what we call emulsions, a stuff that, as we shall see, is abundant in any home.

## Colloidal crystals, gels, and glasses

The characters we turn to now are not simple building blocks, but rather *structures* that the building blocks may form, and which set up the framework of what can be properly called soft matter. Unlike colloidal suspensions or polymer solutions which are liquids (complex, but still liquids), they are *solids*, which keep their shape without spreading around like fluids. In the simplest case of colloidal crystals, they are just a grand version of well-known solids such as ice or diamond, reminding us in a way of the giant pencils or the elephantine baby-bottles once on sale at the fabulous "Think Big" shop in New York. In other cases, however, they are very special solids. While a "real" solid is made of atoms ordered in a simple geometrical arrangement (for instance, the atoms might all lie on the edges of tiny cubes), in gels or glasses the particles or the polymer chains are as randomly placed as molecules in a liquid. Yet, for some reason, everything holds together: our building blocks have produced a strong and stable house, although the architecture is a bit chaotic. The basic difference between glasses and gels is that the former are usually dense and hard, whereas the latter can be "almost empty", standing out as the lightest solids imaginable.

## Liquid crystals and granular matter

In between the Latin mess of liquids and the Prussian order of solids fit other materials, with an even more indecisive behavior: liquid crystals. We shall deal with them only briefly, since they are not usually made of large blocks but of small molecules. But at least we shall see that colloid science can suggest *why* they form. We'll also say a few words about granular materials, stuff like sand, rice, or cereals, but where, unlike with colloids, there is apparently no "suspending medium" like oil or water between the grains apart. Nonetheless, many "packing" problems that we shall address for colloidal suspensions have a counterpart in granular matter, and are actually crucial in understanding their behavior.

## Membranes, biopolymers, and biological machines

Here we are. Approaching what we call "life", we shall reach the climax of our path, a summit where colloids, polymers, micelles, vesicles, and more elaborate structures dance together in a kind of rave party of soft matter. Even though, so far, we have grasped only some of the basic rules of this complex role-play game, one thing is certain: what marks biological structures out from their simpler lifeless forerunners is their *individuality*. For instance, proteins are just polymers, but polymers made of many (about 20) different basic units, the amino acids. Whereas a simple polymer in solution is a randomly coiled chain (so that, as we shall see, all simple polymers with the same length have the same structure, at least "statistically" speaking), every specific amino acid sequence gives a *unique* shape to a protein, and this in turn gives the molecule a single, well-defined function. Moreover, protein chains coil up much more tightly than simple polymers, which make protein coils more akin to rigid colloidal particles. Even DNA, the king of the biological jungle, and RNA, its faithful servant, are polymers, but with such a complex structure that the whole code that uniquely defines each living organism can be written inside them. Finally, the membrane enclosing a cell makes it similar to a simple vesicle, but it is a very special membrane that, aided by proteins, can accurately decide the cell's shape and what it can trade with the outside. So vast a range of biological structures opens up countless ways for them to self-organize into proper "molecular machines", whose operation we are still far from understanding in full. Really and truly, this is the stuff of dreams!

As you see, this is going to be a rather long journey, and we had better take our time to avoid tiring too soon. Let us then begin with the first and simplest item on the list, telling of the deeds and misdemeanors of the eternally suspended life of a colloidal particle.

## 2.2 When it pays to be superficial

For a first encounter with colloids, aiming to learn what is so special about them, let's start with a simple experiment. Imagine I give you a little block, of whatever stuff you like, with a cubic shape and a side of one centimeter. Suppose too that I provide you with a "magic lancet" allowing you, whatever material you chose, to cut it into tiny pieces. Then, begin our experiment by dividing the block in a thousand pieces (you guess how), so as to obtain 1000 smaller equal cubes with a side of one millimeter (if you guessed correctly, this should be obvious). Now repeat the same operation on each of these cubes[1], thus getting $1000 \times 1000 = 1,000,000$ tiny cubes with a side of a tenth of a millimeter. We have not finished yet. Let us repeat everything once more, to obtain one billion ludicrously tiny cubes with a size of a hundredth of a millimeter.

What is the net result of all this effort? If we think of the quantity of matter or, to put it differently, of the block volume, nothing at all. We started with one cubic centimeter of our stuff in a single piece, and now we have a billion tiny cubes, each with a volume of only one billionth of a cubic centimeter. Obviously, we neither gain nor lose anything. But let us think about the *surface* of these tiny cubes. Since a cube has six equal faces, our initial block had a total surface area of six square centimeters, about the size of a stamp. Now the surface of each tiny cube is only $6 \times 0.01 \times 0.01 = 0.0006$ square millimeters, yet, since we have a billion of them, the total area (try it with a pocket calculator, if you do not believe me) is 0.6 square *meters*, which is about the total area of 40 postcards. Every time we divide the cube size by ten, the total area increases by the same factor. Had we repeated our cutting operation twice more, obtaining "nanocubes" with a size of ten millionths of a centimeter, the total area would have increased up to 60 square meters, more or less the living area of a two-roomed apartment!

---

[1] I'd better supply you with a microscope too, to check what you are doing, hoping of course that you have a very steady hand. Which I do, for I thoroughly trust my readers.

Surprised? Welcome to the surface world of colloidal particles or, to use a more fashionable term, to the world of nanoparticles[2]. By "particles" I do not mean electrons, muons, kaons, or any other of those unfamiliar things that (serious) physicists talk about. Rather, as in the original Latin meaning, I am talking about "small parts" of any material, whether they are solid, like dust grains, tiny glass beads, and metal scales, or liquid, like fog droplets, or even gaseous, like bubbles in a champagne flute. What matters is that they are pretty small and therefore have a lot of surface, even if, all together, they fill very little space. When a lot of nanoparticles are suspended in a fluid (usually a liquid such as water or oil, but sometimes a gas such as city air), we speak of a *colloid*[3].

However, "small" is a rather vague term. For instance, I am certainly small if contrasted to Sun Mingming, China's famous top basketball player, but not when confronted with a certain well-known Italian Premier. In physics, anything is big or small *with respect to something else*. To what extent can we call these particles large or small, then? We shall see later what is the maximum size a particle can have and still be "colloidal", for this demands some thought. For the lower limit, we shall only ask that they are much larger than atoms and molecules. It is easy to see that a colloidal particle can be *really* small and still have a large volume compared with a molecule. To make a simple calculation without using too many decimals, let us recall that a thousandth of millimeter is called a *micro*meter, or simply *micron* ($\mu$m), so that 1 $\mu$m = 0.001 mm. In turn, a thousandth of a micron is called a *nano*meter (nm). This is quite a small unit, but not yet small enough, for atoms and simple molecules are no larger than a few *tenths* of a nanometer[4]. For instance, since in water each $H_2O$ molecule takes up a volume of about 0.03 cubic nanometers, it is not difficult to show that a droplet with a radius of just 0.1 $\mu$m still contains almost 140 million $H_2O$ molecules!

A large number of chemists and physicists (including myself) have spent much time and energy in trying to make and characterize colloidal particles and suspensions made of the most diverse materials, and with the most bizarre properties. Why so much interest in these miniature materials? The first and most important reason stems from the huge surface area that is a peculiarity of colloids, together with the observation that it is through the surface that something interacts with its surroundings. This is not too hard to grasp. If you have ever had a camping night in Norway, even in full summer (or

---

[2] A more concise but less precise term that, as already stated, I shall sometimes use *only* because of my laziness.

[3] The word comes from the Greek κολλα, meaning "glue". We shall see later what a colloid has to do with glue (very little indeed), but it is useful to point out that this word stands for the *whole* dispersion, i.e. the particles plus the fluid they are suspended in.

[4] Because of its close relation to the atomic size, a tenth of a nanometer is given the special name of an Ångstrom, where that "dotted A" should be pronounced as a very closed "O" (although few scientists do so).

at least what the locals call summer), you may have realized that, to stay warm, it is worthwhile (and possibly pleasant) to huddle together as much as possible. More precisely, to limit the loss of heat, it helps to reduce the area you expose to the environment. To take another example, contrast the time it takes to dissolve raw sugar in water rather than refined (and, because "time is money", more expensive) sugar. Again, when you notice the walls of a poorly aired bathroom moldering gradually and depressingly in spite of all the protective treatments you have applied, recall that all the biochemical reactions leading to mold growth occur on the wall surface, just like the iron oxidation responsible for rust formation takes place on the surface of balcony railings. The larger the contact surface, the faster that materials and energy can be exchanged with the surroundings, and in this matter colloids are speed aces. In addition, our nano-dwarves are so tiny as to slip into almost any space, hence some of us hope to manipulate these exchanges and reactions so that they take place just where and when it suits us. For example, we may think of inserting anti-cancer drugs into suitable nanoparticles, able to sail down the thinnest capillaries to the sick organ or tissue where the drug will be controllably released.

That is not all. There is a subtler but probably more basic reason to look at colloids with special interest. Until a few decades ago, to build new materials, we had at our disposal only those bricks that Nature or the ingenuity of chemists provided – that is, just molecules. Today things are different: nanoparticles can be regarded as the building blocks of a mesoscopic Lego that, with careful design of the framework and joints, allow us to make materials with totally new and stunning features that do still, in a way, stem from the properties that each single particle has already. So far, this statement may seem a bit mysterious, but in what follows we shall see how to do it.

## 2.3 Colloidal Waterage, an award-winning firm

Before venturing into the world of colloids, let us first face a situation where the huge surface of dispersed particles plays a simple but crucial role. In this "case study", as a bonus, we shall get to know some colloids that Nature places at our disposal with no effort on our side.

The oil crisis (or the recurrent oil crises) has reopened the debate on the timeliness of using nuclear power. Sooner or later, no doubt, we will have used up all the extractable oil, although no one exactly knows *when*. Unfortunately, technologies based on renewable sources, in spite of developments that would have been unimaginable a few years ago, will hardly be adequate to grant our children the same comfortable life we enjoy today. Uranium is neither very abundant (thankfully, in some senses), nor particularly cheap (in the last few years, its price has risen much more than crude oil), but luckily we do not need a lot of it. From the experience obtained with existing nuclear plants we may also be able to draw inspiration for developing fusion power, which would satisfy all our hunger for energy (so far, unfortunately, it remains the Holy Grail of physicists).

The word "nuclear", however, unavoidably raises arcane fears in people, or at least substantial resistance. Having worked for a long time alongside nuclear engineers, I can swear that they are crazy about safety issues, and that nuclear plants, in particular those of the latest generation, are technology masterpieces, equipped with control systems that outshine most oil and carbon plants. So please don't think of enrolling me in the NIMBY ranks which seem to find so many members worldwide.

However, a serious problem does exist: how can we dispose of all that radioactive waste that remains harmful for very long times and that, note carefully, not only includes the exhausted nuclear fuel, but also its container and everything that was contained in it besides the fuel? Actually, those countries (not mine) where reasoning is more common than quibbling have already seriously investigated this matter, outlining what sites are suitable for radioactive waste storage. However, to work this out, we must take into account the role that our colloidal particles may play. Luckily, many engineers are just doing that all over the world.

The story I wish to tell you can be summarized in a few lines. There is a very special place in the United States, the Nevada Test Site, where several (anything but peaceful) nuclear tests were made during the Cold War, which generated a huge amount of waste containing "radionuclides", in other words radioactive atoms such as plutonium, cesium, and cobalt. The Nevada Test Site, like most of Nevada, is a desert zone, but concerns that these infernal dusts will spread around may cross our mind. After all, Las Vegas is less than 100 miles away! Yet, once the waste is stored, radionuclides have just one way to escape, and that is through underground waters. Luckily, these radioactive compounds are practically insoluble in water, so a simple calculation shows that they should stay close to the site for thousands, if not millions of years. Imagine therefore how astonished Annie Kersting and her geologist colleagues were to discover, in 1999, that plutonium had already spread around not by a few meters, but by more than *one kilometer*! What had happened? It's simple: the radionuclides had taken the bus, or perhaps we should say the "colloidobus".

Let me be more specific. A lot of colloidal particles, mostly minerals like clays with a flat particle shape and therefore a large surface, are suspended in underground water. Now, although they cannot be transported directly by water, radionuclides can ride the colloids by *sticking to their surface*, something that they readily do because, with all that surface at their disposal, hitchhiking is really easy for them. Thanks to their huge docking area, colloids can therefore assist in the transfer of stuff which would otherwise be totally insoluble in water. As we already mentioned and shall come back to later, this feature can also be exploited to carry around far safer and more useful substances such as drugs.

Was this the true story? Not all specialists agree. For one thing, under-ground waters flow through porous rocks, with very small pores. We shall see that these rocks, while letting water pass through, should in fact filter out most dispersed clay particles. Yet things could be very different for smaller particles, those made of lignosulfonates derived from lignin, the basic con-stituent of tree pulp. Whatever the truth of the matter, after the alarming observations made in Nevada, nuclear scientists and engineers consider this problem carefully whenever they have to plan a new depository for nuclear waste (which, I should mention, is also safely contained in massive watertight containers). This bears witness once more to their thoroughness and care for public safety. Trust them, for once; they are serious people!

## 2.4 Rock and roll in suspense

Colloids are therefore particles dispersed in a fluid of simple molecules, which we shall call, rather improperly, the "solvent". Let us then see whether there is any relation between these particles and the solvent molecules. Now, par-ticles are quite small on our usual gauge, but still very much larger than the molecules. We might guess that the particles simply ignore the latter, as we usually shrug off midges, at least if they are not too much of a nuisance. Is it really so? Quite the reverse! These "molecular fruit-flies" are bothersome to the point that they leave an indelible mark of their presence on the way the colloidal particles move. In short, solvent molecules have on particles the same effect that several pints of beer have on the drinkers at a bar. To see what I mean by this, let us jump back in time to 1827, and peep through the shutters of Robert Brown's lab. Brown, please note, is not a physicist or a chemist, but an adventurous Scottish botanist who, after traveling all around the world, is observing under the microscope some grains of pollen of *Clarkia pulchella*, a close relative of fuchsias and primroses, suspended in water. The trouble is that these wretched specks do not seem at all keen to keep still un-der observation, but rather seem to suffer from a kind of Saint Vitus' dance, stirring and jiggling madly about before poor Robert's eyes.

At the time, it might have seemed easy to account for this. Most natural-ists, in disagreement with physicists, believed that biological stuff possessed a kind of "spirit of life" that made it superior to inanimate objects, and *Brow-nian motion*, as we shall call the effect observed by Brown, could have been a direct manifestation of this[5]. After all, microscopists had already discov-ered that *animalcula* ("tiny animals") such as sperm cells, though invisible to the naked eye, could move around at will. But Brown, who was no physi-cist, but no fool either, carefully avoided jumping to this conclusion. And he was right, since it was easy for him to show that even specks of humble

---

[5] During the last century, most biologists have changed their mind. Not so certain philosophers, such as Henri Bergson.

and totally inanimate dust shared the same behavior. Nevertheless, Brownian motion remained a puzzle till the beginning of the twentieth century, when a personage you surely will have heard of, Albert Einstein, gave it a full and brilliant explanation[6].

Now, you might regard this result as a trifle, compared with other ideas by the aforementioned Albert, but you would be wide of the mark. The explanation of Brownian motion produced a proof of the molecular nature of matter that could hardly be argued with. Actually, at the beginning of the twentieth century, not all scientists believed in the existence of atoms and molecules. Chemists obviously did, and they already knew a remarkable amount about how to play with them[7]. However, many distinguished physicists held the opposing view that there was no real need of these little things to explain how the world worked. That is to say, they somehow managed to stick to the old idiom "what the eye doesn't see, the heart doesn't grieve over". Einstein's theory, however, which so neatly explains Brownian motion, not only necessarily assumes the existence of molecules, but also allowed scientists to *calculate* how many of them are in a given volume, and this result matched the value used by chemists to make sense of the quantities of chemicals used in the reactions between these hypothetical constituents of matter. Since then, the atomistic view of reality has stood at the foundation of modern science.

At the time, this was anything but obvious, so much so that Svante Arrhenius, Chair of the Nobel Committee, while introducing Einstein as winner of the 1921 price for physics[8], first took time to mention his contribution to the explanation of Brownian motion. So how did Einstein's explanation work? At bottom, it rested on two assumptions:

1. Bodies are made of atoms or molecules (this is what I have already stated);
2. Each molecule has a kinetic energy, called *thermal energy*, which is proportional to the body temperature (and this I shall try to explain now).

If you have any faint memories of school, you may recall that we associate a form of energy to any moving body, dubbed "kinetic" energy, which is equal to half the body mass times the square of its speed. It should be evident why

---

[6] To tell the truth, the same result was simultaneously obtained by a great Polish physicist, Marian Ritter von Smolan Smoluchowski; but he had rather too knotty a name to become popular, and certainly did not have a natural presence for TV shows (at least when compared with Einstein's hanging tongue, as reproduced on a thousand T-shirts).

[7] Just think that Alfred Nobel, to fund the peace prize that bears his name, had already invented dynamite some forty years previously. To make that, he clearly needed to be quite familiar with chemistry, at least if he wanted to save his own skin.

[8] Which, as some of you may know, was not awarded to him for the theory of relativity, but rather for his work on the "photoelectric effect", much less familiar to philosophers and scholars of the abstract, but fundamental for the future of day-to-day things such as solar energy or photography.

this quantity must be proportional to the mass: given that they travel at the same speed, would you prefer to be run over by a small motorcycle or by a trailer truck? To see why this energy is proportional not to speed, but to its *square*, we can observe that crashing into a wall at 50 mph is not just twice as devastating than colliding at 25 mph, but about *four times* (please don't try this at home).

The second assumption tells us, therefore, what it is that we measure when we check for fever with a thermometer: *temperature is a measure of molecular kinetic energy*, or of the speed at which molecules move about. Please note that I am not only referring to the molecules of a gas, where intuitively (or because they told us so) we expect molecules to wander around freely in all directions, unceasingly colliding with other like tiny, mad billiard balls. At a given temperature, the molecules of a liquid, or even of a solid, have the *same* kinetic energy as in a gas. The only difference is that in a liquid something else (we shall see what) keeps them together, forbidding them to escape from the container. In a solid, on the other hand, atoms can "vibrate" around a fixed position, even they cannot get too far from it, and vibration is a kind of motion too.

A little math now – but only a little. To express that kinetic energy is proportional to temperature, we shall write it as $E = k \times T$ or $kT$, where $T$ is temperature measured in degrees kelvin, a scale which is identical in step-size to the common one (common if you are not American, at least) in degrees centigrade, but where "zero degrees" corresponds to the spine-chilling value of $-273.15°C$ (about $-460°F$), and $k$ is a multiplying factor called the *Boltzmann constant*[9]. This is just a rough figure, for some molecules actually move faster and others slower, but *on the average* the kinetic energy is about $kT$, so that we should rather write average energy $E \simeq kT$ [10]. Compared with the figures we usually consider, the thermal energy of a single molecule is ridiculously tiny. To lift me (admittedly no featherweight) from the ground to a height equal to the size of a *single atom*, it would take all the thermal energy possessed at room temperature by something like *sixty thousand billion* molecules[11]. However, the mass of a molecule is even more ludicrously small,

---

[9] In honor of Boltzmann, founding father of statistical physics, this constant is usually indicated in serious scientific literature by $k_B$. For those of you more acquainted with the units of physics, its value is about $1.38 \times 10^{-23}$ joules per kelvin.

[10] The symbol $\simeq$ ("almost equal") means that here, as in other expressions, I am merrily neglecting small factors such as 0.7 or 1.5. I assure you this is not just a sign of the author's laziness, but rather the typical way physicists write formulas at a rough guess, so as to get to the crux of the matter without carrying along too many boring numbers (and to leave something for the engineers to do). Anyway, for the hairsplitter and the know-it-all, mean energy $E = 3/2kT$, at least for monoatomic molecules.

[11] This may seem a lot but, recalling what we said of the volume taken up by a water molecule, all these molecules could comfortably fit in a cube with a side

so that, with a kinetic energy of $kT$, its "thermal" velocity is some *hundreds* of meters per second. In a word, the molecular world is restless.

A grain of pollen, or more generally any colloidal particle, immersed in a solvent is then unceasingly bombarded by these shooting nano-bullets that transfer a little of their own energy to it, so that in a very short time the kinetic energy of the particle will also be equal to $kT$ (neither more, otherwise the particle would give it back to the molecules, nor less, because then it would go on absorbing energy). This means that the particle, too, *must* be moving. But how? Each collision is a kind of "kick" (a feeble one, but still a kick) to the particle, kicks that come from the right, the left, above, below, in front, behind – in other words, from all sides. On average, all these nudges even out, so there is no a preferred direction for the particle to move. Yet, as Einstein realized, if we look at a very short time interval this is not *exactly* true. There will always be a little loss of balance, causing the particle to perform an extremely irregular zigzag motion, similar to the staggering walk of a drunkard.

Let us see whether the latter analogy helps us. For instance, imagine that we have drunk too much and that we get out on the street overlooked by the bar where we have been enjoying ourselves. We have no idea of the right way home, so we make a first move in a random direction, say to the right. Then we stop and think it over, and as a result we resolve to retrace our steps, or to keep going in the same direction, and so on. Where will we be after a certain number of steps? Intuitively, we may expect to find ourselves close to the starting point.

Quite sure? Then let us see whether the computer can give us a hand, asking it to trace the path of an inveterate drunk. To set the scene, suppose that our favorite haunt lies in the open countryside, and that our drunkard tries to set out for home across the fields. At any step (each one of the same length), the computer chooses a direction at random[12]. To be surer, let us try with two heavy drinkers, each one of them taking 100 steps, just to see whether the paths they follow look alike. What comes out from the computer I am working on is shown in figure 2.1.

Surprise! Not only are the two paths extremely irregular (so much so as to be dubbed *random walks*), but they look remarkably unlike each other. Moreover, our first drunkard apparently does not have the slightest wish to come back to the bar (although he moves away from it by much less than 100 steps), and the second is rather reluctant too. What is happening? You may think that my computer has swindled us, and that this result is not to be trusted, but I can assure you that the single Brownian tracks of colloidal particles *really* do look like those shown in the figure.

---

of 0.01 millimeters. To compare it with the sort of energy unit we are used to hear about (and pay for), it would take all the molecules contained in about six gallons of water to get a total energy of one kilowatt-hour.

[12] If we had set this on a street, so that we only had to decide whether to move right or left, we could simply toss a coin. On a surface, the choice is a bit more tricky, but computers know their job!

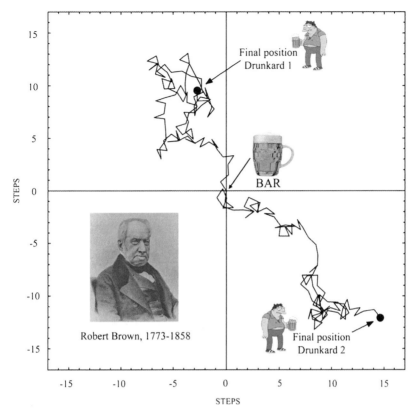

**Fig. 2.1.** Computer simulation of two random walks, each of them made up of 100 steps.

The point is that these paths are *so* random that actually, to grasp what is happening, we must resort to statistics, simulating the motion of a *large* number of drunks. The dotted "spots" in figure 2.2 therefore show, on squares with a side 200 steps long, the final points reached by 1000 drunks after each of them has performed a random walk of 100 steps (panel A), then 400 (panel B), and finally 900 steps (panel C). Note that the spots are actually centered on the drinking place. This means that, statistically speaking, we made a good guess. On average, our random walkers do remain trapped near the pub, although we find a smaller and smaller number of them who have managed to stagger longer and longer distances from its doorway.

There is another feature that is obvious in Figure 2.2. Although we multiplied the number of steps by four, the spot in panel B is about only *twice* as large as the spot in panel A. Similarly, the spot in C is only about three times as wide as the one in A, even though the drunks have taken nine times as many steps. I am sure it would take little effort to convince you that if

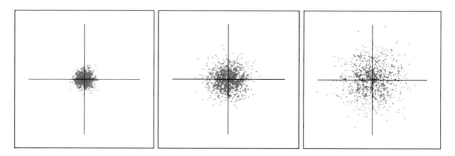

**Fig. 2.2.** Final positions of 1000 drunkards after 100 (A), 400 (B) and 900 (C) steps.

we had simulated random walks of 1600 steps, we would have got a spot four times the size. Now, since 4 is the square root of 16, as 3 is of 9 and 2 of 4, we can conclude that the spot size grows only in proportion to the *square root* of the number of steps. Assuming that our boozers take a step at fairly regular intervals, for instance once per second, this also means that the width of the region they "explore" grows as the square root of time.

It is worth pausing for a while to think about that result, for it is anything but trivial. Suppose for instance that after an hour our spot has a size of 100 meters. What we have found means that the spot would have reached a size of 10 m in under 4 seconds, whereas, to reach a radius of 1 km, we would wait for more than 4 days! Hence, Brownian motion starts like a torpedo, but then becomes slower and slower. It is easy to show that even a snail, no matter how sluggishly it proceeds, will sooner or later overtake even the most go-ahead of the drinkers (provided obviously that the snail is not tipsy too). Things get even more interesting if we note that what I called "spots" may actually be *real* spots. For instance, if we plunge a fountain pen into a glass of still water and release a drop of ink (which, as we shall see, is also a colloid), its size will grow only as the square root of time (which we could write as $\sqrt{t}$). Similarly, if we let a lump of sugar sit in a cup of tea without stirring, the size of the region sweetened by the sugar molecules grows only as $\sqrt{t}$. Phenomena such as these are called in physics *diffusion processes*. And it is not even necessary that there are particles or molecules to diffuse. The same thing happens with heat, the spread of energy from warm to cold bodies, provided that it is transported by diffusion and not by force, as for instance when we use a hairdryer.

This picture of Brownian motion allows us to get an idea of how large particles can be and still be called "colloidal". In abstract terms, there are no limits to the size they could have. When immersed in water, even marbles, the sort we played with as kids, take on an energy equal to $kT$ each. Yet we would hardly think of them as colloidal particles; we know that they would plummet under the effect of their own weight, ending up at the bottom of the container[13], and goodbye to the colloid! If they were much smaller, however, they would settle at a much lower speed. Just to give you an idea, if our marbles had a radius of just one micron (a millionth of a meter, if you remember), their *sedimentation velocity* (the more elegant term for settling rate) would be only 20 cm a day, low enough to allow us to study them while they are still suspended (and after that, a good stir is enough to put us back in business).

But the most surprising news is that, if they were really *very* small, they would not settle at all! Let us try to understand why. You may recall from your basic physics studies that, besides the kinetic part, there is another important contribution to the energy of a body, the *potential* energy. Without racking our brains for formulas or long-forgotten definitions, it is enough to say that it is the amount of energy a body stores when it moves against a force, a kind of investment the same force can give you back later as energy of motion (the kinetic energy). So, for instance, if you throw upwards from the ground a stone with mass $m$, giving it a certain initial speed and therefore kinetic energy $E$, the stone reaches (if we neglect air friction) a height $h$ such that $m \times g \times h = E$, where $g$ is the acceleration of gravity. The quantity $m \times g \times h$, or $mgh$, is exactly the potential energy the stone gains by moving against its own weight that pulls it down. Obviously, once the stone has reached the height $h$, it falls back, losing potential energy and regaining kinetic energy, until the kinetic energy reverts exactly to $E$ when the stone is just touching down. This simple experiment shows that kinetic energy can be converted into potential energy (and the other way around) exactly as we can convert euros into dollars and vice versa. It there is no friction (which, to a small extent, is always present, just as a bank always takes some money off in currency exchanges), the *sum* (that is, the combined total) of the kinetic and potential energy remains constant. This is what we call the "conservation of mechanical energy".

That is all we need, since we also know that a colloidal particle *always* has a kinetic energy $kT$. Therefore, even if it settles under its own weight, it will not necessarily end up at the bottom of the container, for it can harness this small amount of kinetic energy for climbing back to a height $h = kT/mg$. If we consider marbles once again, but this time with a radius of a *hundredth* of micron, it is easy to show that we can always find particles that rise to a height of at least 10 cm, even if we wait forever.[14] Thus, there are always a lot of particles dispersed in the solvent and, in any case, it takes the particles a very long time to reach this final state, since their sedimentation velocity is just

---

[13] Or the top, if they are ping-pong balls. What actually matters is comparing their weight to that of the fluid they displace (Archimedes' principle, if anyone recalls this).

[14] An accurate calculation shows that the particle concentration we would find at this level is about a third of the concentration at the bottom, and that almost 1% of the particles will rise by half a meter!

7 mm *a year*. We may summarize this digression[15] by saying that suspended particles can be regarded as colloidal if they settle slowly enough, or if they are so small that, in practice, they never sediment out[16]. Obviously, it is also a question of viewpoint, since on the Space Station, in the absence of weight, even basketballs would be colloids.

We have spent quite a bit of time in describing how colloidal particles move, but it was worth the effort. As we shall see, all the microscopic world is subject to Brownian motion, including those chemical micro-plants, micro-motors, and micro-repair shops we are made of, which must work in the midst of a restless, jostling molecular crowd. There is really no such thing as a quiet job in the microscopic world.

## 2.5 Osmosis, the breath of a dispersed world

A well-known consequence of the fact that molecules have a kinetic energy is that a gas *presses* on the walls of its container (for instance on the inner tube of a bicycle tyre). Such a pressure is nothing but the overall push that the gas molecules exert on it through their collisions with the walls, divided by the total area of the container (or, as we shall say, the force exerted *per unit area*). If the gas is not too dense, this pressure $P$ is simply equal to the molecular concentration $c$, expressed in molecules in a certain volume of air times (once again) $kT$, or $P = ckT$.

Getting back to our Brownian motion, we have already said that a Brownian particle in a liquid has as much kinetic energy as the solvent molecules. Is there anything paralleling the pressure in a gas that can be ascribed *solely* to the particles? Well, this quantity not only exists, but is also extremely useful for understanding the behavior of a colloid and of many other systems, both simple systems, such as a salt solution, and very complex ones, such as a neuron, one of our nerve cells. It is called *osmotic pressure*, from the Greek *ōsmós*, meaning simply "push", so that, to recall its Hellenic origins, it is usually indicated by the Greek capital letter "pi" ($\Pi$).

What, then, is osmotic pressure and, above all, what visible effects does it have? In a gas, pressure exists because molecules push on the walls, which

---

[15] Which is not totally useless. Remember for instance that many rocks were formed just by particles settling on the seabed. Hence sedimentation certainly does matter to geologists.

[16] Clearly, that "enough" also depends on the density of the particles with respect to the solvent. Checking the results of a blood test, you may find for instance the sedimentation velocity of erythrocytes – your red blood cells, in other words. These are platelets with a size of a few microns, but they have a density comparable to water. Therefore, letting them settle under their own weight requires a ludicrously long time. To avoid a lifelong wait, we get them moving with a centrifuge, similar in principle to a salad spinner or the machines astronauts use to familiarize themselves with what they will feel at takeoff or landing).

in turn push them back[17]. That is, the walls *contain* the gas. If we wish to find a kind of pressure that concerns just the colloidal particles, we should therefore look for walls that confine *only* the latter, while they let the solvent pass through as freely as ghosts through the walls of a Scottish castle. The walls of many animal organs, such as the bladder or the bowels, have exactly this magic property, so that they are called *osmotic membranes* or, more commonly, "dialysis membranes". Basically, the distinguishing trait of these membranes is that they have microscopic pores, large enough to let solvent molecules pass freely, but too small to allow colloidal particles through. Today, chemists and biologists can do much better without eviscerating too many animals. By means of structures we shall call "polymer networks", indeed, we can make artificial membranes with pores of precisely the gauge we want, to hold back only those particles we wish to retain.

Usually, these membranes are shaped like hollow tubes and are rather elastic, a bit like a balloon, a feature useful for a little thought experiment. Let us fill a container with the solvent (no particles), then make a bag with a piece of dialysis tubing, fastening it up at both ends with a thread (like an old-style homemade sausage), and finally fill the bag with a colloidal suspension. If we now dip the bag in the container, we observe the bag swelling progressively as solvent gets inside from the external reservoir. The smaller and more concentrated the particles are, the clearer is this effect. Now, seeing the bag inflating like a balloon tells us that the pressure inside it has *increased*, to the point that, if we fill the bag completely without leaving any air space, the bag may even burst open. Somehow, the colloid "sucks in" solvent until, after some time, the swelling process ends. The osmotic pressure $\Pi$ of the suspension is defined as the *extra pressure* within the bag at this final stage. How large is it? Well, if the colloid is not too concentrated one finds that $\Pi = ckT$, just like the pressure of a gas, but where this time $c$ is the *particle* concentration in the bag (this is called the *van 't Hoff law*). Where does this excess pressure come from? It would be nice to find that it is just due to the "bombardment" of the osmotic membrane by the colloidal particles. In fact (though many colleagues of mine find this hard to believe) it is really so, but the way it works is rather complicated[18].

What is important to stress is that the pressure increase stems only from the trapping of the suspended particles in the bag. To confine particles, an osmotic membrane may not be necessary. For instance, we have seen that

---

[17] Another school memory. This is the notorious "law of reciprocal actions", which you might remember, more prosaically, as the law of "an eye for an eye, a tooth for a tooth".

[18] For the most curious of my readers, the process takes place more or less in this way. To push back particles, the membrane has to apply a force. Since the particles constantly exchange energy with the solvent molecules, they transfer such a force to the solvent through collisions. Yet, when a fluid is subjected to an external force, its pressure must necessarily increase, and this can happen only if some extra solvent molecules pass from the reservoir to the suspension.

colloids settle out (some more, some less), meaning that *gravity* tends to confine them at the bottom of the container. The osmotic pressure must then be larger close to the bottom (where the suspension is more concentrated) than at the top. It was precisely by measuring the concentration profiles of settled colloids that Jean Baptiste Perrin managed to validate Einstein's model of Brownian motion, earning himself the Nobel prize for physics in 1926.

However, osmotic swelling effects are not limited to colloidal suspensions. Biological membranes too are capable of holding back simple substances such as salts or sugars[19], and the osmotic pressure of a solution of small molecules (just because they are small and so, for a given concentration in weight, there are *a lot* of them) can be really gigantic. For instance, the osmotic pressure of salted water suitable for preparing good pasta (slightly less than 10 grams per liter, to my liking) is about *ten atmospheres*, which is the excess pressure you must bear underwater at a depth of 100 meters! This has important repercussions on physiology. For example, to prepare an intravenous shot or drip, we must use a "physiological" (or "isotonic") solution, which is nothing but a salt solution at slightly less than 1% concentration (more or less the same as water for pasta). Why can't distilled water be used? Simply because we need a solution with an osmotic pressure comparable to that of our blood. As we said, the blood contains erythrocytes, small cells with a disk-like shape flattened at the center, and the membrane covering these cells does not allow salts to get through. So, if you add distilled water to a blood drop and look at what happens under a microscope, you will see the red blood cells swelling quickly, and then bursting like balloons. This is the same effect that takes place for our sausage, but strongly (and tragically, for the poor erythrocytes) amplified.

Even without a microscope at our disposal, we can make a simple and less bloody experiment using a couple of eggs. An egg is nothing but a huge cell, which has an outer membrane called the *amnion* (yes, it is really a caul), preventing outside agents from penetrating inside, but which we do not usually notice, for it is hidden under the shell. The eggshell itself is a chalky substance, and dissolves if immersed in an acid liquid like vinegar, or more easily in kettle descaler. The upper picture in 2.3 shows two eggs of similar size which I soaked in a descaler solution. Some hours later, the shells have gone and the eggs have become soft objects, just kept together by the external membranes. I then soaked one of the two eggs in a bowl full of plain tap water, and the other one in water in which I had dissolved as much common table salt as possible. The central picture shows the outcome after one night. The egg that was immersed in the salt solution (to the left in the picture) has slightly deflated, whereas the one in fresh water has swelled hugely. Even their visual appearance is different. While the amnion of the first egg looks whitish, the

---

[19] Here, however, it is not a question of pore size. The reason these molecules cannot get through a membrane is a bit more complicated.

second one is stretched like a balloon, and so translucent we can make out the inside.

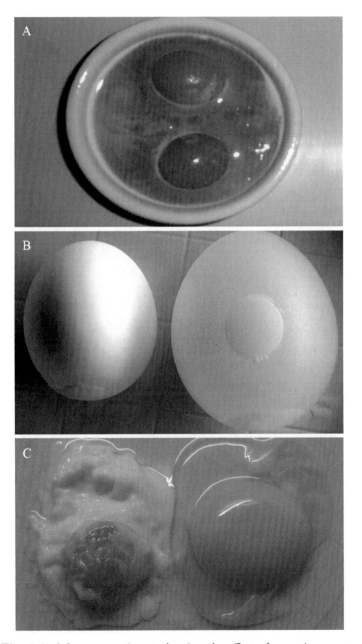

Fig. 2.3. A home experiment showing the effect of osmosis on eggs.

At this stage you should have guessed what has happened. An egg contains many substances like proteins that, for the good of the chick, should not escape from the amnion. Therefore, whereas water gets inside the egg soaked in fresh water by osmosis, diluting its contents, for the other egg the external salt concentration is so high as to favor the *exit* of some water, even if this yields an increase in the internal protein concentration. What is left inside is better appreciated by breaking the membranes (bottom picture)[20]. While the white and the yolk of the first egg have remained liquid and thin, the content of the other egg has shrunk to a semi-solid jelly. Another home example, very familiar to Mediterranean people, is the way we usually remove the "bitter water" from eggplant, by covering slices of it with salt, maybe pressing to improve contact. Here two dialysis processes actually take place. On the one hand, part of the water seeps out from the slices by osmosis; on the other, there is an osmotic exchange between sodium getting into the eggplant, and potassium (which is mostly responsible for the bitter taste) coming out.

In the examples we have seen so far, the solvent flows spontaneously through a dialysis membrane to dilute a colloidal suspension or a salt solution. In some cases, however, it would be useful instead to take the solvent out from the membrane, in other words to make it flow the other way around. For instance, it would be very useful to use an osmotic membrane to extract fresh water from salt water. Because osmosis works in the opposite direction, however, it is clear that this cannot take place *spontaneously*. To do it, we should somehow pay, and in physics money always means energy. For instance, extracting fresh water from a salt solution is feasible provided that *we* apply a pressure (larger than the osmotic excess) to the solution, forcing the solvent through the membrane, whereas salts and other filth are retained inside (obviously this requires work, and therefore energy). This is the principle of inverse osmosis, used in many widely advertised domestic water purification systems[21].

At this point, we should pause to do justice to biological membranes, which are anything but the simple dialysis bags we have been talking about. *Au contraire*, cell membranes, which we will discuss at great length in the final chapter, are masters of direct and inverse osmosis, since they manage to let through what they need when and how they like. Without these osmotic machines, I would not be here writing (nor you reading) this book, since our brains work just because of our nerve cells' wonderful skill in this area.

---

[20] The "balloon" egg actually bursts, squirting all over the kitchen and triggering the rage of the experimenter's spouse.

[21] Technically, the method is a bit more complicated. Usually, pressure is applied *tangentially* to the membrane, which guarantees a much higher efficiency.

## 2.6 Colloidal Lego, matter made to measure

There is a sort of physicist who is perpetually unsatisfied: the condensed matter theorist, whom we shall dub in short a "theomad". Probably the origin of their discontent is that, in their view, God or someone on His behalf has created too messy a world: had they been in His shoes, they claim, they would have made a much better job of it. To tell the truth, theomads are really good at creating ideal worlds and at examining in depth how they work – worlds made of billiard balls, rigid rods, equal coins, but above all worlds made of things that ignore each other or, if they do interact by means of some force between them, do so in a very simple way, as for instance balls connected by springs would. A well-known joke at the expense of theoretical physicists has them wondering how to increase milk production on a farm, and contains the immortal line, "Consider a spherical cow in a vacuum ...". Well, the bucolic world of a theomad is not only a place where all cows are spherical, but one where cows, in a first approximation, never get together at all (clearly there are no bulls around, at least).

Unluckily, the real world is quite different, not only because it is made of molecules, atoms, and "elementary particles" with, deep down, a split personality (somewhat like particles, somewhat like waves, as the ill-famed quantum mechanics says), which are anything but simple balls or sticks, but also because atoms and molecules, far from being lonely hearts, are related to each other by *very* complicated forces. Efforts to explain a world made of *these* things are likely to bring on a chronic headache, at the very least. Okay – you may say – but the real world is this one, and *that* is what you should try to understand, otherwise what do we pay you for? Nonetheless, many colleagues of mine keep on shamelessly designing worlds made of balls and sticks, and, in my opinion, they are perfectly right. Indeed, it's only by studying these ideal worlds that some basic ideas, such as the difference between a fluid and a solid, or between order and disorder, can be fully grasped.

Obviously, till a few decades ago these colleagues of mine led a rather lonely life, for these worlds were only in their minds. Then computers came along, with their rapidly growing skill in simulating virtual worlds, including those all theomads dreamt of. Possibly this suits my kids and their friends, who are virtually PC peripherals, but an old (er... let's say "full grown") generation like mine prefers to see, smell, or touch something substantial (and this is obviously true for *any* kind of material). Luckily for us, colloids can often make our wishful thinking come true, because, using a kind of conjuring trick, we can to some extent *choose* the forces acting between colloidal particles, turning them into the bricks of a "mesoscopic Lego" that allow us to build the worlds the theoreticians long for. You may think that this is totally meaningless for, after all, a colloid is nothing but a bunch of large and complicated clumps of chemicals, immersed in a liquid like water which is anything but simple. Contrasted to stuff such as a salt crystal or a window pane, they must surely be harder to understand! Not so. This magic trick just consists in manipulating the solvent (generally by some additive) and then neglecting it to care *only* about the particles and the forces acting between them. We shall now try to see how this gimmick actually works.

First, note that colloidal particles, compared with most other things around in the molecular world, usually have a much simpler shape. It is true that natural particles, such as clays flowing through water-bearings or dust grains suspended in air, are generally very irregular and varied in shape. Yet, over the past decades, chemists have learnt to create particles with very controlled size, shape, and material properties. Thus, it is not too difficult to produce spheres which are all more or less equal (just like the ideal cows we mentioned), with a diameter that can be as small as a few nanometers, made of glass, metals, and plastic materials. So far, making rods or disks has been less successful, but what we get is adequate for many purposes. Certain chemists and physicist have also been able to produce more elaborate shapes, such as little eggs, pyramids, or dumbbells: a paradise for theomads!

Yet, as we said, controlling the shape is not enough. Let us then see what kind of forces may act between colloidal particles. You might think that the term "colloid" gives us a clue, stemming as it does from the word for glue, but we should not be led astray. When the Scottish chemist Thomas Graham coined this word to set colloids apart from those solutions that were able to pass freely through a membrane (rather oddly, he dubbed these "crystalloids"), he chose it only because the first colloids were obtained from isinglass or sticky jellies. Nonetheless, Graham partly hit the mark, since colloids are often rather "sticky" by nature, and spontaneously tend to clump together. Why is that? All that can be said is summed up in the homeopathic maxim *similia similibus curantur*, or "like cures like" (this is actually the only situation, in my view, where such a claim makes sense). But what is this colloid disease that needs curing? It is precisely their peculiar property of exposing a huge surface area to the solvent. We shall see that building up a surface separating two different materials *costs energy*. After all, for a piece of metal or glass there is no better neighbor than another piece of metal or glass, rather than some solvent molecule that they may like but that is still alien stuff[22]...For a colloidal particle, the easiest way to cut down costs (namely, contact area) is to join to a particle like itself, and then to another, and so on and so forth, till a single lump forms.

What I have tried to summarize in simple (and quite inaccurate) words corresponds, in serious science, to what are called *van der Waals* or *dispersion* forces, a concept that actually lies at the heart of physics[23] and plays a key role in everyday life. Just to give one example, without dispersion forces the

---

[22] For some special particles we shall meet soon, which are quite "friendly" toward the solvent, things are rather different.

[23] That abstruse theory of quantum mechanics shows that dispersion forces magically stem from spontaneous "vacuum fluctuations". In modern physics, a vacuum is not completely empty after all – so Aristotle was right, in a way!

great pyramids would not still be standing, since what we call friction, which is essential for keeping stones together, is strongly related to these forces. The important point is that these forces are *attractive*, namely they tend to make things stick together, and that they are strongest when they act between two objects made of the *same* material. Do you need proof? Just enter a glass-makers shop (although, unluckily, few of them are left) and look at the way glass panes are stored. Between any two of them, the judicious glass-maker inserts a paper sheet. Why? Well, take it out, stack the panes together, and then try to separate them, if you can. You will then realize that van der Waals force do indeed exist![24]

Therefore, when two colloidal particles, tossed about by Brownian motion, come near enough to each other, van der Waals forces make them stick to-gether (or, as we should say, *coagulate*). "Near enough" actually means "very near", since these attractive forces are felt only at the very short range of a few nanometers[25]. Yet, during their random walk, sooner or later two particles do meet (the higher their concentration, the sooner this happens) and...game over.

Why then do colloidal suspensions actually *exist*? The point is that, be-sides dispersion forces, additional *repulsive* forces enter the game, pushing the particles apart and in this way *stabilizing* the colloid. A common origin of this repulsive interaction is that the surface of many colloidal particles is (for chemical reasons we won't go into) electrically *charged*. For instance, glass particles normally have a negative surface charge, whereas certain mineral colloids are positively charged. Do not worry if you don't remember anything about electric (or, more properly, electrostatic) forces. For our purposes, it's enough to keep in mind that:

1. *Like* charges, both with a positive or negative sign, repel each other, whereas charges with opposite sign attract.
2. Charges come by twos like cherries. If there is a certain amount of positive charge, the same amount of negative charge is around. Hence, a solution is always *electrically neutral*.

The second statement, in particular, tells us that if we have a colloidal particle carrying, for instance, 101 negative charges attached to the surface, somewhere there must be an equal number of positive charges. This opposite contribution is made of little charged atoms or molecules (which we call ions) going freely around in the solution. But the first statement tells us that these positive ions cannot be completely free. After all, there are 101 negatives attracting

---

[24] Separating microscope slides with tissue-paper is a good rule too, which unluckily my students have not yet learnt to follow.

[25] At extremely short distances, however, they become really strong, so much so that if the particles were perfect spheres, the forces would become infinite at contact. Perfection does not exist on this planet, but two particles made of a sufficiently soft and plastic material do sizeably buckle when they stick, showing the strength of dispersion forces.

them, so that, like the little Dalmatians, the positive ions are just waiting to... charge! So a little "cloud" of positive charges takes shape around the particle, a cloud of ions that may extend for tens or hundreds of nanometers. How many are there? You won't be too surprised to discover that there are still *exactly* 101, as many as the charges bound to the surface[26]. What happens then when two such particles approach? Before they get close enough to feel the fatal attraction of the van der Waals forces, the clouds surrounding them start to overlap and, since they have the same charge, to repel. As a consequence, particles and associated clouds are pushed away from each other, as if there were a spring between them. Ultimately, the surrounding ion clouds keep the particles so much apart that (sedimentation apart) a charged colloid can unhurriedly remain for years in a bottle, without coagulating and precipitating to the bottom as a dispiriting fine dust.

All this works in water, but surely not in *oil*. Water is what we call a *polar* liquid, where free charges, such as the ions that salts form when they dissolve, can exist. Oil is different. In oil, or in what is in general called a *non*-polar liquid, there is no way to separate a positive from a negative charge (have you ever tried to dissolve salt in oil?). So, colloids cannot be made stable by charge effects, and this is a pity, because for many purposes it would be quite useful to suspend particles in an oil. Moreover, electrostatic forces between charged particles are not so easy to investigate, and our theomads would prefer something *much* simpler, which is easier to obtain, as we shall see, in non-polar solvents. How then can we beat dispersion forces and avoid particle coagulation? The secret lies in coating them in little "hairs", made of molecules in the shape of a rather long chain, but still short compared with the particle size. We shall talk at length about them in the next chapter. So far, it is enough to say that, when two particles approach closely, the "hair forests" covering them do not like to overlap. Once again, this mechanism keeps the particles far enough apart that van der Waals forces are too weak to make them stick.

Charging particles or covering them with hairs are not the only ways to stabilize a colloid. There are special particles we shall soon meet that remain peacefully suspended for ever, even if they are not charged or covered by hairs. The basic reason is that these particles just love to be surrounded by solvent. This may look strange. What happens if we take a lump of salt or sugar, substances that love water, and dip them into a glass containing this liquid? Obviously, they dissolve – in other words, their atoms or molecules slowly scatter around uniformly (by diffusion, remember). Why is it, then, that these special particles *do not* dissolve? There are two main possible reasons, either because they are actually long chains of tightly bound molecules (this is the case for the polymers we shall meet in the next chapter), or because the

---

[26] In this way, positive charges exactly "screen" the negative ones, and another positive ion has no more reason to approach, seeing from the distance the set "particle + cloud" as globally uncharged ("neutral").

molecules they are made of are forced by their special nature to band together to form a particle, which is the case for the surfactants we shall encounter a bit later. In both cases, a thin solvent layer wraps around the particle, a bit like Linus' security blanket. Needless to say, it is just this beloved blanket that keeps particles far enough apart to avoid the ruinous effects of dispersion forces.

Analyzing colloid stability, we have therefore already seen at work those solvent manipulation techniques I mentioned earlier. Van der Waals forces can, for instance, be modified by changing the solvent and making it more favorable for the particles; additives that stick to their surfaces can make them more stable; or the same goal can be reached by charging them. We shall start from this last example to show that in fact we can do much more, adding *more* ions to fine-tune the forces between particles as we might tune a guitar.

## 2.7 Softness without limit: fractal aggregates

In discussing how a charged colloid behaves, I was very vague about the way the ion cloud surrounding a particle is actually made, and in particular about the size of this region, i.e., up to what distance it extends. This crucially depends on the *kind* of water we use to prepare the suspension, something which can be grasped with a simple experiment. Adding ordinary tap water to a suspension of charged particles, we would observe that the particles rapidly settle out as fine dust falling to the bottom of the container. This does not happen if we add distilled water instead. Quite the opposite: the suspension becomes even more stable. Now, we know that the main difference between tap and distilled water is that the former contains salts, mostly of calcium, sodium, magnesium, iron, and often chlorine. The total amount, usually abbreviated TDS (Total Dissolved Solids), is widely variable, but should not exceed a few grams per liter. Hence, it is the addition of salts that gets the colloid into trouble. Let us first get a general idea of why this happens. In the absence of salt, the cloud surrounding a particle, which is swimming in a kind of fresh-water lake, suddenly sees another similar cloud with the same charge. However, if we add a little table salt, mostly composed of sodium chloride containing a sodium and a chlorine atom (NaCl, as you surely know), things drastically change. As we said, salts in water split into electrically charged ions, such as $Na^+$ and $Cl^-$. Even if we add a small amount of NaCl, say a gram per liter, we get a huge number of ions in solution, about ten million $Na^+$ and as many $Cl^-$ ion per cubic *micron*! At this stage, our particle is already swimming among *heaps* of ions. How could any difference ever be made by the additional few ions it finds around a playmate?

To put some numbers on this, theory shows that, by adding salt, both the size of the surrounding ion cloud (which is properly called the *Debye–Hückel length*) and the strength of the electrostatic forces considerably lessen, so that repulsion is weaker and acts at shorter distances. For instance, in the presence of half a milligram of NaCl per liter the Debye–Hückel length is about 0.1 $\mu m$, but it shrinks to just 2 nm if we add 1 gram of salt per liter[27]. At the same time, the repulsion force has decreased a few *thousand*-fold. You may picture the situation by regarding the repulsion due to the overlapping of the ion clouds as a kind of hillock the particle has to climb to get close to another one. If the particle succeeds, on the other side it falls into the bottomless pit of the van der Waals forces, and sticks to its companion. But climbing a hill requires energy, and we know that all the energy our particle has is $kT$. Therefore, if the required energy is much larger than $kT$, the particle will never make it. If we now start adding salt, the hillock shrinks and gets *lower*. Suppose that the energy goes down to just 2 or 3 $kT$. This value is still larger than the *average* particle thermal energy, but some more "gifted" particles may have enough energy to get over the hill, therefore sticking to another one. As we add more salt, the hillock progressively vanishes, and particle aggregation gets faster and faster. As soon as two particles meet they stick, and this happens the more often the higher is the colloid concentration.

To investigate better what happens, let us just consider a case where we have added enough salt to kill the electrostatic repulsion completely. Once two particles have stuck, a third one is more likely to bump into them, because now they are a bigger obstacle, and so on, till a bigger and bigger particle lump forms. What do these aggregates look like?' Well, they have a truly special structure, which is related to one of the most original ideas mathematicians ever had. Although they can hardly be pictured on a flat surface (they are obviously immersed in space), these lumps look rather like I have tried to picture in Fig. 2.4. The first feature that catches the eye is that they are not compact objects, but are full of holes. Why? To understand this, let us see what happens to the particle at the top right, which is approaching the aggregate with its zigzag Brownian motion. There is a lot of room to fill within the aggregate, but the particle will almost never *get* there, for there is a high chance that, before getting into a hole, it bumps into one of those particles in one of the many "arms" protruding outside. And, remember, if it touches it sticks. You can easily see that this is a kind of avalanche process: the more particles attach to the arms, the more the latter branch off outside, the more difficult is for a newcomer to get inside the aggregate. The sketch I made conveys only a rough idea of what is happening, since we should take into account not only that single particles can stick to an aggregate, but also that two of these aggregates can meet and merge, generating even more holes. What results from all this, and can be explored with computer simulation, is really one of the emptiest objects one can imagine.

To see how "open" a colloidal aggregate can be, we have to go on a brief excursion into an abstract mathematical world, one of those only mathe-

---

[27] One can more rigorously show that the Debye–Hückel length is inversely proportional to the square root of salt concentration.

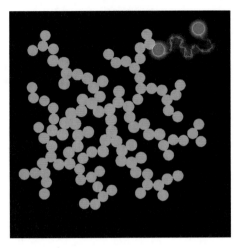

**Fig. 2.4.** Draft of a colloidal aggregate. Note the particle at the top right, which is just sticking to one of the aggregate arms.

maticians can conceive. We are used to regard everyday objects as "three-dimensional" , but what do we really mean by this? Well – you may say – everything, for instance the piece of furniture I just put in my living room, has a length, a width, and a height. True, but in the first decades of the last century a great German mathematician, Felix Hausdorff, thought of looking at this in a slightly different way. Take for instance three balls, made of the same material. The second ball has twice the radius of the first, and the third twice the radius of the second. I think you may agree that the second ball weighs eight times the first, and the third ball eight times the second, that is, 64 times the first[28]. Let us then look at the two series of values to try and find a relation between them:

Radius (compared with the first ball): 1, 2, 4
Weight (compared with the first ball): 1, 8, 64

Not a big issue. The number in the lower series are just the *cubes* of those in the upper one. Thus, the weight (and therefore the quantity of matter) grows like the cube of the radius – or, as mathematicians would say, as the *third power* of the radius. But "three" is just the number of dimensions of our usual space. Hausdorff then had the apparently trivial but actually brilliant idea of stating that an object has three dimensions if, on measuring how much stuff is contained in a sphere of radius $R$ placed inside it, the latter grows as $R^3$. Common things obviously do have this property, but the astonishing fact is that, using this definition, there are objects that, though immersed

---

[28] If you accepted this claim without blinking, you are possibly not very well suited to experimental science, for I did not tell you that all three balls were *full*. Anyway, I'm telling you now.

in a three-dimensional space, have *less* than three dimensions. The colloidal aggregates we have just described are of this kind. Let us first recall that a number can be raised to a power that is not necessarily integer. For instance, we can write the square root of 5, $\sqrt{5}$, as $5^{1/2}$, or similarly $7^{2/3}$ means that we must square 7 and then take the cube root (or vice versa, it does not matter). These two examples may help you understand what we mean by $x^{a/b}$, or in general by $x^c$, where $c$ is not necessarily a whole number. If we now ask a computer to make colloidal particles stick together, so that they generate a fractal aggregate (which is not too difficult), then to draw a sphere inside it, and finally to evaluate how the number $N$ of particles contained in the sphere depends on its radius, we find that $N$ grows approximately as $R^{1.7}$. Hence, not only is the aggregate dimension much lower than three, but it is not even an whole number! These weird objects are called *fractals*, and the power to which we must raise $R$ (here 1.7) is called the *fractal dimension*.

Passed over in silence for almost half a century, fractals were rediscovered by Benoit Mandelbrot (who actually coined the word "fractal") at the end of the sixties, and since then have been recognized by physicists in the shape of an incredible number of things, ranging from snowflakes to coastlines, lightning, river basins, even broccoli. In fact, without pointing it out, we have already encountered an example of a fractal. Consider the trajectory that a particle follows during its Brownian wandering. As we have seen, the radius of the explored region grows as the square root of the number of steps $N$. Turning this relation around, we see that the number of steps contained in a sphere of radius $R$ is proportional to $R^2$, which is like saying that the particle trajectory is so jagged as to be a fractal object with dimension equal to two.

Ultimately, maintaining that aggregates are fractals is a much more precise way of saying that they are full of holes. What's more, it is easy to see that the larger a fractal aggregate gets, the emptier it is. Take for example two aggregates, one about twice the size of the other. The larger aggregate will contain $2^{1.7} \simeq 3.25$ times as many particles as the other (use a calculator to check, if you like). Yet its volume is *eightfold* larger, and therefore its density (the number of particles per unit volume) is almost 2.5 times *lower*. Therefore, the more they grow, the more colloidal aggregates become light, tenuous, made of almost nothing, filling a lot of room with very little stuff. These large objects are cumbersome, and this has noticeable effects on the properties of the solvent they are immersed in. In particular, they make it very viscous, meaning that the solvent finds it harder to flow, an effect that makes these aggregates particularly useful in many technological applications as *thickeners*. For instance, there is a good chance that particles made of silica (silicon dioxide, $SiO_2$, the main component of what we usually call "glass") are present in the toothpaste you use everyday, making it in addition more abrasive (they are very common in "whitening" toothpastes)[29].

"Precipitated silica" (produced differently, but still in the form of aggregates similar to those we have described) is also used as an additive for many other purposes, not just as a thickener. For instance, it's used to prevent clotting of granular materials such as table salt, which is particularly prone to clumping when damp, or to limit the formation of foams. I do not want to scare you too much, but the temptation of adding a bit of precipitated silica to some food to make it creamier is rather strong, and regarded as innocent in many countries[30]. If *this* worries you, regard yourself as lucky in comparison to hundreds of thousands of workers in the mining, construction, ceramic, and obviously glass industry, who inhaled silica particles and have suffered the truly unpleasant side effect of this, silicosis. Luckily, many foods exploit natural colloidal aggregation to get that creamy taste that is so pleasant for the palate. A classic example is yoghurt, which is nothing but the outcome of the aggregation induced by bacteria of a very special colloid, milk, that we shall meet again when we encounter emulsions[31].

Back to our aggregates, what happens if we let them grow more and more, avoiding allowing them to settle too fast (maybe using particles as dense as the solvent)? Well, within a short time they will join and form a *single* huge aggregate filling the whole container. We have thus obtained a very special material, a *gel*, or better a *hydrogel*, i.e., a material that behaves like a solid (if you carefully try to turn the container over, the gel does not spill over the floor), but that, in terms of quantity of matter, is made almost of nothing (once again, it has a fractal structure). These materials, which we shall amply discuss, have a large number of applications precisely because, although there is little stuff inside (apart from the solvent, of course), their internal surface is huge. To tell the truth, gels obtained by colloidal aggregation are very fragile. There are few contact points between the particles, and van der Waals forces are weak, so what we get is really a very "soft" material. In the next section, however, we shall encounter some very special relatives of theirs, which are anything but soft. On the contrary, they are about as hard as could be.

## 2.8 Concrete: united by charge

I have spent quite a bit of time in trying to explain why the presence of charges on the particle surface stabilizes a colloid so much that, to precipitate it, we must add salt to "kill" electrostatic repulsion. Is this always true? No, sometimes the opposite takes place, and particles bearing the same charge strongly

---

[29] If you think you can escape by saying goodbye to toothpaste and exchanging it for certain strongly advertized chewing-gums, you are mistaken. There really is *a lot* of silica in the latter.

[30] It is largely used in animal feeds, where it is officially known as additive E551b.

[31] After all, the Turkish word *yoğurt* has the same root as the adjective "dense".

*attract* each other. Don't worry, I did not lie; it was just a sin of omission. Everything I told you holds true if the surrounding ions carry a single charge or, as we shall say, if they are *monovalent*, for instance simple $Na^+$ or $OH^-$ ions. But what happens when they are *di*valent (or even *tri*valent), meaning that they bear two or more charges each? Well, here even our theomads are partly groping in the dark. While in the former case we have a theory that works fairly well and justifies what we said on stability, here the problem becomes more challenging.

But there is at least one extremely important situation that clearly shows us that things can go the other way around. What I am going to tell you is the short story of a material, anything but soft at first glance, which has deeply marked the story of mankind and, moreover, contributed to the flowering of an empire of which my fellow countrymen and women are the progeny. Let us travel back to the distant past. We have said that friction is a blessing when building anything. Try to imagine how our ancestors could have erected their gigantic monuments if all they had were stones that slid over one another as if there were oil or an air cushion between them. Actually, what permitted them to build the great pyramids or the Mesopotamian *ziqqurat* was their skill in wedging stones carved in masterly fashion or in making bricks of the right shape and solidity. But humans, who are (or at least were) not totally dumb, soon understood that it was worth giving friction a helping hand, by laying first some simple clay between the stones or bricks, and then mortars made of gypsum and lime, skillfully mixed. These "binders" were rather weak, and easily dissolved in water. Therefore, if something dating back to those old times is still standing, we mostly owe it to friction.

Until the Romans came, and with them a magic substance called *puteolana* made of volcanic pumice and ashes, which was abundant in the region around Pozzuoli, close to Naples. Indeed, it did not take them too long to discover that puteolana, mixed with lime, yielded a material that would set and hold firm even in water. Indeed, it actually *needed* water for hardening. This was the birth of the *opus caementicium*, a building technique that revolutionized architecture and made possible masterpieces such as the Pantheon or the great aqueducts. These vast structures used as binder mortar the ancestor of what we now call cement, or more precisely *hydraulic cement*[32].

Then came the Middle Ages, during which the Romans' magic formula was completely forgotten, along with many other things, and Europeans reverted to using techniques that might have looked prehistoric to the Egyptians. A moderate awakening took place only with Humanism, thanks to the rediscovery of Vitruvio's *De Architectura*, and thus hydraulic limes slowly began to

---

[32] Actually, it seems that even before the Roman times there were attempts to use "hydraulic limes", mostly made by the Greeks who (fancy that) also used volcanic ashes dug out of the wonderful island of Santorini. Yet, even if the Greeks may claim priority in many fundamental advances across all fields of knowledge (later duly copied by my old fellow countrymen), here they did not rival the Romans at all...

appear in buildings. Yet it was only at the beginning of the XIX century that, thanks to the development of high-temperature ovens, the advance took place that led an English kiln owner, Joseph Aspidin, to invent *Portland cement*, which is at the roots of most modern concrete binders. A full book would not be long enough to describe how modern concrete is made, or its properties and many different formulations and uses. Yet, bizarrely enough, even though more and more sophisticated concretes had already invaded the planet and gigantic skyscrapers already soared over many cities, until the end of the twentieth century no one had any sound way to explain *why* cement works so well. And this is because not only the chemistry, but also the physics of this wonderful material is extremely complicated.

What on earth have cement and concrete to do with colloids? Well, water (at least at the beginning) is there; it *must* be there. And then there are particles. Besides the large sand and gravel grains that make up the mixed matrix of concrete, the cement keeping it together is mostly made up of particles with a size between tens and hundreds of microns (a micron being a mere thousandth of a millimeter, remember), mainly (but not exclusively) composed of calcium silicate, a mineral containing silica, calcium, and oxygen. At the beginning cement is therefore a highly concentrated suspension, containing more than 40% particles. But the high adhesive strength of cement is due to smaller particles still. These are formed in a second stage of the setting process, as some of the original calcium silicates are dissolved and then crystallize back out of the liquid. These new particles are "platelets" with a thickness of about 5 nanometers and a diameter of a few tens of nanometers, made of hydrated calcium silicate hydrate (CSH for short), that, besides sticking to the original grains, rapidly form a gel similar to those we already met, but very much sturdier.

Why? It certainly cannot be van der Waals forces that keep them together so firmly; they would never manage it. Only electric forces can fill this role. Yet, since both calcium silicate grains and CSH particles are negatively charged, how can they attract each other? The point is that our ion clouds this time are mostly calcium ions $Ca^{++}$, which, as the double plus sign might suggest, are *di*valent – doubly charged. As I already mentioned, the reason why divalent ions can induce attractive forces between like-charged particles has only started to become clear in the past few years. To fully understand it requires very sophisticated concepts such as that of "ion–ion correlations", but I shall try to give you at least a hint of a reason. Just picture a $Ca^{++}$ ion as if it had two "little arms", each one carrying a positive charge. These ions are obviously attracted by the CSH particles, having an opposite sign, but it may happen that these two limbs are attracted by two *different* particles. So, the ion becomes a kind of "little bridge" that holds two particles together at a very short distance. This is a very crude model, dangerously and superficially appealing to intuition (which is almost always misleading, in these tricky problems), but not totally alien to reality. On much firmer ground, in the past few years many proofs have accumulated in favor of a cement theory based on colloidal forces, and not on real chemical bonds, as many formerly believed.

In fact, colloidal systems where forces of this kind are in action existed well before the invention of cement. These are the clays present in aquifers that were used as binders by our ancestors. Even more, in this case platelet particles (of a different kind) are already present, and do not require a precipitation process as in cement. The main differences are, first, that the particle charge is much lower and, second, that a lot of singly charged ions are there too. As a result, particle adhesion is much weaker and, moreover, water is a nuisance, since it just increases the normal electrostatic repulsion between the particles.

## 2.9 Particles spreading waves: colloidal light and colors

If you ever happened to stroll (not to drive, I hope) on a winter day through lowlands such as those surrounding my native town of Milano, you are surely familiar with the topic with which I want to start discussing the rich relation between colloids and light: fog. For on foggy days, when there is actually no cloud in the sky, nothing can be seen. I do not mean just houses, trees, or (worse) cars, but sometimes even your feet. Where does this weird (and dangerous) atmospheric effect stem from?

Let us start with saying that fog is a particle dispersion too, and to be precise a dispersion of water droplets in a "solvent" or medium which is not a liquid, but air, so it is an example of those suspensions called *aerosols* we shall deal with later. Actually, all fog does is to raise to a fever pitch what all colloids do, namely to scatter light. When illuminated, indeed, all kinds of particles pick up a little light and, instead of letting it pass straight on, scatter it in all directions. But they do not do this "democratically". Instead, they prefer scattering blue rather than red light. To be more precise, every kind of "electromagnetic radiation", which includes also forms of "light" that we cannot see, such as infrared, ultraviolet, X-rays, and gamma-rays[33], has a well-defined *wavelength*, usually indicated by the Greek letter $\lambda$ (lambda).

Just picture a wavelength as the distance between two neighboring crests of the sea waves you see from the beach. Here, something more complicated than water is forming wave patterns, or "oscillating", but the idea is the same, although for visible light these waves are very short, because $\lambda$ ranges between 0.4 $\mu$m for the blue–violet, to 0.7 $\mu$m for deep red. We can therefore re-express what we have just said by stating that particles diffuse shorter wavelengths better than longer wavelengths. More precisely, a wavelength twice as long is scattered sixteen times (i.e., $2^4$) less. To tell the whole story, it is not only

---

[33] Concerning the latter, we had better avoid staring at them for, even if they cannot be seen, we can certainly *feel* them. In any case, if there are a lot of gamma-rays around, you are probably witnessing an atomic explosion, and this could be the last of your problems. In all senses.

colloids that scatter light. Gas molecules scatter too, although to a much lesser extent. Indeed, it is atmospheric scattering that renders the sky blue and the sun red at sunset (this is the color that manages to reach our eyes, just because blue is scattered more). A crucial point is that, at least when the particles are small compared with the wavelength $\lambda$, the amount of light they scatter grows terrifically with their size: if we double its radius, a spherical particle scatters 64-fold more! When the particles becomes much larger than the wavelength, the growth with size slows down. Moreover, the particles tend to scatter preferentially in the forward direction and, to a lesser extend, backward, namely, in both cases along the direction of propagation of the light falling on them. A physical property on which scattering also depends, one you may be less familiar with, is the *index of refraction* of both particles and solvent. Without going into details, let me just say that this quantity is related to how fast the light moves in a material (the higher the refractive index, the lower the light speed), and also to how much the light is bent from its course (or better, *refracted*) when it encounters that material (the wonderful light effects in diamonds are just due to their high refractive index). For our purposes, it is enough to know that the closer the refractive indices of particles and solvent, the weaker is the light scattering. In particular, if a particle has the same refractive index as the solvent (a condition known as *index matching*), its scattering totally *vanishes*, and it becomes fully invisible[34].

Let us see some of the effects of these observations on light scattering by colloidal particles. Unless it contains very special particles (we'll come to these in a minute), a suspension of very small particles, for instance with a size of a few nanometers, looks pale blue for the same reason the sky is blue, and it is the more bluish the more it is concentrated. Yet, if we increase the particle size, all wavelengths get scattered more and more, and the suspension becomes whitish, like fog or milk. On the subject of milk, in the introduction I suggested you try a simple experiment. In Fig. 2.5B you can see the outcome I obtained with no special effort in my kitchen. It should not be too hard for you to reproduce it, provided that you use:

1. *Well-skimmed* milk, so to avoid the presence of large fat drops that, even at high dilution, will yield a whitish shade (we shall see later what the smaller "droplets" are that remain after milk skimming).
2. A flashlight producing a *really white* light, for instance made by an LED. In my case, it was my daughter's pig-shaped keyring flashlight, whose light-emitting nostrils are visible as a reflection in panel 2A (which shows a comparison with plain tap water, just to assure you I did not cheat). If you use a regular lamp, emitting a yellowish light, you can forget about the blue. . .

The weak blue halo in Fig. 2.5D shows that, as we suspected, a small amount of particles with non-negligible size is also dispersed in a solution of common

---

[34] This is basically the trick on which is based the famous novel *The Invisible Man* by H. G. Wells, which has been the subject of more than one movie.

table salt (which should not display any difference with pure water). This does not mean that additives such as silica particles are necessarily present (I checked on the packet that additives to avoid clotting are indeed present, but they are not colloidal particles), for these could simply be left over from the refining process. Anyway, something big is there, and we do not need any chemical analysis to discover it.

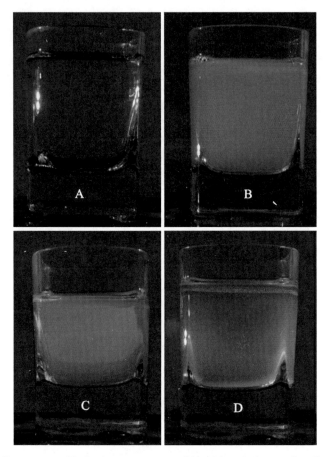

**Fig. 2.5.** Scattering of light from tap water (A), diluted skimmed milk suspension of colloidal particles with a size of 180 nm (C), and a solution of table salt (D). The light source is to the left.

We mentioned that large particles such as fog droplets, which have a size ranging between a few microns and some tens of microns, mostly tend to scatter forward, but also *backward*. This is the reason why using headlights on a foggy day is definitely self-defeating[35]. A related annoying effect takes places when the headlights of a car coming from the opposite direction fall on your windscreen covered, if the wipers are not working properly, by raindrops. There are not many drops, and they are much larger than fog droplets, but they scatter practically all the light forward (towards your eyes), appearing as bothersome shining little stars.

All in all, if colloids were just white or bluish, they would be rather boring chromatically. Actually, however, certain colloids are a riot of color, for colloids are a large fraction of the colors on an artist's palette – the *pigments*. Most of the colors appearing on the frescoes and canvases of the late Middle Ages and Renaissance, from ochre and verdigris to the wonderful ultramarine blue, were obtained by skillfully grinding soils ("earths") and then dispersing them in a "binder", generally egg yolk for frescoes and temperas, then linseed or poppy-seed oil with the revolution started by van Eyck in the XV century, which led to the predominance of oil painting. Hence, pigments are colloidal suspensions in every respect. Here, however, the colors do not arise from scattering, but rather because the particles chiefly *absorb* some specific wavelengths, reflecting just those we actually see. This is a very different way of producing color. Scattered light is only diverted from its original path, but in absorption, which for pigments usually comes from ions of metals such as copper, lead, or cobalt, the light loses energy that is turned into thermal motion of the molecules (so, on dog days, people prefer dressing in light colors, not in black). Anyway, even without calling in at a decorator's shop, we still bump many times a day into colloids that absorb light. Coffee and tea, for instance, are colloidal dispersions too. Just take a look at the ring that forms at the bottom of a cup of tea (green tea, for preference), caused by particles that clump together and settle out, or at the curious appearance of a coffee stain on clothing where, because of the surface tension we shall talk about later, particles gather together at the rim of the stain[36].

Coming back to pigments, you may have got the impression that their color is just a result of the chemicals they are made of. If so, you are on the wrong track. The colloidal state of the pigments has important effects on their color, for at least two reasons. First of all, particle size *matters*, as the Tuscan painter Cennino Cennini guessed, back in the early XV century. On pigment grinding, Cennini writes in his celebrated masterpiece *Libro dell'Arte*:

---

[35] Fog lamps, low-mounted because fog is generally weaker close to the ground, emitting a yellow light without any blue component, may help (you should now understand why), but only partially.

[36] By the way, both these drinks show interesting effects related to suspension stability. For instance, coffee with milk (or, if you prefer it, "espresso macchiato") generally contains smaller particles, because some components of milk reduce aggregation, whereas lemon reduces the formation of scum on cooling tea, besides changing its optical properties.

"E tanto le macina, quanto hai sofferenza di poter macinare, ché mai non possono essere troppo; ché quanto più le macini, più perfetta tinta vienne" [*And grind them as much as you can stand. For it is never too much. For the more you grind it, the more a perfect hue comes out*]

So much that, for vermilion:

"se il macinassi ogni dì persino a venti anni, sempre sarebbe migliore e più perfetto" [*Had I gone on grinding every day for twenty years, it would still improve and become flawless*]

Grinding brightens pigments mainly because, when we reduce the size of the dispersed particles, we decrease the scattering at all other wavelengths, which blurs the natural color. When a pigment is heavily ground, as we can do with modern techniques, our beloved "colloidal blue" peeps in once again. This happens for some hues such as "Mussini Transparent White", which actually shows a light blue shade. But the appearance of a pigment also depends on the refractive index of the binder (i.e., of the solvent) used for dispersing it. For instance, when distempers for painting on canvas were progressively replaced by oil, which has a much higher refractive index than water, painters realized that some colors lose most of their gorgeousness. So the magnificent and expensive ultramarine blue becomes sadly dark, vermilion does not coat any more, whitewash looks almost transparent. When the goal is to obtain particularly brilliant and reflective white surfaces, the particle refractive index takes on a vital role. For instance, using pigments made of titanium oxide, which has a refractive index only 30% larger than zinc oxide, the fraction of light reflected by a painted surface grows from 20% to more than 80%.

Actually, scattering and absorption by dispersed particles are closely related. A suspension looks very different from a chunk of the same material, even when the latter is dipped into the same solvent. This difference may become sensational when the particles are *very* small (say, with a diameter of a few nanometers) and made of a pure metal, such as silver or gold. Obviously, particles of this size cannot be obtained by grinding (grinding a metal is not easy in any case). However, these particles can be made to grow *directly* from a solution of metallic salts. A pioneer of this technique was Michael Faraday, probably the greatest experimental physicist ever. When he wasn't providing the basis for a coherent understanding of electric and magnetic phenomena – allowing James Clerk Maxwell to explain what light really is – Faraday dabbled in his spare time with the synthesis of gold particles by precipitation of a gold salt. Faraday was totally baffled by observing how the color of the suspension progressively changed from light yellow to orange and to ruby red as the chemical reaction went on. Stranger still, if the particles were aggregated by the addition of salt (Faraday did not know why, but he made a clever guess), the suspension could even turn blue.

Today we know that these effects arise because not only the purity and brilliance, but the *color* itself of the absorbed light depends on the size of the particles, when they are very small compared with the wavelength[37]. For instance, while growing, the color of silver particles shifts from yellow, to brown, to red, until they become grey and then completely black. Obviously, as in any scattering effect, the make-up of the solvent has a role too. Using some tricks of the trade, one can obtain mauve, purple, or violet colloidal gold, practically any shade between red and blue. Changing the optical properties of a suspension by controlling the particle size or solvent has its practical uses. Even more interesting, in the field of the so-called "nano-biotechnologies", is the chance of deducing the particle size from the color of the suspension, as a technique, for example, to follow the progress of a biochemical reaction.

Optical properties apart, typical colloidal aspects of pigments are their stability and the way they disperse. Quite often, a pigment is made of particles that do not like water, or conversely oil, a problem making it hard to disperse them in these solvents without them aggregating. To do this, additives must be introduced to alter the particle surface and their interaction with the solvent. In ancient temperas, the egg yolk, besides being a very good binder with the painting surface, contained substances (similar, as we shall see, to soaps) that had exactly this protective action. Today, the formulation of industrial pigments is crucially related to the physical chemistry of colloids and surfaces.

I chose painting as a starting point to emphasize the relation between pigments and colloids, but what we said clearly applies to all the varnishes and enamels that the industrial world spreads on thousands of different surfaces, from building walls to car bodies. Just think of how many colloidal stability problems crop up in the production of a water-repellent paint, which must contain particles that "hate" water, but that, at the same time, must be dispersed in water-based solvents to be spread on a surface. And then there are most inks, which, since the first recipes based on soot particles, or made by precipitating iron salt with the tannin contained in oak galls, have been colloidal dispersions stabilized by the addition of gum arabic, a polymer that sticks to the particle surface.

Quite often, in particular in printing, it is important to spray an ink with high speed and accuracy though a nozzle, without the latter getting clogged. We shall later see that the presence of particles or other "big" stuff such as polymers or surfactant aggregates alters the way a liquid flows in quite remarkable ways (these effects are actually essential for paints to work properly). Here I will just mention a problem related to flow through a nozzle that has a marked practical importance. Suppose you fill a syringe with a colloidal suspension, but insert between the syringe body and the needle a filter containing a membrane with holes that are smaller than the dispersed particles (biologists use these a lot). If you now press on the plunger, you may expect the solvent to come out, while the particles remain inside, but this is not

---

[37] This is because absorption is due to the generation of a kind of collective wave effect in the electrons within the metal the particles are made of, whose frequency depends on the particle size.

the case. Unless the colloid is very dilute, the filter gets clogged in moments, and nothing more comes out the syringe. Now, this is not *too* strange. The particles collecting close to the membrane form a dense layer that blocks the filter[38]. What is very surprising is that quite often the same thing happens, unless you press the plunger very gently, even if you use a membrane that has pores ten or twenty times *larger* than the particles you need filtering out.[39] Why and how particles, crowding to get through the membrane, hinder each other until they clog the pores (like a panic-stricken crowd trying to escape from a burning building) is still not completely clear, but you will appreciate that devising solutions that limit this effect is of great practical importance.

## 2.10 A very particular particulate ink

Pigments and inks are and will always be one of the most important technological uses of colloidal systems, but, in the Internet Age, information is more and more frequently spread electronically, and acquired by means of TV, computer screens, palmtops, and mobile phones. Just to put a figure on it, all around the world more than a square meter of liquid crystal displays (LCD) is produced every *second*. Will paper disappear? Probably not completely, for I believe and hope that none of us will give up leafing through the pages of a beloved book, I might almost say smelling them. But in many other cases using paper is really a waste. It does not make sense, for instance, that so much newspaper is sent every day for pulping, after a brief and partial fruition.

However, the reasons we are still so fond of newspapers are many. They can be read in any illumination, from full sunlight to dim candlelight; they are made of very thin paper, so much that they can be folded and pocketed, and above all no plug-in or battery is needed! Unluckily, from all these points of view, television or computer screens perform very poorly. Firstly, they require an *internal light source*, both when they are "back-lighted" as in liquid crystal display (LCD) screens, and when the elements that make up the image themselves emit light, as in the "organic LED" (OLED) screens which are coming onto the market (more efficient, and promising, but still a bit too dear). Keeping these light sources on obviously costs energy (and thus eats up batteries) and, moreover, the screen is generally not bright enough to be read in full sunlight. Things will be a bit better with OLEDs, but not too much. What probably will still improve is the maximum *acceptation angle* for reading, which for LCD screens is rather limited (but not for paper).

---

[38] This is the main reason why, in inverse osmosis, pressure is applied *tangentially* to the membrane.

[39] A practical consequence is that generally syringe filters are "disposable", meaning that you get a little filtrate, and then you have to throw them away these rather expensive devices.

How nice it would be if instead we could make use of what Nature gives us for free, using a screen based on *reflection* of sunlight, as newspapers do all the time! A screen of this kind would do without internal light sources, and would be all the brighter in a better-lit environment. This really would be an ingenious application of solar energy. Moreover, if the screen uses energy only when it has to be "refreshed", namely for *changing* type and not for *keeping* it, and were also thin and flexible, we could almost hail it as a miracle. Till a few years ago, all this seemed to be just a dream of science-fiction writers, but things are changing fast and, once again, it's because of colloids.

To tell the truth, the first attempts to make electronic paper, or more accurately an electronic ink (*e-ink*) date back to the end of the 1970s. Yet, before a suitable and cost-effective material was obtained, we had to wait until the end of the millennium, when a group at the Massachusetts Institute of Technology, after duly creating a spin-off company ("E-ink"), started marketing the first e-paper displays. How do they work? As in any screen, they are made of pixels, each one containing a small capsule about a hundred microns across, filled with a colloidal suspension in oil of two kinds of particles. The first, made of titanium dioxide ($TiO_2$) and a few microns in size, scatter light strongly because of their refractive index. The others are made of the same material, but dyed with a black pigment. The gimmick devised by MIT researchers consisted in using a pigment that, by sticking to the particle surface, gives them a positive charge, whereas the original $TiO_2$ has an *opposite* sign. Thus, by applying a voltage between the upper (towards the reader) and lower side of the pixel using two transparent electrodes (electrical conducting plates), the white $TiO_2$ particles move to the top or to the bottom, depending on where the positive electrode is, whereas the black ones go the other way (see Fig. 2.6). What is very important is that the oil surrounding the particles does not conduct current. Thus, once a white or black pixel has been created, it stays that way with no need to supply further energy[40].

By dissolving a black dye directly in the solvent, the same trick was later possible using only non-modified $TiO_2$ particles. Using this method, it has been possible, from the first years of this century, to make high-contrast screens with a high resolution (thousands of pixels per square centimeter) and a thickness of a few tenths of a millimeter, very flexible (although not as bendy as paper yet) and, above all, with very low energy consumption. These so-called *electrophoretic displays*[41] lie at the roots of those electronic book readers which are starting to be in high demand. Countries such as the Netherlands are already experimenting with these as substitutes for standard textbooks, something that will soon substantially lighten our children's schoolbags.

---

[40] In physics, a system showing two "equilibrium states" of this kind is called *bistable*.

[41] Electrophoresis is particle transport driven by an electric field. We shall come back to it when we deal with proteins.

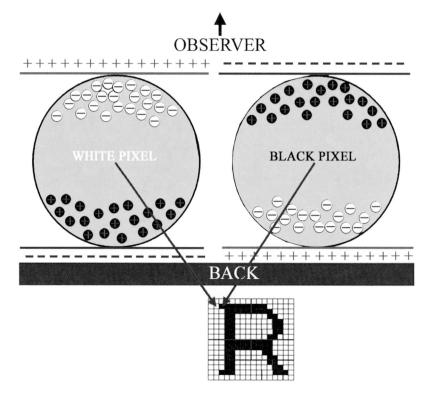

**Fig. 2.6.** Layout of an e-paper display.

Electronic displays still show some drawbacks when compared with LCD screens. In particular, since moving large colloidal particles is not as easy as aligning tiny molecular liquid crystals, they are quite slow. Refreshing a full page may take as much as a second. Moreover, black-and-white displays clearly do not meet the demands of the general public, by now accustomed to a profusion of colors that are easy to obtain on paper, but not so easy on a screen. To comply with this requirement, each single pixel in an LCD screen (or sensor in a digital camera ) is actually made up of (at least) three elements, respectively fronted by red, blue, and green filters[42], which can be turned on or off independently (each pixel on an OLED screen instead uses three different emitters with these colors). The same layout can be used for "electrophoretic" screens, but this means that the amount of sunlight reaching each colored pixel

---

[42] Any other hue is obtained by "summing" these three base colors; red + blue gives magenta (a kind of fuchsia), blue + green gives cyan (a greenish-blue), and red + green, perhaps surprisingly, gives us yellow. Note that this method, which is called *additive synthesis*, is very different from that exploited with printing inks, where the apparent color is all that is left after "subtracting" what the ink absorbs (*subtractive* synthesis), and where primary colors are yellow, magenta, and cyan.

is reduced to one-third of the original, and that the brilliance, the main feature making e-paper so attractive, is noticeably penalized. Luckily, new methods to make electronic inks are currently being studied. These are no longer based on colloid transport, but rather on electrically altering the "wettability" of a surface, a concept we shall discuss in Chapter 4. Although still in its infancy, this technology promises both to render e-paper screens much faster, and to yield brilliant colors through a very different strategy, more like that used for generating them on paper with pigments.

## 2.11 Flying colloids: deceptive beauty of the aerosols

We close this chapter with some words on particle dispersions in a gas, in other words on aerosols. This is worthwhile, both because they have many technological uses, and since they have a lot to do with our health. We have already described fog as an extreme situation of airborne water droplets (clouds, obviously, are similar), but the air we breathe is always *full* of particles in any case. Just look at a sunray pouring through a loophole into a dark room. Those restless sparkling dots you see are nothing but scattering from suspended particles. Carefully note, however, that their frantic agitation is not due to Brownian motion (though it is there), but to unceasing air microcurrents dragging the particles. It is only thanks to these currents that the particles, which are *far* denser than air (and are not small at all), remain suspended and do not settle on the floor. To give some idea of this, the air in large urban areas typically contains tens to hundreds of micrograms of aerosol particles per cubic meter, with a size from a few tenths of micron up, which corresponds to tens of millions of suspended particles for each cubic meter of air we breathe.

There are obviously natural causes for the presence of aerosols in the atmosphere. Besides the noticeable uplift of sand from the desert and of water droplets from oceans due to wind and wave motion, the main ones are volcanic eruptions and forest fires (though these, in most cases, are not natural). Particles may be given off directly, or form by the "condensation" of gases into liquid droplets. For instance, volcanic eruptions produce a large amount of sulfur dioxide ($SO_2$), which is ejected to stratospheric height. In the stratosphere, the part of the atmosphere at a height between 10 and 50 km, sulfur dioxide condenses as droplets of sulfuric acid, which is the main source of cloud formation at these altitudes. The great eruptions, such as that of Mount Pinatubo in the Philippines in 1991, may actually increase the concentration of $SO_2$ in the stratosphere more than a *hundredfold*, with noticeable effects on the whole planet's climate that can last for a very long time. Worse still, in polar regions, the surface of the droplets in stratospheric clouds is the site where chemical reactions occur that lead to the reduction of the ozone layer (the notorious "ozone hole", caused by compounds such as freon used until the 1990s in refrigeration and many other industrial applications[43]).

In contrast, particles formed "at source", such as powders, dust, water droplets, combustion products, and also "bio-aerosols" like bacteria, viruses, and fragments of organic matter, are the main components of the aerosols in the troposphere, the part of the lower atmosphere which extends up to a height of about 10 kilometers. On the whole, in normal conditions, the natural production of aerosols is a few *billion* tons a year, whereas those due to human activity are typically ten times less. But while the planet is used to coexisting with the natural aerosols, the manmade sort are a bit more troublesome. This stems both from their type and the fact that they are concentrated around urban areas. There are notorious events such as the "killer smog" of December 1952 that, owing to the wicked combination of "thermal inversion", which generates fog, and the use of coal as major fuel in heating plants, caused more than 4000 deaths in a few days in London and surroundings. Yet anyone who lives (or at least survives) in a polluted city like Milano, as I do, does not need extreme examples of this kind to realize that particulate emission is one of the most pernicious side effects of the industrial society. The main components of aerosols deriving either from natural sources or human activity are summed up in Table 2.1. The last column is particularly meaningful.

|  | Oceans | Soil | Volcanoes | Fires | Transport | Industry |
|---|---|---|---|---|---|---|
| Carbon |  |  | X | X | X | X |
| Silica |  | X | X |  | X | X |
| Sodium | X | X | X | X |  | X |
| Chlorine | X | X | X | X |  | X |
| Calcium | X | X | X | X |  | X |
| Magnesium | X | X | X | X |  | X |
| Potassium | X | X | X | X |  | X |
| Aluminum |  | X | X |  | X | X |
| Iron |  | X | X | X | X | X |
| Sulfates | X | X | X | X | X | X |
| Nitrates |  |  |  | X |  | X |

**Table 2.1.** Main components of aerosols, either natural or from human activity.

In urban areas, aerosols can be roughly divided into three groups, differing in both origins and particle size. First, there are the so-called "nuclei", coming

---

[43] By themselves, these compounds are chemically inactive, but this is actually bad news. Since they do not react with anything, they rise without problems up to the stratosphere, where they are chemically split by "hard" ultraviolet rays (those that do not reach the ground), apparently triggering a devastating "chain reaction" that kills off the protective ozone layer (without which, I would not be here writing, nor you reading). Not all the experts agree on this mechanism, but the ban on freon and similar chemicals does not hurt, at least.

directly from combustion processes (industrial processes and domestic heating, but today mostly from vehicle emissions) or generated by gas condensation in a similar way to what we have seen for volcanic emissions. These particles have a very small size (lower than a tenth of a micron, 0.1 $\mu$m) and reach very high concentrations close to highways, for instance. Just because they are so concentrated, however, the nuclei have a short lifespan, and tend to combine rapidly into larger particles that make up the so-called "accumulated fraction" of urban aerosols, made of particles with a size ranging between 0.1 and $2 - 3$ $\mu$m. The latter grow much more slowly and are small enough to remain suspended for a long time, so they make up the main cause of poor visibility in big cities. Just to give an idea, a few thousand particles per cubic centimeter cm$^3$ with a size of half a micron (0.5 $\mu$m) will reduce visibility to less than a mile, whereas, owing to the strong dependence of scattering on size, a hundredfold larger concentration of 0.1 $\mu$m particles has negligible effects. Coarse-grain particles with a diameter larger than a few microns (besides natural sources, these could be from construction or industrial activity or, in rural ares, from agriculture) settle much faster. An approximate distribution of the single components of urban aerosols is shown in Fig. 2.7.

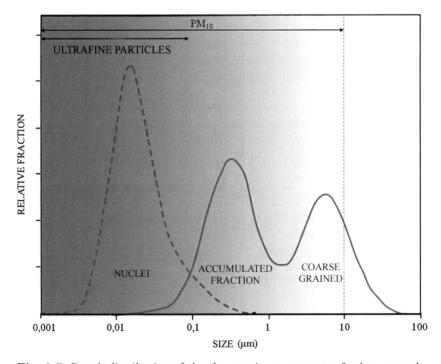

**Fig. 2.7.** Rough distribution of the three main components of urban aerosols.

For what concerns our health, the size of aerosol particles matters. It determines their power to reach different regions of the respiratory tract, and consequently their potential harmfulness. While larger particles, those more than about 5 $\mu$m across, settle into oral or nasal cavities, where they are rather easily removed, those with a size between 5 and about 1 $\mu$m penetrate deeper and deeper in the bronchi and, even though the mucus wetting the bronchial walls aids in getting rid of them, they can cause breathing troubles, asthma, and, in the long term, chronic bronchitis. Particles with a size smaller than about one micron are probably the most injurious, since they reach the lung alveoli, those tiny bags where oxygen is exchanged with the circulatory system, which have an overall internal surface comparable to a *tennis court.* There they remain trapped, because there is no "washing" mucus like in the upper bronchial tract[44]. So far, we have not looked in detail at the processes leading to particle trapping (which are similar to colloid filtering), nor do we have a clear view of the ways in which they induce disease or trigger immuno-reactions. However, surveys by the World Health Organization suggest that these fine particles may be responsible for at least one in every 200 deaths in large cities.

But this is not the whole story yet, because *tinier* particles (say, around 100 nm), known as "ultra-fine" particles, can pass from the alveoli *to the blood*, directly entering the body circulation. Note that this is the typical size of the particles, dubbed DPM (diesel particulate matter), emitted by modern diesel motors, which is strongly suspected of having cancerous effects. As I have said, there is still a lot to be understood concerning the effects of fine particles on health. What should be already clear to you, though, is that the simple "$PM_{10}$" denomination, which gathers together all particles with a size smaller than 10 $\mu$m to fix their highest legal concentration, is totally inadequate. The effect of ultra-fine particles (so far partially obscure) could be quite different from those with a diameter of $5 - 10$ $\mu$m[45].

Before leaving the not-so-pleasant topic of fine dusts, I just wish to point out that in some conditions, dusts can give rise to curious and aesthetically fascinating optical effects. Gorgeous sunsets where green, yellow, red, and violet mix in a chromatic fancy often arise from the presence of aerosols, which can generate countless color shades[46]. Actually, chimneys, oil refineries, or Beijing buildings often provide a foreground for these gorgeous scenes. Personally, I am content with the more traditional colors that fabulous and pollution-free places such as Santorini island may offer. And my bronchi thank me for it.

---

[44] As we shall see, this is also, and not by chance, the typical size of viruses.

[45] On the pleasanter side for our health, suitable aerosols can be used, as you surely know, to carry drugs. Even in this case, determining the size of the droplets produced by an atomizer is useful to establish the region of the respiratory tract in which they will probably settle.

[46] Volcanic eruptions that send particles high into the stratosphere are the masters of this art.

The colors of aerosols can be due to the absorption of specific wavelengths by the suspended compounds. For instance, they are mainly nitrogen compounds which intensify "sunset red" or yield yellow and brown nuances. Yet, even in the absence of absorption, colors may simply stem, once again, from light scattering. Let us see how. As we said, for non-absorbing particles the intensity of scattered light soars as we decrease the wavelength $\lambda$, but this is true only if the particles are small compared with the wavelength[47]. For larger particles, things are much more complicated. In particular, when the particle size is about the same as the wavelengths of visible light, say between a few tenths and a few microns, the scattered intensity may even *grow* with wavelength, so red is scattered more than blue.

This can generate very curious phenomena. Although the English idiom "once in a blue moon" originally meant one more full moon than the usual twelve in one year (something happening every two and a half years, approximately), in current language it is used to indicate something very rare[48]. Yet, in very special conditions, one can *literally* see a blue moon (not necessarily full), in particular after volcanic eruptions or great forest fires. For instance, after the gigantic outburst of Krakatoa volcano in Indonesia in 1883, it was apparently possible to see the moon turning blue (and also greenish) for almost two years. This is because these events can inject tons of particles into the atmosphere[49] with a size and a refractive index that happen to reverse the usual scattering trend with $\lambda$. However, just because this effect is so sensitive to size, it also requires the particles to have a similar size. This is the reason it is so hard to see a blue moon, possibly more so than glimpsing the famous "green flash" at sunset.

Without looking for a blue moon, however, there is a wonderful atmospheric effect that has a lot to do with aerosols, and which I think any of you will have appreciated after a rainstorm, looking at waterfalls, or simply observing a water sprinkler. In all these (and in other) conditions, there is a large number of airborne water droplets that are large enough for their optical properties to be explained without resorting to the complicated theory of scattering. It is enough to use the simple laws of light reflection and refraction, which some of you may remember from schooldays. The light rays bumping into a droplet are partly reflected backwards, but a certain amount is refracted, penetrating into the droplet to be partly reflected again from its back surface, so that, after a further refraction, they come out again in the

---

[47] And if their refractive index is not too high. You actually need to take the difference between the particle's refractive index and that of the suspending medium, multiply the result by the particle radius, and compare the answer with the wavelength.

[48] The Italian equivalent means "once in a bishop's death", which is rather unfair to the church hierarchies!

[49] As I am preparing this English edition, we are all realizing what silica particles emitted by a formerly unfamiliar volcano with the almost unpronounceable name of Eyjafjallajokull can do.

backward direction. The curious outcome of this "ping-pong" of reflections and refractions is that, eventually, much of the light is reflected only within a narrow cone forming an angle of $40 - 42°$ with the direction of the original rays. Since water's refractive index depends on wavelength, the exit angle is slightly larger for blue than for green, and for green than for yellow or red.

The outcome, of course, is what we call a *rainbow*. When we stand with the sun behind us and look at the clouds still building up near the skyline after a storm, the overall effect we see is an arc of a circle centered in the opposite direction to the sun (the *anti-solar point*), which is therefore always *below* the horizon and has an angular aperture of $40 - 42°$ on the vault of heaven (a couple of hands' breadth at arm's length). Obviously, therefore, unless you are on a mountain top or on a plane, the rainbow can be observed only if that the sun is less than $42°$ over the horizon, since otherwise it is completely below the skyline (this is why rainbows are usually seen near sunset). The more observant of you may also have noticed that, in some cases, a *double* rainbow forms, with the colors of the additional outer arc *reversed* compared with the usual ones. This secondary arc is due to rays suffering a *double* reflection before escaping the droplets[50]. The rainbow is only the most spectacular of many atmospheric phenomena such as "glories" and "halos" that are due to suspended water droplets or ice crystals, and which I recommend you to look for. However, the moment has come for us to temporarily leave our simple colloidal particles, either solid or liquid, and to devote ourselves to other systems, which are still in some sense "colloidal", but quite different and more complex.

---

[50] If you are really observant, you may also have noticed that, between the two arcs, the sky is considerably darker. This effect, due to the fact that back-reflection is very low for these angles, is called "Alexander's band", after the Greek philosopher from Aphrodisia who first described it.

# 3

# Freedom in chains

*Polymers, Italians in miniature: messy, but very flexible – Rubber soles: a brief history of tennis shoes, from rubberwood to sneakers – Chains for all tastes: from granny's phone to microfluidics, from baby colic to space ships – Stretching and bending without breaking: the surprising mechanical properties of rubber and plastic – Snake dance: polymer coils, nets and molasses – Elastic by chance: entropy tricks and superheroes – Panta rei: even the mountains flow before God – Indiana Jones' nightmares: the rheology of reptiles – Polymers to the charge: the polyelectrolytes.*

No moment in a scientific meeting is as thrilling as the opening session, which the organizers always await with bated breath. Even if they have done their very best to take care of every detail, it always happens that some presentation was prepared using software incompatible with the version installed on the conference PC (obviously, the speaker will just have come with a USB stick, not his own computer). Or, if this does not happen, the projector lamp will burn out within the first three minutes. Or, more simply, some participants have no hotel room, while their luggage is performing a Brownian walk around the planet's airports. Therefore, while opening a meeting I had organized some time ago in the wonderful venue of Varenna, on Lake Como, I tried to forestall any problems, reassuring my colleagues that, when it comes to organization, Italians are a bit like polymers: rather messy, but very, very flexible. In the remainder of this chapter, I shall try to make it clear to you what I meant (my colleagues obviously needed no explanation).

## 3.1 Long and disordered queues

Packagings, bags, and food containers. Synthetic fibers, sunglasses, and pantyhose. Furniture, mattresses, and tubing. Videotapes, records, and DVDs. Glues, sealants, and varnishes. Switches, electric cables, and optical fibers.

R. Piazza, *Soft Matter*, DOI 10.1007/978-94-007-0585-2_3,
© Springer Science+Business Media B.V. 2011

Car tires, bumpers, and dashboard. Keyboard, mouse, and screen of the computer I am using... And so on and so forth. From children's toys to space probes, the world as we see it today could not exist without the essential contribution of what we call plastic materials. To give some idea of this, every year Americans alone (who, to tell the truth, are not too careful about their consumption) "devour" more than thirty billion tons of these materials, which have countless applications, reasonable costs, and practically unlimited duration[1]. Yet, no more than a century ago, all these things did not exist. Our great-grandfathers would have struggled to understand what a large part of the stuff we use in our everyday life is made of.

Even more surprising is that, at least in simple cases, all these materials are just variations on a common theme, namely that of making "molecular chains", called polymers, from a long sequence of chemically identical "links" known as *monomers*. Unlike the particles we discussed in Chpt. 2, polymers are therefore single molecules, but really huge, to the point that they can be made of tens of hundreds of thousands of atoms. For good reason, then, they are called *macromolecules*. Although the properties of interest for chemistry are set by the specific chemical nature of the monomers, most of the physical properties that make polymers so different from simple molecules are very general, and stem from their being so long and flexible. Flexibility and freedom of motion are the keywords of this chapter, since these give polymers the versatility that make them unique and invaluable materials. For once, freedom and chains go hand in hand.

To understand how this remarkable stuff can be made, we have to dwell upon the properties of a wonderful chemical element, carbon, so "creative" that we had to invent a branch of science entirely devoted to its countless compounds: organic chemistry. Even alone, carbon shows its restless imagination. Probably you already know that the charcoal you use to grill sausages, the lead of the pencil you write with, the splendid diamond shining on your (or your wife's) engagement ring are essentially just different forms of pure carbon. As you can see, then, carbon binds quite well even to other carbon atoms. But its imagination can be fully unleashed when it binds to *other* kinds of atoms. This happens with most chemical elements, but with some preference for nitrogen, oxygen, chlorine, fluorine, sulfur, silicon, and, above all, hydrogen. In fact, the number of new forms of organic compound invented every year is greater than the total number of natural inorganic compounds. What's more, it is inconceivable that structures as complex as living beings could have emerged in a carbon-free world[2]. The final part of this book will

---

[1] Unfortunately, exactly this last feature makes me doubt that the world as it is will last forever. Every day, more than sixty million plastic bottles end up in US garbage dumps or incinerators.

[2] Some science-fiction writers have speculated on life forms based on silicon, a chemical relative of carbon, but to our knowledge, this really is just science fiction.

be entirely devoted to the complex organic macromolecules that make life possible, or at least life as we know it.

Each carbon atom usually binds to four other atoms (thus, carbon is said to be *tetravalent*). Some of these four can be carbon atoms too, and this is how the repetitive chains we want to talk about may originate. To see how, consider two simple organic compounds, ethane ($C_2H_6$) and ethylene, also known as ethene ($C_2H_4$). The chemical formulas do not tell us much[3]: both compounds contain two carbon atoms, and the rest is just hydrogen. In other words, they are what in chemistry is called a *hydrocarbon*. Yet, if I point out more explicitly the way atoms are bound to each other

$$\text{ETHANE}: \ \underset{\displaystyle \overset{|}{H}}{\overset{\displaystyle \overset{H}{|}}{H - C}} - \underset{\displaystyle \overset{|}{H}}{\overset{\displaystyle \overset{H}{|}}{C}} - H \qquad \text{ETHYLENE}: \ \underset{\displaystyle \overset{|}{H}}{\overset{\displaystyle \overset{H}{|}}{C}} = \underset{\displaystyle \overset{|}{H}}{\overset{\displaystyle \overset{H}{|}}{C}}$$

you can see an important difference, for ethylene contains a *double* bond between the two carbons. If we then manage to break one of these two bonds, each one of them gets a "free arm" to bind an additional carbon atom. If this occurs, we face a situation like this:

$$-\underset{\displaystyle \overset{|}{H}}{\overset{\displaystyle \overset{H}{|}}{C}} - \underset{\displaystyle \overset{|}{H}}{\overset{\displaystyle \overset{H}{|}}{C}} - \underset{\displaystyle \overset{|}{H}}{\overset{\displaystyle \overset{H}{|}}{C}} - \underset{\displaystyle \overset{|}{H}}{\overset{\displaystyle \overset{H}{|}}{C}} - ,$$

which we could simply write $[-(CH_2)_4-]$, where two ethylene molecules have joined to form a compound with four carbons.

This does not occur spontaneously, because the double bond, although not the favorite condition for the joined atoms, is rather stable. However, using suitable chemical compounds called *catalysts*[4] that allow the reaction to start, we find that the process goes on further, for the end carbons *still* have two free arms. This chain reaction can proceed pretty much for ever, or at least until two hydrogen atoms get bound to the ends of the chain, effectively capping the ends, as in ethane. Hence we obtain a chain molecule $CH_3 - (CH_2 - CH_2)_n - CH_3$ where, apart from the end groups $CH_3$, the number $n$ of ethylene links can be really huge. This conga line is properly called **polyethylene** or polythene, the simplest of polymers; it is used to make,

---

[3] Don't worry if you have forgotten all your high school chemistry; we will start gently!

[4] Often a chemical reaction does not take place spontaneously. Between the starting and the final state, even if the latter has a lower energy (namely, if in terms of energy the reaction as a whole proceeds "downhill"), in between there may be a hillock that the molecules, with their low thermal energy, fail to climb over. Just think of a catalyst as a kind of "bulldozer" boring a tunnel through this hill and linking the start and end points directly.

for instance, common plastic bags, as the acronym PE they are marked with suggests[5].

Being made of a lot of repeated units, a macromolecule like polyethylene has a very large "molecular weight". Roughly speaking, this term means the molecule's total weight measured with respect to a hydrogen atom, the lightest element. Thus, for instance, since an oxygen atom weighs as much as 16 hydrogens, the molecular weight of water, $H_2O$, is $2 \times 1 + 16 = 18$. Because on this scale carbon "weighs" 12 units, the molecular weight of a polyethylene monomer is just 28, but since a polymer can be made of tens of thousands of these monomer "links", it is easy to get chains with a molecular weight ranging into the *millions*.

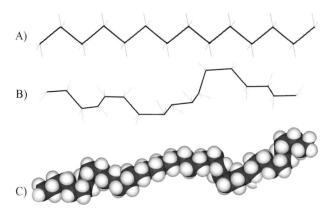

**Fig. 3.1.** Sketch of a (short) polyethylene chain, either perfectly ordered (A) or slightly bent and twisted (chemists would say it has "conformational defects") (B). A more realistic drawing, where carbons are painted in black and hydrogens in light gray, is shown in (C).

More than the polymer's chemical formula, it is helpful to look at the "conformation" of polyethylene, meaning how the monomers organize in space. The fully ordered shape is that of a *zigzag chain*, where the chain is straight as shown in Fig. 3.1 A. However, the macromolecule takes on this structure only when it is in a crystal state (and not perfectly, even then). When the polymer is dissolved in a solution, the case we shall mainly investigate, this zigzag trend is often interrupted by defects (see Fig. 3.1 B), making the chain kinked and twisted. Before dealing with this topic and with some other aspects of polymer behavior, however, I wish to tell you the curious story of a natural macromolecule which has irreversibly changed the world we live in.

---

[5] As a dabbler in polymer chemistry, I am considerably simplifying the matter. Actually, the properties of the outcome (for instance, the chain length and regularity, or whether it is linear or "branched") depend on the way polymerization occurs and on the kind of catalyst we use.

## 3.2 A tale of cross-links and double-crosses

Our fondness for running after a ball is timeless. Therefore, when Europeans landed on the shore of Haiti, during Columbus' second voyage to the New World, they took with them some hempen balls for their leisure time. Yet they were amazed to discover that the natives already had much better balls, made of some unheard of material, which bounced wonderfully. For the first time, Westerners had set eyes on rubber. Both the Mayas and the Aztecs considered rubber a sacred material, to the point that they used these spheres in a ritual game, like a kind of basketball played with the feet, where the punishment for the losers was rather severe, since it generally consisted of beheading[6].

However, rubber was already used for less bloody purposes in Amazonia, where local populations such as the Tsachali extracted it by tapping the bark of *Hevea brasiliensis*, today known as rubber tree, but called by them *cahuchu* or "weeping tree". The Tsachalis used the stuff we call caoutchouc for many purposes, such as making canoes waterproof, or manufacturing objects by heating caoutchouc in wooden molds. It is even said to have been used for homemade shoes, simply made by dipping the feet in it and letting them dry. These were not long-lasting shoes, and probably smelt rather bad too, but they certainly would have fitted well enough to comply with modern orthopaedic prescriptions. Also, it would surely have been much better to walk through the forest in these do-it-yourself shoes than barefoot.

Many different plants and trees, among them *Castilla elastica* (precisely the one used by the Aztecs), gutta-percha, guayule, some kinds of ficus, and even common flowers such as the dandelion, secrete sticky latex, but the rubber tree really produces a lot of very high quality juice. This latex is actually one of those substances we shall call *emulsions*, where a natural polymer, **polyisoprene**, is stably dispersed in water as droplets by means of special biomolecules called phospholipids[7]. Evaporating the water makes the latex clump, or coalesce, in a pretty similar way to what we have seen for colloids. In this case, though, the droplets merge into a single polymer mass behaving as an *elastomer*, in other words a material that is much stretchier and springier than common solids.

Rubber became known in Europe only two centuries later, thanks to the description made by Charles-Marie de Condamine who, after reaching

---

[6] Because the Geneva Convention had not yet been drawn up (and the Aztecs would hardly have adhered to it in any case), the players were usually war prisoners.

[7] For those of you who really cannot do without chemical formulas (which I shall carefully try to avoid in what follows), the monomer of polyisoprene:

$$- C = CH - CH_2 - CH_2 -$$
$$\underset{CH_3}{|}$$

is not much more complex than in polyethylene, apart from having a *double* bond between two carbons, but it's this double bond that is of primary importance to obtain rubber.

Ecuador with the aim of measuring the length of a degree of longitude, was so struck by its properties as to beg the Tsachalis for a waterproof bag made out of it. But rubber was slow to take off, mainly because the latex slowly coagulates even in the absence of evaporation. Thus, the fluid shipped in barrels from the New World constantly landed in Europe as a useless dense paste. The first practical use of rubber is due to the insight of the great British scientist Joseph Priestley[8] who, in 1770, noticed that, when rubbed on paper, it erased his pencil marks. Then came the first truly successful idea for exploiting a key property of rubber, its waterproof nature. The trick lay in dissolving the latex in a suitable solvent, first turpentine (used for instance by the Montgolfier brothers for their balloons) and then naphtha. The latter made the fortune of Charles Macintosh, who in 1823 used it to make the first kind of raincoat, which still bears his name.

A true fever for rubber started at the beginning of the nineteenth century, mostly in America, with the birth of companies trying to make gloves, boots, gaskets, and other technical items from latex. Unfortunately, the outcomes were anything but encouraging. Natural rubber stiffens in cold weather, and it rapidly softens at not very high temperatures. As a result, the first boots made of rubberized canvas used to melt in summer, sticking to the skin, whereas they literally went to pieces during the harsh American winters. Moreover, after a short time and in the absence of anti-bacterial preservers, rubberized items begin to stink, because natural latex, beside coalescing, undergoes a progressive and unpleasant rotting process. Consequently, the rubber mirage ended in rapid and disastrous commercial failures.

Failures and bankruptcy were familiar to Charles Goodyear, over whose head already hung the debts run up by his father, also an inventor, who had attempted to launch pearl buttons on the US market with little luck. Nevertheless, our stubborn Charles, unmoved by the miserable end to which the newborn rubber industry seemed to be heading, strongly believed in the potential of this material. Indeed, he felt, by his own account, "called by God" to give rubber a future. Imprisoned in Philadelphia for debt, he persuaded his wife to bring in raw materials and started, even in jail, his strenuous fight to obtain the "ideal" rubber. It was a fight that, through exciting moments, partial successes, failures and further imprisonments, lasted for more than five years, driving him to misery as his friends refused to finance him further. In 1839, however, one of the most famous accidents in the history of technology took place. While Goodyear was trying to improve rubber by processing it with sulfur, he unintentionally let a sample fall on the stove where his wife was cooking. The sample suddenly became a piece of solid, heat-resistant rubber, much more similar to the stuff we are used to than to caoutchouc. Heat and sulfur together had worked the miracle, causing *vulcanization*, an effect that

---

[8] Primarily known as the discoverer of oxygen, though he had been preceded by a Swedish chemist, Karl Wilhelm Scheele, who was, however, slower to publish his results.

marked a turning point in the history of rubber[9] and basically consists, as we shall further discuss, in inducing cross-linking between the polyisoprene chains to form a sort of mesh. As a further bonus, vulcanized rubber is what is known as a *thermosetting* material – that is, it hardens irreversibly when heated. Hence, once molded into the desired shape, it withstands heat, and intense cold, too, much better than natural rubber. In a short time, vulcanized rubber spread on the market through countless applications. One of the riskier, but with hindsight the most ingenious of them all, was spotted by the Michelin brothers, who ran the Paris–Bordeaux–Paris race after equipping their car with those strange rubber tubes formerly used only for bicycle wheels.

Because the monopoly of latex production was in the hands of Brazil, the amazing boom in rubber seemed to herald rapid economic growth, if not a true "Amazonian renaissance". Thousands of immigrants poured into towns such as Belém and Manaus, which filled with banks and company offices, but also with telephones, artificial lights, restaurants and theaters. Yet the British, by illegally purloining some rubber tree seeds from the Amazon in 1876 and planting them in London's Kew Gardens, managed to double-cross the Brazilians. New, more resistant crossbreeds were shipped and sown in Far East colonies, in particular Malaysia, Singapore, and Ceylon, where they grew luxuriantly. Whereas in the Amazonian forest growers had often to walk for miles to move from one rubber tree to another, in southeast Asia the trees could be planted very close to each other. As a result, Brazil had soon to face an economic crisis of vast proportions.

When World War One broke out, the English cut the Germans out from the rubber market, by then fully in their hands, exactly when they badly needed it (just think of gas masks). But the Germans had no intention of being stabbed in the back as the Brazilians had, and, thanks to their already strong chemical industry, undertook colossal efforts to develop a synthetic alternative to natural rubber. Their most successful result, called *nitric rubber*, made from a monomer similar to isoprene obtained from coal and lime, admittedly lacked the elasticity and durability of natural rubber, but at least allowed Germany to hold out and, above all, to develop new synthesis methods. One of the reasons it was so difficult to synthesize artificial rubber is that no one actually knew what polymers were, or even that they *existed*. It was Hermann Staudinger who first speculated in 1920 about the existence of what he called macromolecules. For this brilliant intuition, he was awarded (obviously, I would say) the Nobel prize in 1953[10].

---

[9] But not in Goodyear's life. He let other people (Macintosh, above all) exploit his unpatented results. Much less able as an investor than as an inventor, he ended up with a debt of $200,000 and a ruined family, nonetheless proud of having fulfilled his mission (enough is as good as a feast).

[10] Even though he was a German, Staudinger, who lived in Switzerland during World War One, joined Einstein in signing a document condemning the belligerent actions of Germany. His fellow countrymen never forgave him for it. Back in his homeland, he was treated in the postwar period as a poor incoherent fool. Given

A crucial step towards synthetic rubber was taken by Julius Nieuwland, a Belgian priest settled in Indiana and a first class chemist too, who, by experimenting in secret during the war (as a scrupulous Christian, Nieuwland did not want his results to be used for war purposes), found for the first time a suitable monomer. Together with Wallace Carothers from DuPont, who later invented nylon, Nieuwland obtained *neoprene*, an elastomer showing properties even superior to natural rubber. In the meantime the Germans, remembering the lesson from the previous war, were making strenuous efforts to reach the same goal. Thus, they first synthesized **polybutadiene**, a very strong elastomer which was particularly useful to make tires but, at that time, rather expensive, and later cheaper, but still good quality polymers, made from two different monomers together (these are what we shall call *copolymers*), butadiene and styrene[11].

At the beginning of 1942, it was the turn of the allied forces to be threatened, for the rapid occupation of Malaysia and the Dutch Indies by the Japanese put more than 95% of the world production of rubber in the hands of the Axis. Because a single Sherman tank contained more than half a ton of rubber (not to speak of ships, planes, and vehicle tires), while the production of synthetic rubber was just 8000 tons a year, no more than 1% of what was needed, the USA was left with just over a year before it would have to surrender for exhaustion of war material. The American reaction was the most extraordinary recycling collection in history. Countless rubber items (whose retail sale was immediately forbidden, except for absolutely essential uses) were given by US citizens to thousands of depositories spread around the nation, which ensured troops an adequate supply until the end of 1942. Moreover, the great effort made to develop new synthetic elastomers shot US scientists to the top of polymer chemistry, and also led to some great basic advances. A large part of what I shall tell you about the properties of polymer solutions is due to work by Paul Flory, Nobel laureate for chemistry in 1964, who, first in DuPont and then in Cornell University[12], laid the foundations for the modern science of macromolecules.

The years after the Second World War witnessed a true flowering of polymer chemistry, mostly made possible by the invention of new synthesis methods. The latter have enormously broadened our ability to make macro-

---

the developments made in elastomers in Nazi times, despite German chemists not understanding what they were, we may wonder what might have happened had Staudinger had the praise he deserved. So much the better. . .

[11] Again for chemistry fans, the monomer of polybutadiene:

$$-CH = CH - CH_2 - CH_2 -$$

is very similar to that of natural rubber, and contains a double bond as rubber does.

[12] Where he worked under the direction of Peter Debye, a great Dutch scientist, Nobel laureate too in 1936.

molecules with properties that could never have been foreseen in the pre-war period. Many synthetic elastomers too have progressively replaced natural rubber in more and more advanced applications. But in the modern war zone of the global market, the "mother of all battles" is fought on the ground of the consumer market. For instance, since my children and their friends refuse to wear shoes other than trainers (by whatever name), the big producers are battling to impose on the market new or revamped models of these post-modern versions of the Tsachali rubber shoes. Judging from their price, or at least the ratio of price to production costs, I wonder whether, in this long-lasting fight about rubber, it is not our turn to be stitched up.

## 3.3 Necklaces for all tastes

The need for synthetic rubber as good as or even better than caoutchouc prompted a real hunt for macromolecules that, besides elastomers, has also begotten *plastic* materials of even greater interest. Starting from the beginning of the twentieth century, when **bakelite**, the first synthetic polymer, was invented and used for most of the "plastic" stuff made for several decades, the number of synthesized polymers has rocketed. Indeed, given the wide range of properties of these compounds, polymer science has become an independent subject, to the point that today entire books are devoted to a *single* polymer. Since our aim is just to grasp some general physical aspects of macromolecules (and in any case, I would feel like a mere dabbler thrown in at the deep end) I shall not dwell on either the chemical properties of the different polymer classes, or their structure. Thus, I shall not fill your head with mysterious chemical formulas of no use for our purposes. Nonetheless, I wish at least to present you with a potpourri of those polymers used to make many everyday objects, hoping to give you a glimpse of their uses.

Let us start from plastic bottles, which have unfortunately become an integral part of many charming landscapes. They are almost always made of PET, which is an acronym for the terrifying name of **polyethylene terephthalate** (clearly some sort of relative of polyethylene). PET, like PE, has the advantage of being a *thermoplastic* polymer; namely, unlike bakelite and vulcanized polyisoprene, it can be melted and molded as a liquid. PET can be easily recycled at low cost, provided that trash separation and collection is done properly: on the environmental side, at least, this is a good point. Another basic use of PET, which actually takes up almost two-thirds of its global production, is easily guessed if I mention that it is more commonly known as **polyester** (although PET is just one kind of polyester among many, for this is a wide polymer class). As such, it is the synthetic fiber most widely used to make garments, but also supports for magnetic tape and, because it is very light, even sails. A PET by-product, **Mylar**, can be metallized by evaporating a thin aluminum layer onto a polymer film. As such, since it is non-toxic and impermeable to gases, it lends itself together with polyethylene to making flexible food containers[13], but it is also used as an electric insulator, and even in space suits.

Back to the kitchen, when I was a kid dishes and tableware were mostly made of **melamine resin**, a polymer that, being thermosetting, could be easily shaped in a mold. The same polymer makes what we call Formica, a common laminated plastic for kitchen furniture. Melamine, however, has the defect of scratching easily, so in kitchenware it has mostly been replaced by **polystyrene** (PS), another thermoplastic polymer with countless industrial uses, from toy making to biomedicine. Moreover, as you surely know, in the expanded form called styrofoam (which can still be regarded as a colloid, but where this time the particles are air bubbles, and the solvent a solid polymer matrix) polystyrene's lightness makes it the most common filling material for packing. In addition, being mostly made of air, which does not conduct much heat, it is a very good thermal insulator. Yet it is not particularly suited (in particular as styrofoam) to use in microwave ovens, and the same goes for most plastic materials. For this purpose, specific thermoplastic polymers such as **polysulfone** (PSU) are required, although the research is shifting towards polymers "reinforced" with glass or carbon fibers.

Two other basic thermoplastic polymers are **polypropylene** (PP) and **polyvinylchloride** (PVC), respectively the second and third most common polymers in terms of global production after polyethylene, which are so widespread that we can give only a hint of their countless applications. Because of its toughness and resistance to solvents, PP is used to make plastic furniture, synthetic carpets, ropes, adhesive tapes, stationery, car components, loudspeakers, high-quality food containers, even plastic banknotes which some countries are already using. PVC is the most widespread polymer for biomedical components such as containers for blood or urine, catheters, inhaler masks, surgery gloves, and cardiac bypasses, and in construction building to make tubing, electric insulators, or window and roll-up shutter frames. Moreover, when properly "loaded" by incorporating other materials, it turns into the most suitable polymer for products such as gloves and boots (where it is known as "imitation leather"), car parts such as bumpers or driving belts, acoustic barriers, insulation, and better (but dearer) adhesive tapes than those made of polypropylene.

Current photo and video cameras provide digital images collected by electronic sensors. Nevertheless, most of you surely remember the old films made of **cellulose acetate** (CA), a by-product of cellulose, which constitutes the basic component of tree pulp, and also of paper[14]. Close relatives of CA,

---

[13] Looking at the laminated plastic with which these containers are made, the external matt side is made of PE, whereas the glossy internal one is in Mylar.

[14] Like polyisoprene, cellulose is an example of a *biopolymer*, meaning one of the many polymers naturally present in living beings that we shall meet in the last chapter. As a matter of fact, cellulose acetate has replaced another formerly used polymer, bearing the well-known name of **celluloid** (cellulose nitrate), which had the great defect of being highly inflammable (some readers may remember

also obtained from cellulose, are **Rayon**, one of the first synthetic fibers, also known as *viscose* or *art silk*, and **cellophane**, which I guess I don't need to comment on further. Under the difficult name of **polymethylmethacrylate** (PMMA) is hidden a polymer surely better known by its commercial names of Plexiglass$^{TM}$ and Perspex$^{TM}$, which shows very good optical properties, for it is even more transparent than glass. In contrast to the latter, it is moreover shatterproof (and lighter), and therefore replaces glass in the manufacture of safety transparent surfaces. However, it scratches easily, and thus is not particularly suited to making lenses. Much better for this purpose is polycarbonate (PC), a transparent polymer showing superior strength and high resistance to temperature, because it softens only above 150–200 °C, or about 300–400 °F (although it cannot withstand boiling water). Besides eyeglass fabrication, PC is used to make DVDs, MP3 players, high-quality laptops, and particularly strong and durable suitcases.

When extreme transparency is needed, such as that required to make optical fibers, glass is better replaced by **silicones** (SI), or polysiloxanes, which are for once not polymers of carbon, but of silicon bound to oxygen (thus they are akin to glass, which, we recall, is basically made of $SiO_2$). I guess you know, however, that silicones have many more uses, mostly as adhesives and sealants, for making soft toys, and as additives in cosmetics. This last application suggests that they are particularly bio-compatible, and in fact they are extensively used in the medical world to make implants (and not just those which many of my readers may be thinking about). A specific silicone, PDMS, is also having a noticeable impact in basic scientific research, because it is an important material for *microfluidics*. This is a new technology allowing researchers to make *labs-on-a-chip*, true miniaturized chemical or biomedical labs taking up the space of a stamp, which are built with the same methods developed for electronic circuits. Curiously, however, the same polymer mixed with silica particles makes simethicone, a drug probably known to young moms, for it is supposed to soothe baby colic (not that it worked with my kids). A final important use of silicones is to make elastomers, with properties which are often superior to ordinary rubber O-rings.

When mechanical strength is more important than elasticity, other outstanding polymers such as **polyamides** (PA) hold sway. Among them, it is useful to mention **nylon**, another synthetic fabric which, arriving with the American liberation troops, delighted so many Italian women (and, probably, some allied soldiers too), but which is also well-known to sailors, for it can be used to make robust sails such as spinnakers. Close relatives are **aramides**, which are used to make motorbike helmets, ropes for sailing and climbing, and high-tech sport garments. Aramides hit their top level with **Kevlar**, so impact-resistant as to be the main component of bullet-proof jackets.

---

the tragic experience of Alfredo, played by Philippe Noiret, in the great movie *Cinema Paradiso* directed by Giuseppe Tornatore).

All the polymers we have considered so far are mostly made of carbon (or silicon) and hydrogen. If most or all of the hydrogen atoms are exchanged for fluorine, which is also monovalent (like hydrogen, it forms a single bond) and strongly binds to carbon, we obtain polymers with very distinctive properties, known as *fluoropolymers*. For example, substituting fluorine for all the hydrogens of polyethylene, we get **polytetrafluoroethylene** (PTFE), which you probably know under the commercial name of Teflon$^{TM}$, and whose chemical formula is $CF_3(CF_2 - CF_2)_nCF_3$. The story of PTFE is a very peculiar one. It was discovered by chance in 1938 by Roy Plunkett, a chemist from Kinetic Chemicals, a subsidiary of DuPont, as a whitish stuff settling in fluorinated gas cylinders used to make new coolants (those *chlorofluorocarbons* which were later banned because of their possible effects on the ozone layer), which apparently defied any attempt to be dissolved. The chemical hallmarks of PTFE are almost incredible, since it resists practically *all* solvents. Thanks to this property, it is an ideal material in all those applications requiring very high chemical resistance, such those related to aerospace technology. Without PTFE, probably we would have neither reached the moon, nor made the Space Shuttle[15]. Today, the applications of PTFE as a corrosion-free material are widespread, ranging from aircraft components to the production of gaskets, special cables, and technical fabrics[16]. Moreover, because it scarcely interacts with other materials, PTFE sticks to nothing. Indeed, its friction with any other stuff is so feeble that it has earned its place in the Guinness Book of Records, for it is equivalent to sliding wet ice on wet ice (because of this, PFTE is a very good *lubricant*). The only way to make PTFE adhere to something is to trap tiny particles of it on the very slightly roughened surface of a material such as aluminum by depositing a film from a colloidal dispersion, the form in which PTFE is made. This method was discovered in 1954 by Marc Grégoire, a French engineer and keen fisherman, who first thought of using it to coat reels and tackles. Yet it was his wife who had the brilliant idea of trying instead to coat aluminum *pans*, to make them non-stick. This stroke of genius has made many cooks happy, and obviously the Grégoires too. Back to rubbers, **fluoroelastomers**, which have a chemical resistance comparable to PTFE, are today crucial for making many components of rockets and spacecraft. An additional gasket made of Viton, one of the major fluorinated elastomers, would probably have prevented the tragedy of the Space Shuttle *Challenger*.

With all these acronyms, I suspect you are already risking a poly-headache. Alas, we are not yet done. When I told you that polymers are long sequences of equal monomers bound to each other, I partially lied. As we have seen when

---

[15] The first technological use of PTFE was actually within the Manhattan Project which led to the first atomic bombs, mainly as a material suitable to contain uranium hexafluoride, a very toxic substance.

[16] The well-known Gore-Tex$^{TM}$ is made by stretching PTFE film to make it porous, a key feature in fully waterproof, but still breathable, textiles.

discussing synthetic rubber, there is no reason why all the monomers should be the same. It is perfectly feasible to synthesize chains, called *copolymers*, made of two or more kinds of monomers that have some "affinity", some attraction to each other. To simplify the matter, consider a copolymer made of just two monomers, A and B, and suppose that A is attracted to B as much as itself, and vice versa. Even in this simple case, there are many ways to make a "mixed" chain. For instance, we may choose to make a well-ordered chain such as

$$A - B - A - B - A - B - A - B - A - B - A - B\ldots,$$

or choose instead to make

$$A - A - A - B - A - A - A - B - A - A - A - B - \ldots$$

which contains less *B* (only one-third as much as A), but is still ordered. This may be too hard to make, and usually this is the case, so B monomers might rather be placed randomly within the chain:

$$A - B - A - A - B - B - A - A - A - B - A - B - \ldots$$

making what is called a random copolymer. There is a last, very special way to order the monomers:

$$A - A - A - A - A - A - B - B - B - B - B - B - \ldots$$

which builds what is dubbed a *block copolymer*, but we shall deal with this later.

Nature is a master at making copolymers, because biological machines work only thanks to special polymers, the proteins, which are made of twenty different monomers placed, as we shall see, in a specific sequence for each different macromolecule. Learning from Nature is never a mistake, and these mixed chains do indeed increase the range of possible uses. For instance, by mixing what are called epoxy resins with a "hardener" made of a polyamine[17], we obtain copolymers known as **epoxy glues** with exceptional strength, so much so that they are the adhesives universally used in car, bicycle, boat, or ski production, and even to attach the wings of a fighter plane to the fuselage. Similarly, copolymers of acrylonitrile with several other monomers yield **acrylic fibers**, used for example to make synthetic sweaters or camping tents. Finally, adding an extra type of monomer to the mix may help in processing plastics. For instance, although PFTE has good thermal stability, it cannot be processed as a melt, that is, in liquid form. This hinders the use of some methods that are common in the production of polymeric materials, such

---

[17] To avoid nauseating you, I confine to this footnote the names of the most common polyamines, whose derivation should be obvious: putrescine, cadaverine, spermine...

as extrusion, where pipes and shapes are molded by driving the melt through a duct which gives it the desired size and form. Adding small amounts of co-monomers, themselves fluorinated, leads to the formation of copolymers such as **PFA** which, although it keeps all the chemical resistance of PTFE, can be extruded more easily.

## 3.4 Plastics: false solids with a biddable disposition

Starting from the long list of macromolecules and of their applications we have just made, let us try to understand why polymers are so special. The key question is how these long chains can form a solid, at least in the broadest sense of this word. For a physicist, the main difference between liquids and real solids is not so much that the latter keep their shape, with no need for a container to prevent them from flowing. On the contrary, in this book we shall meet materials that are basically solid even if they can be poured from a glass, or liquids that, conversely, flow only over a period of years. The true hallmark of a *real* solid is that atoms or molecules are organized in a crystal, meaning a structure that regularly repeats in space, much as floor tiles are a recurrent motif on a plane[18]. For instance, in table salt $Na^+$ and $Cl^-$ ions are regularly arranged, on the edges and on the sides of little cubes. It is the presence of this spatial regularity that makes a solid a totally different phase of matter compared with a liquid. To tell the truth, we do not fully understand yet what drives atoms and molecules, randomly scattered in a liquid, to order spontaneously into a crystalline structure, for nothing in the structure of a liquid close to freezing point suggests that this will occur.

Ordering into a solid is a far harder problem for polymeric chains than for simple sodium chloride. Things still go fairly smoothly for a simple polymer such as polyethylene. As we said, the happiest arrangement, in terms of energy, for this macromolecule is a linear zigzag chain, where all carbon atoms lie on the same plane. Lengths of chain so ordered can lie side by side, forming regular structures with all chains aligned along the same direction. This is what actually happens for short hydrocarbon chains $CH_3 - (CH_2 - CH_2)_n - CH_3$ with no more than 20 to 30 carbon atoms, the *paraffins* that constitute common wax. But for long macromolecules such as polyethylene, the story is different. Those defects we talked about, which twist the chain out of its flat, "planar" shape, are not too expensive in terms of energy, namely, they do not cost very much *compared with thermal energy* $kT$. Therefore, at room temperature, there will always be a few defects around to break the orderliness of the zigzag chain. As a consequence, only a certain fraction of the $CH_2$ groups manages to arrange itself neatly in crystalline form, whereas the remainder

---

[18] We shall sometimes retain the common sense of the word and describe as solid even amorphous substances such as glass. I would certainly never try to walk through a window on the grounds that it was a liquid.

stays in a disordered state. Therefore, solid polyethylene is always *partially* crystalline, and is composed of little crystallites, often in the shape of tiny platelets, embedded in a disordered (or "amorphous") mass of chains.

We might think that all "linear" polymers similar to polyethylene share the same behavior, forming planar chains that organize themselves into a crystal, but it is not necessarily so. For example, fluorine atoms in PTFE are quite a bit bigger than hydrogens, so that, in a planar zigzag configuration, they would constantly get in each other's way. To avoid this, the chain has to twist, forming a *helix* with a specific pitch (like the distance between neighboring turns on a screw thread). Curiously, these helices find it easier to organize into a crystal than planar chains do, so that polymers like PTFE may have a larger crystalline fraction than PE. The same happens for a polyamide such as nylon, which can therefore form very rigid fibers. In practice, however, the presence of a randomly tangled amorphous fraction is unavoidable, and even desirable, since high crystallinity does lead to strength, but, as we shall see, also to a certain fragility.

Things definitely become worse for those polymers – and that means most of them – that have *side groups* bristling out from the main carbon frame. Often, these groups of atoms are exactly the parts of the molecule that give the polymer its distinctive features. For instance, the monomer of polypropylene:

$$-(CH_2 - \underset{\underset{CH_3}{|}}{CH}\,)-$$

differs from that of polyethylene just because of a $CH_3$ group, quite a bit bulkier than a simple hydrogen, attached to one of the two carbons. If PP is polymerized with no special care, these groups lie randomly with respect to the chain plane, some over and some below it,

$$-CH_2 - \underset{\underset{CH_3}{|}}{CH} - CH_2 - \overset{\overset{CH_3}{|}}{CH} - CH_2 - \overset{\overset{CH_3}{|}}{CH} - CH_2 - \underset{\underset{CH_3}{|}}{CH} -$$

making crystalline order impossible. Solids do form, but they are those *fully amorphous* structures we shall deal with soon. The problem is that, to have mechanical strength, a plastic material must contain a non-negligible crystalline fraction. For instance, the mechanical properties of the kind of polypropylene we just considered, which is called *atactic*, are rather poor. To get (partially) crystalline polymers, controlling the *tacticity*, or orderliness, is therefore a must. This has been feasible since the second half of the last century, thanks to the genius of Karl Ziegler, from the Max Planck Institute, and Giulio Natta, from Politecnico of Milan, joint Nobel laureates in 1963. They developed new catalysts allowing them to make, for instance, fully *isotactic* polymers, where lateral groups are all placed on the same side of the chain:

$$\begin{array}{ccccccc} CH_3 & & CH_3 & & CH_3 & & CH_3 \\ | & & | & & | & & | \\ -CH_2 - & CH - CH_2 - & CH - CH_2 - & CH - CH_2 - & CH - \end{array}$$

or *syndiotactic*, where the groups regularly alternate with respect to the chain axis:

$$\begin{array}{ccccccc} & & CH_3 & & & & CH_3 \\ & & | & & & & | \\ -CH_2 - & CH - CH_2 - & CH - CH_2 - & CH - CH_2 - & CH - \\ & | & & & | & & \\ & CH_3 & & & CH_3 & & \end{array}$$

Back to amorphous polymers, the materials they form, although they look like solids, do not show any crystalline structure and are fully disordered, much more like liquids than crystals. Later we shall call these "substitutional solids" glasses, although they apparently bear little resemblance to window glass. A manifest clue that a glass is not a "true" solid is that a crystal *melts* above a definite temperature, becoming a liquid, whereas a glass just *softens* progressively when brought above its *glass transition temperature* $(T_g)$, as butter does when heated in a pan[19]. We shall shortly see that it is exactly this "glassy" nature that gives polymers those distinctive mechanical properties that justify the name "plastics" for these materials. When the temperature is raised above the glass transition temperature, a simple glass, such as that of a window, progressively turns to liquid. Yet many polymeric materials behave rather differently. Even when their glass transition temperature is much lower than room temperature, they don't appear to be liquid at all, but apparently keep together as a solid, although they show properties that are quite unusual for ordinary solids, in particular an incredible elasticity. These are the *elastomers*, whose prototype is natural rubber.

But what do we exactly mean by "plastic" and "elastic"? To understand it, let us sketch some general aspects of the mechanical behavior of a material. Suppose we pull a wire, fixed at one end, with a force $F$, for instance by hanging a weight on it, and that we measure how much it stretches with respect to its initial length (this is called its *strain*), by increasing the load we apply. Obviously, the bigger the cross-section $S$ of the wire, the larger $F$ must be to get the same stretch. Therefore, we should express the strain as a function of what we shall call the applied *stress*, given by force divided by this cross-sectional area, $F/S$, as shown in Fig. 3.2.

A first rather trivial thing we can see is that, sooner or later, the wire breaks. Thus, there is a maximum possible value of the load that depends on the material the wire is made of. A weight of about 40 lb is enough to snap a copper wire with a section of one square millimeter (1 mm$^2$), whereas for a tough metal such as tungsten 300 lb would not be enough. Obviously, most plastics have a lower breaking load than metals. For example, a nylon fiber with the same section can bear about 15 lb at most, whereas for a rubber

---

[19] Actually, semicrystalline polymers also do not have a definite temperature but rather a temperature interval where they melt.

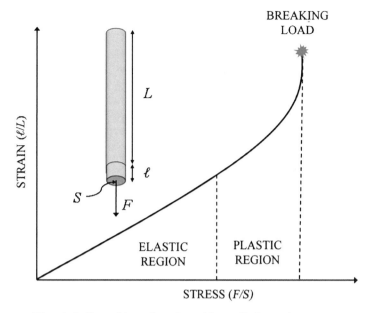

**Fig. 3.2.** Stretching of a wire with applied traction stress.

thread 5 lb would be too much. Special fibers such as aramides, however, may outclass even the strongest metallic alloys. A similar wire made of Kevlar, for instance, could withstand more than 600 lb.

However, there is much more to it than that. First, the behavior of the wire before breakage shows two distinct regions. For small loads, the graph is linear – that is to say, the strain is proportional to the applied stress. This corresponds to what is called an elastic behavior, similar to that of a stretched or compressed spring. The shallower the slope of the line, the more rigid is the wire, or, in technical jargon, the higher is its *elastic modulus* (or *Young modulus*), given by the ratio between stress and strain. All materials are more or less elastic, but the Young modulus can vary across a wide range. Plastic polymers are consistently less rigid than metals, which are already quite elastic materials. A steel wire, again 1 mm$^2$ in section, loaded with a weight of 4 lb, stretches by only one part in ten thousand of its initial length, but with the same load the relative stretching of a nylon fiber can be 1 or 2%, which is hundreds of times larger. When it comes to stretching, however, elastomers are king. In the same conditions a rubber thread would *double* its original length!

Second, in the elastic region the wire *reverts* to its original length if we stop pulling. This no longer happens once the wire gets over its *yield point*, beyond which its behavior turns *plastic*. What this familiar term means, in the technical sense, is that the strain grows rapidly with the applied stress and becomes *irreversible*, partly remaining even if the load is removed[20]. Not all materials display a plastic behavior; in fact, many of them reach the breaking point still behaving elastically, with no appreciable plastic region. Among crystalline solids, the plastic behavior is typical of metals, and specifically of those metals named *ductile*. Ductility and malleability (the ability to be, respectively, pulled into wires or hammered into plates) are exactly those features making metals so pliable, and therefore so fundamental for technology[21].

Moreover, it is plasticity that provides the *toughness* of metals, which in material science means something quite distinct from *hardness*. Materials such as diamond can be hard, because they have a large maximum stress, and at the same time *fragile*, since they are barely elastic and not at all plastic, so that, as soon as they are deformed even a trifle, they break. A tough material, conversely, does not at all endorse the old Italian motto (which does not seem typical of my fellow countrymen) that goes, "I break, but I do not bend". Hence, it can withstand considerable loads, deforming a great deal before breaking. You may recall from school that an applied force times the distance by which a body is moved is equal to the work done on a body. Then, noticing that a deformation is nothing but a relative displacement of the different parts of an object, it is not difficult to grasp that a material is tough if it manages to stockpile a lot of energy before breaking.

For engineers, toughness is much more important than hardness (just think about the construction of an earthquake-proof building, where these two properties can play diametrically opposite roles!), and therefore plasticity is quite often a desired property. Obviously those we commonly call "plastics" have it in plenty, and are moreover lighter and cheaper than metals, which is one of the main reasons for their success. This holds true both for amorphous and for semicrystalline polymers. The main effect of the presence of crystalline regions is indeed to increase hardness noticeably, making plastics more similar to metals, whereas their disorderly amorphous regions ensure plasticity. When it comes to their ability to stretch without breaking, plastic materials such as polypropylene, PET, or nylon can more than double in length, and carry a peak load up to one-third that of copper, whereas the polyethylene used to make plastic bags can reach much larger stretches, although with a lower breaking load.

Quite often, "internal stresses" build up in a material solidified into a given shape. That is to say, the stuff is not fully uniform, as if some inner regions were more stretched or compressed along specific directions than the others.

---

[20] For the plastic region, the trend shown in the figure is purely a rough guide. Some materials, such as concrete, have much more complex behavior.

[21] It may seem paradoxical that metals, crystalline materials *par excellence*, are so plastic. Actually, their plasticity stems from the presence of specific defects, called dislocations, consisting of crystalline planes that are "out of register". In metals, and only in metals, these defects are extremely mobile, and their motion and reorganization prevents excessive stress from concentrating in small regions, where they might lead to fractures.

This is particularly true for glassy materials such as plastics, where an uneven cooling of the melt yields these kinds of inhomogeneities. In a transparent polymer, the presence of internal stresses can be easily detected, because they modify the *polarization* of the light passing through the material, namely, the direction along which an electromagnetic wave vibrates. In Fig. 3.3 you can see what happens if a rectangular polystyrene container is placed between two crossed polarizers, which are optical components allowing us to detect changes in the direction of the light's polarization, and thus the areas of different stress in the material, as these show up as different colors. This technique, called polarimetry, allows engineers to spot those "critical" regions where cracks are likely to appear in a mechanical part.

**Fig. 3.3.** A transparent polystyrene container observed in natural light (left) and between two polarizers (right).

For the moment, we shall suspend this brief survey of plastic materials and of their properties, although we shall later come back to elastomers. To find the physical origin of their exceptional elasticity, however, we need to start from the properties of macromolecules in solution: that is, to investigate what polymeric chains do when they are dispersed in a solvent. In doing so, we shall find a unexpected affinity between these and those colloidal particles we formerly met.

## 3.5 Snake dance

> *This is the dance of the snake*
> *coming down from the mountains*
> *to find its own tail*
> *that it had lost one day...*

This is an nursery rhyme that most Italian moms and dads know by heart; what we shall do now is to discover exactly what kind of dance a "polymeric snake" performs in solution. But let us first have a short intermezzo, to help us to grasp what we really mean by saying that a polymer is "solubilized".

In the first chapter we considered colloidal particles. These, as far as we have seen, do not aggregate so long as there are repulsive forces keeping them apart. Therefore, the particles are not truly dissolved (or, as we shall say from now on, *solubilized*), but simply *suspended*. On the other hand, if the stuff they are made of were really soluble, like sugar in water, the particles themselves would, almost by definition, *dissolve*, and the colloid would be no more. In the technical jargon, the colloids we talked about are *lyophobic*, meaning that the particles "fear the solvent". I stress again that, if we are to have particles, this has to be so, otherwise the material would separate into individual molecules scattered throughout the solution.

With polymers, we are dealing with an apparently opposite situation. The fluid they are dispersed in is a "good solvent" for the plastic, meaning that, if particles of the plastic are poured into it in solid form, they dissolve. But whereas sugar dissolves in water into the form of small molecules, a plastic dipped in a good solvent, for instance polystyrene in acetone, dissolves into polymeric chains, which are huge molecules. In what follows, we shall see that these chains, far from remaining ramrod straight, coil to form globular "particles", though very special ones. In the end, therefore, we shall discover a new kind of colloids which we shall call *lyophilic*, or "solvent lovers". Their main difference from those we already know is that they are naturally stable, with no need to fend off the van der Waals forces. But where have the latter gone? Actually, they are still there, but the affinity between the polymer and the solvent ensures that a thin solvent layer forms around the chain. This thin layer (technically known as the *solvation layer*) forms a kind of shield that prevents two molecules from approaching closely enough to stick because of dispersion forces. As for the rest, much of what we said about lyophobic colloids still holds true[22].

---

[22] Some confusion may arise from the fact that in Chapter 2 we also talked about "polymeric" colloidal particles, made for example of polystyrene or PMMA. How-

*Which* solvents a specific polymer prefers obviously depends on the monomer chemistry. A very special solvent is water, whose importance for our purposes is evident, and whose rather peculiar properties we shall discuss. Most of the polymers we talked about do not love water at all, preferring other liquids such as hydrocarbons or, more generally, "oily" substances. This is for instance the case for polyethylene, which after all is a very long hydrocarbon itself. The same happens for natural rubber, which, as we stated, dissolves quite well in naphtha. Polymers of this kind are called *hydrophobic*, meaning again that they are afraid of water. Polyoxyethylene, also known as polyethylene glycol (PEG), whose chemical formula is $HO - (CH_2 - CH_2 - O-)_n - H$, is conversely a *hydrophilic* polymer, which loves to plunge into water. As you can see, PEG's main difference from polyethylene is the presence of an oxygen atom in the monomer. This atom can form very special ties with water molecules, the *hydrogen bonds* we shall discuss later, making PEG highly water-soluble. A more familiar example is starch, whose monomers are molecules of glucose, the sugar produced by plant photosynthesis (whereas table sugar is saccharose) which, like all sugars, is also water-soluble because it forms hydrogen bonds.

To simplify the matter, let us now leave aside the complex chemistry of polymers (I imagine you will not regret this) and move to the much coarser, but not totally pointless, view a physicist has of this stuff. I have insisted on drawing a parallel with a chain because this is just what physicists notice about a polymer molecule – namely, a long chain where the links are made of monomers, and where each link can be arranged anyhow, provided that it remains bound to the preceding and following ones. Understanding the structure, or, more properly, the conformation of a polymer in a good solvent then amounts to studying the shape taken on by a chain when its links are randomly oriented. If this would help you to grasp the problem better, think of those construction toys that surely most of my few readers, depending on their age, have either played with or given to their children, the sort made of rods that link to each other by magnetic balls attached to both ends. It is just the spherical shape of the magnets that enables each segment to orient in any direction.

Clearly, this is not totally correct. In polyethylene, for instance, each monomer segment can be placed only at specific angles with respect to the preceding one, so the chain cannot bend at will. If the polymer is long enough, however, this problem can be sidestepped. Suppose that it takes a certain number of monomers, say $n$, to allow the last monomer to be placed at any angle to the first. Then, we may just group together these $n$ monomers and consider them as a single block, which is called the *Kuhn segment*, playing

---

ever, in that case we were referring to particles made of *many* polymer chains that keep together. All in all, therefore, they are just tiny plastic bits that *do not* dissolve in the surrounding liquid. This corresponds to having a little ball of polymer inside what is called a *bad* solvent.

the role of a new "effective" monomer. Obviously, this strategy works only provided that the length of the Kuhn segment, called the polymer *persistence length*, is still short compared with the extended chain length. Otherwise, we are dealing with a so-called "semirigid" polymer, whose conformation can be very different, being more akin to a rigid stick than to the coils we are soon to meet. Biopolymers, which we shall discuss in the final chapter, are often of this kind. When the chain is really *very* long and flexible (i.e., with a tiny persistence length), picturing the chain becomes even easier. The rings of the thin chain that I imagine many of my female readers (and possibly some of the males) wear as a necklace are so small with respect to its length that, all in all, it can be regarded as a string. Studying the conformation of a polymer in solution is therefore almost equivalent to studying how a string coils on itself.

Walking out of the complicated world of chemistry, in which they are rarely expert, to embrace that of geometry, where they are a bit more confident, physicists can indulge in the business they are more fit for: drawing with the mind (with the hands, they may be less adept). Since my own mind is currently rather fatigued, I have tried to get my computer to do it. Figure 3.4 shows the shapes that are obtained for three distinct simulations of a polymer made of 1000 monomers, which are pictured as dots connected by sticks. As you see, each chain is entangled, forming a kind of coil.

**Fig. 3.4.** Different simulated configurations of a polymer made of 100 monomers.

Of course, these configurations are not "frozen", since monomers move around unceasingly, so much that the three images could refer to the *same* polymer pictured at different times. On examining them carefully, these intricate shapes look somewhat similar to the drunken paths we have met in Chpt. 2. If you think about it, the two problems have much in common; just imagine that the sticks connecting two monomers each represent a single step, and the dots the positions attained by our drunkard after each step. As the drunk randomly chooses what direction to move at each step, making steps of equal length, each monomer is placed in a random direction relative to the previous one. Hence the shape of a polymer is nothing but a random walk, or, if you prefer, the trace left by a particle during its Brownian motion. We have already learnt a lot of things about the latter. For instance, we may guess that, as for the drunken walk, the coil radius will be proportional both to the monomer size (the step length) and to the square root of the number $N$ or monomers, that is, to the square root of the polymer molecular weight.

Nothing thrills a physicist more than seeing two apparently unrelated problems described by the same model – so much so that often, owing to excessive enthusiasm, he or she loses sight of subtle but crucial differences. For a difference does exist: while the drunken man can retrace his steps, passing through a place where he has already been, this cannot happen for a polymer, because two monomers *cannot* overlap. Does this seem a minor detail? Quite the reverse! Whereas the problem of a simple random walk could be tackled even by a high-school student (a bright one, I must admit), the problem of a random walk where passing twice over the same place is not allowed (namely, where all paths must be "self-avoiding") is so challenging to have causes terrible headaches to several Nobel laureates. Better still, no one believes that it can be solved *rigorously*, and even approximate solutions require quite advanced methods in theoretical physics. However, we can grasp the main consequence of being forced to avoid step-retracing. To avoid overlapping with the others, a monomer should prefer to "move out" from the coil core. This means that the coil will be more expanded than it is when no region is "excluded". Hence, on increasing the number of monomers, its size increases *faster* than in the ideal random walk case. While for Brownian motion the radius of the explored region is given by $R = LN^{1/2}$, where $L$ is the step length, an approximate but rather good model of the self-avoiding random walk yields $R = aN^{3/5}$, where $a$ is the monomer size. Since $3/5 > 1/2$, this agrees with our rough guess.

In a good solvent, polymers therefore roll up to form particles that have a rather irregular but still globular shape, so that a polymer solution is after all a colloid, but with some important differences. First, it is a *stable* suspension, so that there is no need to introduce repulsion forces to keep the particles apart. Second, these globules, although much bigger than the solvent molecules, are quite small compared with the particles we formerly considered, having a typical size of tens of nanometers[23]. Last but not least, the coils are mostly *empty*. As for the path traced by a Brownian particle, polymer coils are fractal objects too. In fact, reversing the relation we have just mentioned, we get $N = (1/a)R^{5/3}$, so that the fractal dimension of a polymer chain in solution is $5/3$, which is even *lower* (just because the coils are expanded) than for an ideal random walk. Thus the coils are very "soft" when compared with the hard spheres we formerly met. Indeed, their density $d$, given by the ratio between $N$, which grows only as $R^{5/3}$, and the coil volume, which is instead proportional to $R^3$, gets *smaller and smaller* by increasing $R$.

---

[23] Clearly it is hard to define their exact size, and the value I stated is for the so-called *gyration radius* of a polymer chain. In analogy to Fig. 2.2, this is roughly speaking the region including the largest part of the monomers (for the ideal random walk, about 2/3 of the drunkards).

Everything we have said refers to a single chain or to a small number of chains in solution. Each chain will have a globular shape only if it is sufficiently far from the others not to be "troubled" by their presence. But what happens when we increase the polymer concentration? Just because they are mostly empty, the coils quite soon begin to overlap, and the concentration at which this takes place can be *extremely* low. If you manage to follow the not-too-hard estimate below, you will more easily realize why (if not, try to convince yourself using some intuition). Let us call $n$ the total number of monomers in a volume $V$, and therefore $c = n/V$ their concentration, i.e., the number of monomers per unit volume. If each chain is made of $N$ monomers, the total number of chains will obviously be $n/N$, each one with a volume $V/(n/N) = N/c$ at its disposal. The volume filled by a chain is, however, of the order of its radius cubed, and therefore of $a^3 N^{9/5}$. When this quantity becomes comparable to $N/c$, that is, to the free volume per chain, the chains start to overlap. Comparing these two values, it is easy to show that this takes place for a monomer concentration which we'll call $c^*$, given by

$$c^* \simeq \frac{N^{-4/5}}{a^3},$$

which is called the *overlap concentration*. As you see, $c^*$ decreases quite fast with the number of monomers in the chain. For values of $N$ that in practical application may reach the tens of thousands, $c^*$ may be as low as 0.1 percent!

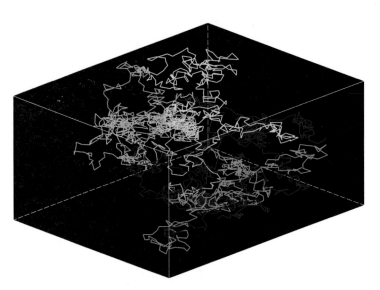

**Fig. 3.5.** Sketch of a semidilute polymer solution in a good solvent.

A polymer solution where the chain widely overlap, which is called *semidilute*, cannot be regarded as composed of separate coils. What does it look like, then? Once again, let us ask the computer. Fig. 3.5 shows that the solution looks like a tangled net, with a mesh size becoming smaller and smaller as the polymer concentration grows beyond $c^*$. Merging together, the coils have created something quite different from the colloidal suspensions we are used to. Much of what we shall say in the next sections has to do with the surprising properties of these networks. However, some features of the colloidal suspension we have formerly discussed are shared by semidilute polymer solutions. In particular, the latter have a high osmotic pressure. We shall later see that a polymer network can be strengthened by forming cross-links between distinct chains, so making what is called a *polymer gel*. Because of osmosis, a polymer gel craves solvent, which it can suck in by swelling. Thanks to what are called super-adsorbent polymers, and in particular to **polyacrylic acid** (PAA), it has been possible to make the diapers and sanitary napkins of today. Unlike PEG, which is electrically neutral, PAA is water-soluble because it is charged: it is a *polyelectrolyte*, a polymer class we shall soon encounter. The most interesting property of polymers in solution, in particular when semidilute, is their effect on the fluid viscosity, namely how much they alter the solvent flow, even at low concentration. At the end of this chapter, I shall try to give you some simple ideas about what we shall call the *rheological behavior* of a solution.

If we go on concentrating a polymer solution, reducing the amount of solvent until it vanishes, we end up with a material that, for an amorphous polymer, will be a glassy polymer or a polymer melt like natural rubber, depending on whether we are below or above the glass transition temperature $T_g$. In a simple picture, a semidilute polymer solution reminds us of spaghetti boiling in a pot, whereas an elastomer looks like just-strained spaghetti. Nothing binds the strands together, so they can in principle move as freely as molecules in a liquid. Yet, since they are so entangled, they have to move in a rather complex and definitely slow motion, so that the spaghetti ultimately stays together more or less as in a solid. That is to say, if we stick a fork in, we can pull up many of the strands at the same time, at least if we do it fast enough. If we then let the spaghetti plate rest for a full day, without adding a trickle of oil as any good cook would do, they stick together hopelessly and irreversibly. In this way we obtain a true disordered solid, similar to glassy plastic materials.

What is the polymer configuration in these materials, taking into account that at this stage, without solvent, a chain sees around it only the likes of itself? Since a monomer must make itself room among the other chains, there are no longer any "external" regions to be preferred to those already occupied. As a curious result, the conformation taken on by a polymer chain, instead of corresponding to a self-avoiding path, just becomes that of an *ideal* random walk. This means that the number of monomers enclosed in a sphere of radius $R$, centered anywhere in the medium, grows only as $R^{1/2}$ (obviously, these monomers belong to different chains, entangled even more than in a semidilute solution). Studying the properties of polymer networks will enable us to grasp many aspects of elastomer behavior. Before dealing with them, however, we have to make an important, although not easy, digression. In the next section, indeed, I will introduce you to one of the tougher, but also more fundamental, characters in physics.

## 3.6 Entropy: disorder or freedom?

There is a rude word used on purpose to scare young students in physics, but that, on the other hand, philosophers, artists, architects, and recently some politicians[24] like so much: *entropy*.

That it *sounds* like an insult is unquestionable. Which one of my lady readers, hearing this term called after her, would take it as a compliment? Yet, name apart, it is probably the most important concept in physics together with energy. For that matter, when Rudolf Clausius (who was a great physicist, but would probably not be a great adman today) dubbed entropy with this Greek expression, meaning something like "internal turning", he actually chose it mainly because it sounded a bit like "energy".

What, then, is entropy? Readers with a high-school education may remember that it has something to do with heat, motors, and the end of the Universe. Those of you who attended scientific courses at college, are mad about Star Trek, or simply pay attention to the said philosophers and architects, may have learnt to associate it with the word "disorder" and to the fact that the latter always grows. If so, forget it immediately. I am not claiming that these concepts are alien to each other. On the contrary, quite often entropy and disorder walk hand in hand. But it is not *necessarily* so. In Chapter 5 we shall indeed encounter situations where a system spontaneously orders *because* of entropy. Moreover, what is disorder, truly? *This* is really a hard question.

The real meaning of entropy is definitely much easier: entropy is the *freedom of motion* that any physical system tends to maximize. As such, this seems to be more a vague New Age allegation than a real scientific statement, but simple examples will make everything clearer. For instance, if we have a gas enclosed in a box placed in an evacuated room, and we open a valve allowing the molecules to get out, they spread all around the room because this allows each molecule to get more free space. Therefore, the gas entropy

---

[24] The relationship between politicians and the words of science is curious. Now and then they enthusiastically embrace one of them, using it in a much wider and fuzzier sense than its original meaning implies (have you ever heard of "synergy" between ministers?). In other situations, conversely, they are outraged by simple mathematical concepts, which a scientist needs using to avoid talking haphazardly exactly as musicians need notes to compose. For instance, I remember a talk-show where a well-known Italian politician (at present, one of the highest rank) railed at a mathematician who, in asking him a question, took the liberty of using the word "derivative", which just means how fast something is changing. After that, he answered the question using many times the word "trend" which, apart from being more trendy, just means the same thing expressed more vaguely.

increases. Similarly, when we put sugar in water, we know that it dissolves, namely, it diffuses throughout the whole container. Diffusion takes place because, in this way, the sugar molecules get more freedom of motion.

What we have said holds for *insulated* systems, that is, systems that cannot exchange energy with the surroundings. What happens if they are not? Put simply, a system can "buy" or "sell" entropy by getting or giving up heat. Indeed, when this trading is made *slowly*, the entropy gain or loss is just equal to the heat absorbed or yielded, divided by the temperature (in Kelvin degrees) at which this transfer takes place (here is the simple relation with heat you may have found hard to grasp in high school!). For instance, if we compress a gas in a cylinder with a piston, the available volume per molecule becomes smaller, and therefore entropy should *decrease*. Yet entropy does not get lost, but is simply transferred to the environment as heat.

These examples, however, do not fully capture the idea of freedom of motion associated with entropy, which does not necessarily mean "having more room" in the day-to-day meaning of this expression. For example, suppose once again that we compress a gas, but that this time its container is a thermos bottle, which does not allow heat to get out. Since here the volume decreases too, where does the entropy go if it cannot escape? This time, unlike the former trial (where we supposed the cylinder to be kept at constant room temperature), the experiment shows that the gas *heats up*. If you remember, in the previous chapter we have seen that a higher temperature is associated with a larger average kinetic energy of the molecules. The latter, therefore, have less room to move around, but move faster. Thus, they gain freedom in a different "space", which is not measured in length, breath, and depth, but rather in terms of the values that the molecular speed[25] may attain. Imagining this "velocity space" is a bit hard, but when calculations are duly made, one can see that the volume the gas loses in ordinary space is regained as possible values of the speed. In short, everyone has their own notion of freedom; one person likes to travel all around the world, no matter how, whereas another is content to stay within Marion County, preferring the freedom to shoot along the Indianapolis track in a Ferrari. We shall shortly see that, for polymers, this freedom is to be meant in yet another way. Nevertheless, one is never mistaken in regarding entropy as the total number of ways a physical system has to move about or rearrange. In a nutshell, entropy is basically related to the total number of distinct *configurations* (as we physicists call them) that a system can take on, although setting them out may be, in practice, rather difficult. After all, Clausius did not choose such an abstruse name as it seemed!

---

[25] Better, the three "components" of their velocity. To describe how a molecule moves, we really need to state how much it shifts up/down, left/right, and forward/backward. That is, to specify its motion in full, we need *three* directions. Along each of these directions, the molecule in general will have a different speed, so we also need *three* numbers (the three values, or components, of the velocity along these directions) to state the molecular velocity.

Before moving to polymers, let us try to fix some ideas that are useful for what follows. A system which is insulated from the surroundings has a fixed energy, which cannot change because heat exchanges are impossible. What it tends to do spontaneously is just to maximize its entropy in ways that are compatible with the amount of energy it has at disposal. But what happens if the system is *not* insulated? In the previous chapter we saw that the system approaches an equilibrium state of minimum energy, like a ball that spontaneously rolls down a hill[26]. Yet, when energy is lost as heat, there is an entropy loss, which the system does not like. The only way out is a kind of compromise, where the demands for energy decrease and entropy gain somehow balance. Indeed, calling $E$ the system energy and $S$ its entropy, what should *always* be minimal for a system exchanging heat with the environment is the so-called *free energy* $F = E - TS$. The name "free" energy comes from the observation that, if we try to make a machine work (which is rather advisable, if the machine is for instance a motor) by supplying it a given amount of energy $E$, maybe by burning fuel, the maximum *work* it can do is not $E$ but just $F$, namely, what is left after we pay "pizzo" (an Italian word that has became almost as widespread as "pizza", but with the rather nastier meaning of extortion money) to entropy. Therefore, $F$ is the maximum *useful* energy, which is more relevant to engineers than plain energy. For our purposes, it is useful to point out that $E$ and $S$ do not stand on equal footing in the definition of $F$. Entropy is *multiplied by temperature* (in Kelvins), and therefore counts for very little at low $T$, whereas it becomes more and more important as the latter increases. That's it, for the moment, since you may already have a fairly severe headache. Anyway, what matters is that you have some idea of the relation between entropy and freedom of motion. We shall profit from this in what follows.

## 3.7 Elastic by chance

We have seen that all solid materials are more or less elastic; that is, if we try to stretch, compress, or bend them, they resist with a force proportional to the deformation. This is because stretching, compression, and bending require gentle deformations of their crystalline structure, which are barely detectable, but cost a lot of energy. A spring works pretty well both because of its easily deformable helical shape, and since it is made of particularly elastic metals such as "harmonic" steel. Actually, the idea of elastic force is among the most important in physics, and allows us to explain many effects, ranging from the way waves move on the sea surface, to why those mysterious objects called "quarks" stay tied together in a proton, never able to wander freely

---

[26] You may ask yourself: "Why? Didn't he tell me before that potential energy does *not* get lost, but simply becomes kinetic energy?" That is true, but in practice every real ball, little by little, comes to a halt, because, through friction, it *transfers* kinetic energy as heat to the surroundings.

apart. It seems natural to think that a rubber band works the same way a spring does (after all, why do we speak of *elastic* forces?), but this is wrong: between a spring and a foam rubber mattress there is a subtle but deep difference (for that matter, we have seen that, compared with a metal, rubber has truly exceptional elastic properties). The basic reason is that rubber's elastic response does not stem from energetic, but from *entropic* grounds. To see what I mean, let us first consider a single polymer chain.

Suppose we keep the two ends of a polymer chain at a fixed distance. Provided that such a distance is shorter than the fully extended chain, this does not cost any effort, for we are not stretching or bending any chemical bond, but just forcing the chain to start and end at two set points. Yet, between the two situations shown in Fig. 3.6, *there is* a difference, since in B, where the ends are further apart, there are many fewer chain configurations joining the extremes. When it is stretched, a chain loses a lot of freedom of motion, thus losing entropy, something it does not like at all. As a result, the chain tries to pull back and, when calculations are properly done, one finds that its reaction force is just proportional to the end-to-end distance. Hence, the chain behaves exactly like a stretched spring not because it feels strained, but rather because it wants to keep the freedom to rearrange itself.

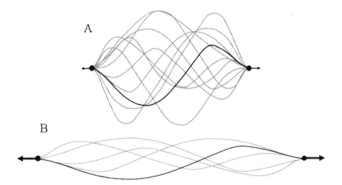

**Fig. 3.6.** Reduction of the allowed configurations for a stretched polymer chain.

Curious and far from our daily experience as it may seem, this is exactly the mechanism yielding the spectacular elasticity of polymers. Better still, looking at it carefully, we can see the real connection between entropy and temperature. Every particle in a solvent has a thermal energy, but, whereas for rigid colloidal particle this is just the source of Brownian motion, a flexible polymer chain also exploits thermal energy to "toss about" internally (for instance, to move from one to another of the different arrangements in Fig. 3.4). Because of this, the reduction in entropy due to chain extension costs more at high temperature, where the thermal energy is larger (so that the amount of free energy to be paid is larger). Therefore, unlike what happens for a spring, the restoring force exerted by the chain *grows* with temperature, a rather surprising result which is nevertheless correct for polymeric materials.

But what has a single polymer chain got to do with a rubber band, a mattress, a tennis ball, or the little duck your kids play with in the bathtub? And how can this desire for entropy account for the stunning power of Reed Richards, the leader of the Fantastic Four? In a nutshell, what is rubber from a physical point of view?

## 3.8 The secret of Mr Fantastic

We have seen that natural rubber is basically polyisoprene well above its glass transition temperature. Thus, it is a polymer melt, a liquid. Even if at room temperature it is so viscous as to behave almost as a solid, it is only useful to waterproof fabrics, certainly not to make rigid parts. There are empirical methods to make caoutchouc harder (the Aztecs surely knew some of them) but, as we have said, the successful way to turn this sticky liquid into real rubber is vulcanization, serendipitously discovered by Goodyear. What, then,

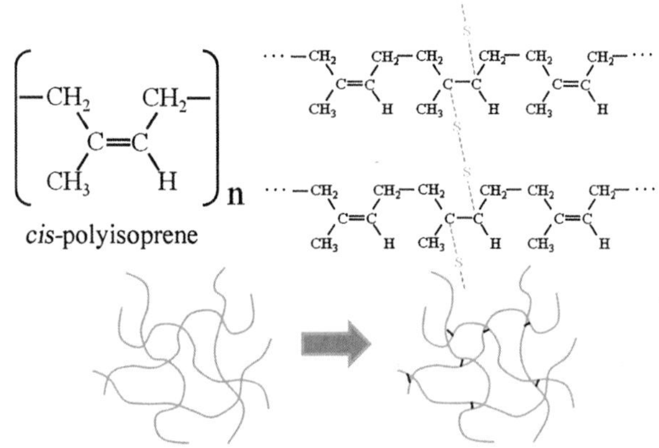

**Fig. 3.7.** Formation of sulfur bridges between isoprene monomers, as a result of the vulcanization process. The transformation of natural rubber into a vulcanized network, highlighting the cross-links between the chains, is shown at the bottom.

is vulcanization? In Fig. 3.7 you may notice how the polyisoprene monomer[27] contains a double bond between two carbon atoms. When the temperature is high enough, sulfur can interpose between the two bonds, binding to two carbon atoms and forming bridges (made by several sulfur atoms, since sulfur

---

[27] More precisely, it is *cis*-polyisoprene, corresponding to a specific orientation of the monomers with respect to the chain.

itself tends to form short chains) that link *distinct* polymer chains. The poly-mer network is then tied up by "knots", called *cross-links*, and grows much stronger. It is as if the cross-links have "frozen" the polymer melt into a glob-ally solid structure, yet the polymer segments connecting the knots are still totally free to explore a large number of configurations (all those compatible with having given start and end points). Therefore, what we have said for a single polymer chain is still true, provided that we apply it to the segments in between the cross-links. When we stretch a piece of rubber, the cross-links get farther along the stretching direction. But this reduces the number of possible configurations of the segments, and therefore their entropy. As a consequence, we again have many "entropic springs" trying to bring the piece back to its original shape. As I said, one of the most evident indications of the entropic nature of rubber elasticity is that the elastic modulus grows with tempera-ture. This means that if we heat a rubber band kept under constant tension, it must *shrink*. This effect is hard to appreciate by eye. After all, if for in-stance we heat the band from 20 to 50 °C (about 70 to 120 °F), its *absolute* temperature, which is what really matters, varies by only 10%, corresponding to an increase of about 10% of the rubber's *extension* (and not of its *total* length).

However, are the elastomer properties entirely identical to those of a single polymer chain? Actually no, for we have neglected an important aspect; if we stretch a rubber band in a given direction, for instance by pulling it along its length, it must *shrink* in thickness, because its volume should stay more or less constant[28]. Across the thickness, the cross-links get closer and the chains connecting them gain new allowable configurations, so their entropy *grows*. This lowers the cost of stretching the band, and consequently the restoring force is smaller. It also explains why rubber reacts elastically to squeezing as well, because here the chains get more freedom along the line of compression, but lose more of it along the other directions. Unlike a spring, however, rubber behavior is not the same when squeezed or pulled; it stretches more easily than it is squashed. A fairly simple evaluation yields the behavior in Fig. 3.8, showing in particular that, unlike an ideal spring, the elastic modulus of an elastomer is anything but constant. Actually, the plot refers only to moderate compression or extension. For higher strain (in particular when the stretching becomes so large that the chains are almost fully extended) the elastomer behavior is much more complicated.

To make a good rubber, the number of cross-links need not be very large. On the contrary, excellent elastomers can be made by co-polymerizing a monomer without double bonds (which therefore cannot be vulcanized) with 2–3% isoprene. This is the case for silicone and butyl rubber, the latter of which is so gas-proof as to be used for coating the inside of tube-free tyres.

---

[28] This is by no means trivial and, in fact, is true only for entropic elasticity. Indeed, the total volume of a metallic sheet pulled along a given direction would slightly *increase*.

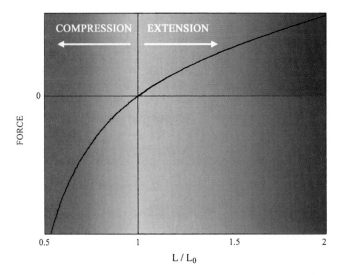

**Fig. 3.8.** Length $L$ of an elastic material, relative to its "unperturbed" value $L_0$, when it is stretched or squashed by a stress along this direction.

The only essentials to get a rubber are having a small fraction of cross-linked monomers, and a glass transition temperature for the polymeric matrix at least 30–40 °C (a hundred or so degrees Fahrenheit lower than its working temperature. Having a very low $T_g$ is what makes silicone rubber, for instance, so soft and elastic even at low temperature. Conversely, too many cross-links yield rigid elastomers; this is actually the way to obtain melamine resins or epoxy glues.

We have discussed the mechanical properties of polymeric materials, discovering in particular that they may show both a solid-like elasticity and a plastic behavior, that is, the property of deforming irreversibly under stress like pliable metals. In the latter condition, the stuff somehow flows, being in this regard more similar to a liquid than to a solid. What about polymer solutions then, which are true liquids? Are there situations where a liquid behaves like a solid? More generally, what makes liquids so different from one another, in terms both of the stress required to set them in motion and of the friction they exert on a body moving within them? In short, why is swimming in honey is much harder (I guess, for I've never tried it) than doing it in a regular pool? To see that, we must give a precise meaning to the word "viscosity".

### 3.9 Panta rei

Some readers may dig out of their high-school memories the famous saying by Heraclitus of Ephesus, stating that "everything flows", which gives the title to this section. This is probably one of the deepest physical intuitions of ancient philosophy, so much so that it is the basic motivation of *rheology*, a scientific subject covering the flow of those complex fluids we are dealing with. Compared with simple liquids, the latter have many surprises in store for us. Usually, we can tell a solid from a fluid by considering how they behave under stress: a liquid flows freely, whereas a solid reacts elastically, or at most deforms plastically. However, in the Middle-earth of colloids, macromolecules, and the surfactants we shall meet soon, things go differently. The presence of particles, polymers, or supramolecular aggregates can provide the solvent with properties that are in between those of a solid and a liquid, which are extremely useful for applications ranging from the food industry to oil recovery.

Let us begin by asking ourselves what we really mean by "viscosity". This concept has first of all to do with the difficulty a body finds in moving through a fluid. The speed a marble can reach while falling through a bottle of syrup is much lower than it can attain if the bottle is full of water, which we sum up by saying that syrup is much more viscous than water. The marble is subjected to friction because it has to set the surrounding fluid in motion, since the fluid layer touching a solid body is always *at rest* with respect to the latter. This is exactly what physicists mean by saying that a liquid "wets" a solid surface. It is not difficult to find common situations confirming this statement. For instance, dust settles on the blades of a fan, even one kept in constant motion, something that would be impossible if the air close to the blades moved with respect to them. Similarly, try to remove the fine layer of chalk deposited on a blackboard by blowing, instead of using an eraser; even the lungs of a racing cyclist will not suffice, since, no matter how hard you blow, the air film in contact with the board is always at rest[29].

The friction force is therefore related to the effort required to set in motion some regions of the fluid (those closer to the body) *with respect to others* (in our example, those closer to the bottle wall, which must remain at rest). What viscosity does is actually trying to hinder the establishment of velocity *differences* in a fluid. Now consider water flowing through a hose. I guess it is evident to all of you that, for constant pressure at the water intake, the narrower the pipe, the less is the exit flow. This is a strong effect, since on halving the pipe diameter, the flow decreases *sixteenfold*. On the basis of what we have said, we can conclude that water has a maximum speed in the middle of the pipe, whereas in contact with the pipe wall it must be still. This tells us that, for a given viscosity, what really matters is how *fast* (namely, over what distance, here related to the pipe diameter) the speed changes.

We are now ready to define viscosity rigorously. Consider a solid block, as in the top panel of Fig. 3.9, and suppose we push only the upper face to the right, while keeping the lower one fixed. Similar to what happens for pulling, the block bends until it reaches a given deformation angle, which depends

---

[29] Exceptions do exist, but they are so rare and peculiar as to be an active (and controversial) research topic in fluid physics.

again on the applied stress and on the elasticity of the material it is made of[30]. Similarly, let us now think of confining a fluid layer between two parallel sheets, which we suppose to be so large that we do not have to care about what happens at the borders. If we pull the upper face with a force $F$, the fluid at contact must yield, whereas the liquid close to the lower plate, which we keep still, remains at rest. Thus the fluid layer deforms, but, unlike what happens for the solid block, it flows, or, in other words, it *goes on* deforming. In fact, what happens is that the whole fluid layer starts moving with a velocity profile that, after a short time, becomes linear (see the bottom panel in Fig. 3.9). What is the force $F$ required to set the upper plate in motion with a velocity $V$? Obviously, the wider the surface $S$ of the plate, the larger $F$ should be, and therefore we had better once again consider the stress, in other words the force per unit surface $F/S$. We have seen that the latter should be the larger the faster the velocity changes, that is, the bigger is the ratio $V/d$. The simplest guess is then to assume that the required stress is just *proportional* to $V/d$, so that we can write

$$\frac{F}{S} = \eta \frac{V}{d},$$

where the proportionality constant we have written with the Greek letter $\eta$ (eta) is exactly the fluid viscosity. This simple conjecture, originally made by Newton himself, works extremely well for all simple fluids; not only those with low viscosity like water, but also liquids such as glycerol, which have a viscosity coefficient thousands of times larger.

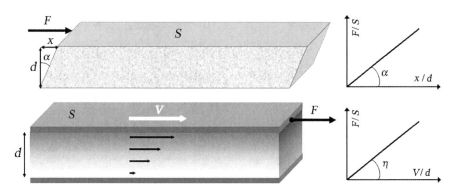

**Fig. 3.9.** Behavior of a solid (top) and of a fluid (bottom) subjected to a shear stress.

Things are very different for supramolecular fluids, since their rheological behavior strongly depends on *how large* the applied stress is. On increasing the stress, some of them apparently become less viscous, whereas others

---

[30] In this case, the factor relating stress to strain is called the "shear modulus".

show the opposite effect. We could say that the former are fluids that pliably "yield" when sufficiently stressed, whereas the latter become obstinate, rebelling against our wish to move them. The rheological behavior of these two kinds of fluids, which are respectively called *shear thinning* and *shear thickening*, is very different from what was predicted by Newton. Altogether, they are therefore known as "non-Newtonian", or also as *viscoelastic*, because the general features of their behavior can be grasped using a simple model, developed by Maxwell, that combines the properties of a viscous fluid with those of an elastic solid. Let us then examine some examples of viscoelastic fluids, keeping one eye on Fig. 3.10, where they are compared with simple Newtonian fluids on a graph showing the stress required to get a given strain rate (in our experiment, a given value of $V$).

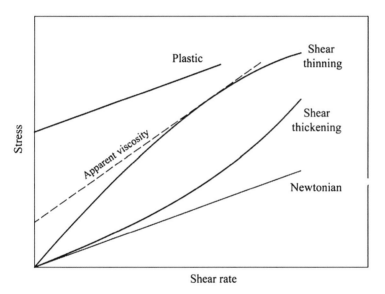

**Fig. 3.10.** Rheological behavior of non-Newtonian fluids.

## Shear thinning

Finding daily examples of shear thinning fluids is rather easy. Indeed, to work properly, many common products, such as paint, toothpaste, or nail varnish, must be reluctant to flow when they are strained weakly or not at all, but should yield when forced to do so. For example, we wish a paint to spread easily when brushed onto a surface, but we also want to avoid having it slide to the floor under its own weight before drying! Similarly, it is useful to be able to squeeze toothpaste out of the tube, but we would not be happy to see it slip spontaneously down from the brush. From Fig. 3.10 we can understand why they behave so, since the larger the strain rate the *easier* it becomes to increase it further. In other words, to do this requires little additional strength, so that the *apparent viscosity* of the fluid, which is given by the *slope* of the curve in the figure, becomes smaller as we increase the stress or the velocity of the fluid[31].

Polymer solutions and polymer melts often show a shear thinning behavior, for reasons we shall soon discuss. Polysaccharides, which are long water-soluble polymers whose monomers are sugars, are for instance extensively used in the food industry, where they play the role of "thickeners". They enormously increase the viscosity of a product, but still allow it to be easily spread, blended, or even chewed. A widely used polysaccharide is **xanthan gum**, which is extracted from cultures of the bacterium *Xanthomonas campestris*, a common parasite of crops. Less than 0.5% of this additive is sufficient to make a sauce very thick or an ice cream particularly creamy. But the same biopolymer is commonly used in widely different applications, for instance to increase the viscosity of drilling mud in the oil industry, or as an additive to reduce the erosion of concrete by water. For many foods, however, introducing additives is not necessary, since natural polysaccharides of fruits and vegetables such as the **pectins** already behave as thickeners. For instance, pectins in tomato sauce provide ketchup with its distinctive rheological properties, so that we have to shake the bottle vigorously, or tap its base, to get this thick sauce out. Certainly the most important biological fluid showing shear thinning properties is blood. It would never be able to flow along capillaries unless its viscosity decreased when it was pushed through these narrow vessels. Here the main source of the non-Newtonian behavior is that the red blood cells, which swarm through the blood and are largely responsible for its viscosity, elongate when they enter the capillaries, so making the blood much thinner. If this were not the case, since the typical capillary diameter is not much larger than the red blood cells, we would surely come up against the "filtration catastrophe" we met for colloidal suspensions.

### Shear thickening

Unlike the polymer solutions above, concentrated suspensions of rigid colloidal particles can show the opposite trend of increasing their viscosity when we increase the shear rate or the applied stress. Besides these, these effects are particularly relevant for materials such as wet sand. Here, water manages to "lubricate" the grains only when they flow slowly, whereas it does not

---

[31] An ideal plastic material should behave as a solid, not flowing at all until a certain value of the applied stress is reached, and then become a liquid in every respect, according to the line shown in Fig. 3.10. This is the reason these fluids are also called *pseudoplastic*. Indeed, if we consider their apparent viscosity, that is, the slope of the curve, and we trace a line with the same slope, the line cuts across the stress axis above zero (we say it has a "non-zero intercept"), as if it were in actual fact a plastic material.

have enough time to fill the gaps between them when they move fast. These effects are related to the peculiar property of granular materials to be *dilatant*, meaning that they *expand* under pressure[32].

A colloidal, and at the same time polymeric, system that shows a surprising shear thickening effect is obtained from corn starch, itself a polysaccharide. By jumping vigorously, it is not hard to walk across a pool filled with a suitable mixture of starch and water, basically a colloidal suspension of globules, whereas one rapidly sinks when standing still (if you don't believe me, just check on YouTube). Even more spectacular is the behavior of a material that has cheered the days of many kids, including, I'm sure, many of my readers: Silly Putty. When lying on a table, a ball of this stuff behaves as a very soft plastic material, collapsing and spreading across the surface; yet, if you *throw* the ball at the table, subjecting it to a stress much larger than its own weight, it bounces back like a rubber ball, with an elasticity that is even larger if you first cool it down in a freezer. These two wildly different behaviors are shown in Fig. 3.11. Even the story of this material is curious. It was discovered by chance, during the frantic search for substitutes for natural rubber prompted by the Japanese occupation of caoutchouc producing areas, by letting a little boric acid drop into silicone oil. One of the main ingredients of Silly Putty is, once again, PDMS, but its surprising properties are also due to other components, such as siloxanes and quartz silica, which are present in the secret recipe that has had such incredible commercial success. Although it is mostly used for play, Silly Putty has also been useful in hand rehabilitation therapy and, since it has good adhesive properties, by the Apollo astronauts to stick instruments to the walls of the command module.

**Thixotropy and rheopecty**

We have seen that the viscosity of some liquids changes on varying the applied stress or the shear rate. Yet our survey of complex fluid rheology does not end here. The viscosity of some polymer solutions or colloidal suspensions can also vary *with time*, if we apply a constant stress for longer and longer periods. These effects are even more intriguing and intricate, since they are related to the *history* of the material. That is to say, the measured viscosity depends on how we have handled the sample at earlier stages. There are, however, two common situation that somehow parallel what we have just discussed. In some fluids, called *thixotropic*, the viscosity under fixed stress decreases with time, whereas for others, dubbed *rheopectic*, the opposite takes place. Thixotropy and rheopecty are therefore the analogies of, respectively, shear thinning and

---

[32] For a long time, a similar mechanism was held to be acting in concentrated colloids, so that shear thickening materials are also called "dilatant". Actually, the reason seems to be more complicated, and colloid thickening seems rather to be related to complex aggregation processes induced by flow. These effects are still only partly understood.

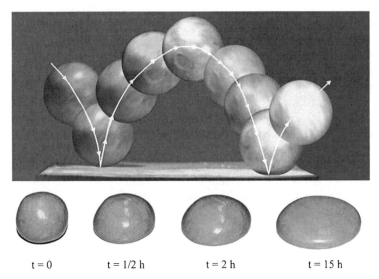

$t = 0$       $t = 1/2$ h       $t = 2$ h       $t = 15$ h

**Fig. 3.11.** Upper panel: bouncing of a Silly Putty ball on a table, reconstructed from a frame sequence obtained with a fast camera. Lower panel: the same ball allowed to rest on the table, at the start of the experiment, half an hour later, two hours later, and after one night.

thickening, but we had better avoid mixing up these two conditions. Here, what causes the non-Newtonian behavior is not *how much*, but rather *how long* we strain the material[33].

Whereas rheopectic liquids are quite rare (with the noticeable exception of gypsum paste, which progressively hardens when kneaded), many colloidal suspensions, in particular when made of particles that attract each other, are thixotropic[34]. This behavior is often observed for suspensions of platelets such as clays, and seems to play an important role in geophysical phenomena such as "soil liquefaction", taking place during lengthy earthquake shocks and *lahars*, the mudflows of water and pyroclastic rocks flowing down the slopes of a volcano during an eruption.

Many biological fluids are also thixotropic, in particular the synovial liquid lubricating bone joints: the more we move a joint, the smoother it gets. Reproducing this effect is the main problem to solve for making durable artificial hip replacements (as you may know, these conversely tend to "seize up"). Just as a curiosity, it has been suggested that the "miraculous" event of the annual liquefaction of the blood of Saint Januarius in Naples takes place

---

[33] For example, corn starch has strong shear thickening properties, but does not show any dependence of viscosity on time, whereas toothpastes are shear thinning, but hardly thixotropic (and it better be so!).

[34] Actually, this is a common feature of the "physical gels" we shall deal with in a later chapter.

because the latter may not be real human blood, but a mixture carefully made to be thixotropic[35]. Clearly, since direct chemical checks are not allowed, any final verdict is impossible. All we can say is that everything that is observed is fully compatible with this hypothesis and, moreover, that solutions with these properties can be easily obtained starting from materials that were fully at the disposal of medieval people.

The rheological properties of supramolecular systems can be very complex, bringing together several of the features of non-Newtonian fluids I have highlighted. Basically, there are two different causes of such a complicated behavior. First, when they flow, these systems may *restructure*, because their building blocks change size or shape. For example, in a colloidal suspension, aggregates may form, and then redissolve when the fluid comes back to rest. Similarly, the surfactant aggregates we shall discuss in the next chapter can totally change shape, turning for instance from flat sheets into coiled "onions". Second, this "restructuring" takes place *over long timescales* compared with those proper of simple liquids, which actually restructure too. In fact, any liquid behaves either as a fluid or elastically, depending on whether the strain last much longer or much less than its spontaneous restructuring time. Even water, for instance, if subjected to a stress for very short times, around a *millionth of a millionth* of a second, acts like a piece of rubber! Within Middle-earth, these times are simply prolonged to values that are easy to detect experimentally. However, grasping the basic nature of these restructuring effects is rather hard, and is an active research topic. At least we may have learnt something about polymers, which is the topic we discuss in the next section.

## 3.10 Nightmares for Indiana Jones

Most of you will remember that Indiana Jones, the intrepid hero of many movies by Spielberg, had a secret Achilles' heel: he was scared stiff by snakes, in particular when they slithered all around him. Now, since I have always envied, like many colleagues of mine, the adventurer/professor Harrison Ford has masterfully given a face and a voice to, I cannot resist the temptation to beat him at least in this field. Therefore, we shall just talk about the rheology of our (admittedly less dangerous) little snakes, the polymers.

Let us start with a dilute polymer solution, well below the overlap concentration, where the chains are well-separated coils. We shall discover that, even in this conditions, they can noticeably alter the viscosity of the solvent. To see this, I have to provide you with one of the few exact results in rheology, obtained in 1906 by Einstein (fancy that, him again!), just after having ex-

---

[35] Actually, these events are just regarded by the Catholic Church as "prodigious", a sensible precaution, since many other relics as prodigious as Saint Januarius' blood, as far as concerns their rheology, are spread throughout Italy.

plained Brownian motion[36]. Einstein's model shows that the viscosity $\eta$ of a suspension differs from its value for the pure solvent $\eta_s$ just by a term proportional to the fraction of the total volume occupied by the suspended particles, usually denoted by the Greek letter $\Phi$ (Phi). More precisely, we have

$$\eta = \eta_s \left(1 + \frac{5}{2}\Phi\right).$$

This result is sufficient for understanding why polymers increase the viscosity so much, even at low concentration. Because they are basically "empty", the coils fill a very large space, much more than solid particles would at the same concentration.

But the story really changes drastically when we cross the overlap concentration and the chains start to form a polymer network. Do you still remember our plate of spaghetti? How can polymers move around this tangle? Getting in each other's way, they make life hard for each other, and hard also for the solvent, which has to flow with them. The easiest way to picture how chains can move in these conditions is to look at a snake creeping among a host of its fellows. Actually, the basic mechanism of chain motion in a concentrated solution or in a melt, suggested with great insight by Pierre-Gilles de Gennes, Nobel laureate in 1991 and true "founding father" of soft matter physics, and later developed by Sam Edwards and Masao Doi, is named *reptation*, after the creeping motion of a reptile. The idea is to consider the motion of a chain as an "obstacle race", where the handicaps are the other polymers it must avoid. I have tried to sketch this in Fig. 3.12, where you should regard the dots as the cross-sections of other chains in the plane of the drawing.

As you can see, the surrounding macromolecules define a sort of "tube" through which the chain has to crawl. Although narrow (it gets narrower the higher is the polymer concentration), this tube nevertheless leaves the chain enough freedom to twist about inside it. It is just this "twisting" that enables the polymer to crawl, for it can advance essentially like a worm, by progressively pushing forward a "kink" as shown in the figure. Clearly, this tube is not a lifeless structure, since the other chains are moving too, but rather a "mobile tunnel", which closes at the back of the chain when it advances, whereas the chain has somehow to thread its way through the crowd, digging a new tunnel while it moves forward. This may seem a trivial picture, but it can be restated in rigorous terms, enabling polymer scientists to work out, for instance, how viscosity depends on concentration or on the polymer's molecular weight. The model also allows them to understand why polymeric solutions show a viscoelastic behavior, and to account, at least qualitatively, for their shear-thinning properties.

---

[36] Please note: the basic law of viscous flow was proposed by Newton; the first model of a viscoelastic fluid was made by Maxwell; one of the most important results in rheology was obtained by Einstein. Three names making up a veritable pantheon of physics, in the teeth of those who regard soft matter as a second-class subject compared with the "big themes" of physics.

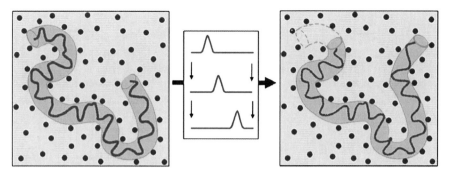

**Fig. 3.12.** Reptation model for polymers. The middle panel shows the motion of an "undulation", or kink, along the chain.

## 3.11 Charged polymers: polyelectrolytes

The large number of different polymers we have already met could make you think that we have used up all the possibilities offered by these intriguing chains. No, we have not. So far we have totally neglected a very different class of macromolecules, which, by the way, make up a large fraction of water-soluble polymers and by far the majority of biopolymers: charged polymers, or *polyelectrolytes*, whose monomers become ionized in water. A simple example is **sodium polystyrene sulfonate** (NaPSS), used for instance as an additive for cements, to fix dyes in the textile industry, and more recently also to make membranes for fuel cells (those to be used in future, or futuristic, hydrogen cars), where a $NaSO_3$ group is attached to the hydrophobic polystyrene monomer. In the case of NaPSS, sodium is released as a $Na^+$ ion, leaving the chain negatively charged. In the final chapter, however, we shall encounter proteins, which are polyelectrolytes showing *both* negative and positive charges on the chain at the same time. Because of this "dual nature" a protein can be neutral overall, but this does not means that charge effects do not matter.

We have seen that the presence of charge on a colloidal particle induces new forces, owing to the presence of a cloud of ions (known as counterions) around it. This happens also for polyelectrolytes, because, as for simple colloids, each coil is surrounded by a little cloud with opposite sign which repels the other chains. But in polyelectrolyte solutions, charges do not only yield new forces between *different* chains. In fact, their major effect is on the conformation of a *single* chain. Because equal charges repel, trying to be as far apart as possible, the chain gets "stretched". Thus, the polymer persistence length $L_p$ *grows*, and the chain becomes much more rigid[37]. As for the forces between colloids, these effects are very strong in the absence of added salt, so that

---

[37] In this case, we speak of the *electrostatic* persistence length, to contrast it with the "intrinsic" rigidity of the chain.

a polymer that would be flexible if we neglected charge effects may become a rigid stick, whereas they become less relevant the more electrolyte we add to the solution. If we assume for simplicity that the chain is almost ideal, which is true for concentrated solutions, the gyration radius is then given by $R = \sqrt{N}L_p$, where this time $N$ is no longer the number of monomers, but the ratio $L/L_p$ of the total extended length of a polymer chain to the persistence length[38].

What is the maximum charge of a polyelectrolyte? We might believe this to be related only to the chemistry of the macromolecule, so that, by placing the charged groups very close to each other, the charge can be increased at will, but it is not so. Detaching from the chain and wandering freely in the solvent, a counterion surely gains entropy, but as it gets further from the chain, it loses electrostatic energy. One can show that, to balance these effects, a fraction of the counterions has to "recondense" on the chain, partly neutralizing the charge of opposite sign. This process goes on until the free charges on the chain are at least a *Bjerrum length* apart, corresponding to the distance at which their electrostatic interaction energy is equal to $kT$ (in water, about 0.7 nm). This effect, called *Manning condensation*, therefore sets an upper limit to the number of charges that can actually be present on a polyelectrolyte.

Even more interesting is the behavior of a semidilute or concentrated solution of polyelectrolytes, which form what is called a *polyelectrolyte gel*. The main difference between these gels and simple polymer networks is that the large number of counterions surrounding the chains increase the osmotic pressure a lot. As a consequence, the gel swells by sucking inside as much water as it can, behaving similarly to an osmotic membrane that does not let the small ions escape. This is the working principle of those super-absorbing polymer gels we already mentioned, but also of our bone joints, which have to sustain large compression stresses. The presence of charge therefore makes polyelectrolytes more complex than simple polymers, generating many other effects which are still under intensive investigation. Among them, as we shall see, are the mechanisms allowing our DNA, a polyelectrolyte about two meters long, to fit inside the narrow space of a cell nucleus, which has a diameter of a few *microns*.

To avoid wearing you out completely, here I shall end this brief introduction to polyelectrolytes which, because of their solubility in water, play a major role in many areas. But please don't breathe a sigh of relief, for in a different manner we shall meet them again. Anyway, by looking at the complex world of polymers, we have learnt something about the mechanical and rheological properties of soft materials. Above all, we now have a better grasp of why we call them "soft", meaning by that a material that very easily changes shape at almost constant volume, because it simultaneously has characteristics of

---

[38] It can easily be shown that this corresponds to writing $N = \sqrt{L_p L}$, namely, the gyration radius is the geometric mean between the persistence and the extended chain length.

both solids and liquids. It is now time to turn our attention to systems that are even more amazing, systems that *spontaneously* organize themselves into mesoscopic structures.

# 4

# Double-faced Janus molecules

*A pretty elastic skin: surface tension – Molecular schizophrenia: the surfactants – A multitude of multicolored bubbles – Foams: when a blind man teaches us to see – Micelles: group therapy for schizophrenic molecules – Whiter than snow: the science of detergency. Pastis, pomade and petroleum: the multiform world of emulsions – Chronicle of a death foretold (epitaph for a suntan cream) – Smart drugs, white diesel, and art restoration: microemulsions, when small is beautiful.*

"These guys are like oil and water." You may have used this idiom to refer to people who really don't get along. After all, could any two things be more hostile to each other than these liquids, which do not like to mix at all? In fact, all common substances can apparently be split into two factions: stuff like alcohol, which dissolves in water but not in oil, and stuff like fats, behaving totally the other way around. At first glance, there seems to be no way to make these different worlds talk to each other.

If you think so, you have been misled. In this chapter, we shall see that "water" and "oil" (where the quotation marks indicate that we are not necessarily talking about tap water and cooking oil) often live together quite happily, in stuff ranging from milk to anisette, from suntan creams to pesticides, from crude oil to drugs. To do it, however, they need special "mediators". These are neither mesoscopic particles like those that form colloids, nor long chains like polymers that coil up to form soft particles. Instead, they are rather simple molecules, but with the wonderful property of *self-organizing spontaneously* into very special particles when dissolved. Moreover, because of their nature, these molecules can build a bridge between water and oil lovers. Their wonders are due to their "split personality" or, if you prefer, to their affectionate nature, driving them to like *both* water *and* oil. These *amphiphilic* molecules, as we shall call them[1], are the main subject of this chapter. Before introducing them, however, I wish to dwell briefly upon a liquid that is really very special

---

[1] Judging on what we have just said about their loving preferences, I have no objection if you prefer to dub them "sexually ambiguous compounds."

R. Piazza, *Soft Matter*, DOI 10.1007/978-94-007-0585-2_4,
© Springer Science+Business Media B.V. 2011

to me, not least because it makes up more than 90% of each of my readers –
a unique liquid we call water.

## 4.1 Striding on water: the physics of Jesus bugs

As far as I know, there is only one person who managed to walk on water,
and He was a very special person. However, in this respect, He would look
much less unusual if compared not only to us lesser mortals, but to the whole
animal kingdom. Indeed, many bugs not only walk, but spend a large part of
their brief life and carry out all its basic tasks (see Fig. 4.1) floating on water,
where they are able to stand thanks to very long legs. These are the water
striders, members of the Gerridae family, also known as pond skaters or Jesus
bugs (I don't know if you have ever happened to see them, while looking at a
clear stretch of water)[2]. How do these strange little creatures manage to float

**Fig. 4.1.** Superficial life of pond skaters, suspended over the water surface. The right
panel shows that the surface tension can bear at least *two* bugs. [*Source: picture by
Markus Gayda, from Wikimedia Commons*]

so effectively? It certainly can't be because they are lightweight since, albeit
small, they are almost as dense as we are, so they should sink, at least partially.
Quite the contrary, they are literally suspended *over* the water, which actually
they do not even touch. The reason is that the tips of their legs are covered
by a fuzz of microscopic hairs, trapping microscopic air bubbles. This "air
cushion" keeps them afloat because it *cannot* penetrate water[3].

---

[2] They are not the only ones. Some of their close relatives like the marsh treader
also "skate" on water. Other insect grubs, moreover, grow attached to the water
surface, but *below* it.

[3] This is similar to the system ducks use to stay dry when they swim. Until a few
years ago, when a group in the Massachusetts Institute of Technology pointed out

Where does this impenetrability of water by air stem from? In liquids, the molecules attract each other with rather strong forces (that is why, unlike gases, they have a fixed volume). In water, these forces are mainly due to what is called a *hydrogen bond*. A water molecule is made of an oxygen atom bound to two hydrogens, but in fact each of these hydrogen atoms is also partly bound to the oxygen atom of another molecule. In other words, hydrogens are somehow shared by the water molecules, hence forming a network of bonds, each one with an energy of about 8 $kT$ (thus quite a bit larger than the thermal energy), which keeps the liquid together[4]. However, looking at Fig. 4.2, where water molecules are pictured with their approximate "Mickey Mouse head" shape with the hydrogens as ears, we see that the situation is quite different for a molecule lying *on the surface* compared with another one inside (in the *bulk*). While molecule $A$ is equally pulled in all directions by the surrounding molecules, molecule $B$ feels only those molecules which are below or to its side, whereas it has no particular attraction to the air molecules above. To lie on the surface, a molecule must therefore give up half of the attractions, and this is why forming a contact surface between water and air has an *energy cost*, as we shall shortly see.

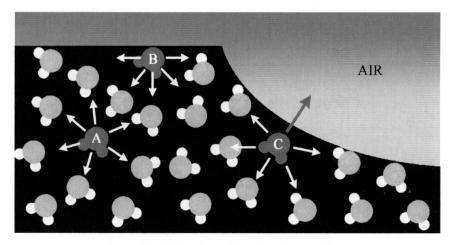

**Fig. 4.2.** Molecular origin of surface tension.

Let us first try to see why this prevents our bugs sinking. As long as the surface is horizontal, the molecules lying on it, though interacting with only half of the usual number, do not feel too bad. After all, the net downward force due to all the molecules beneath is just what keeps it from leaving the

---

the presence of these micro-bubbles, people believed that the tips of the skater's legs secreted a kind of water-repellent wax. If this were the case things would not change too much, as we shall see, but using an air cushion is more effective.

[4] This structure is even more pronounced in ice, namely in solid water.

surface and wandering off into the air (in other words, these attractive forces are what keeps water from evaporating). But suppose we make a "dent" in water with an air bubble, which could be one of those little ones attached to the insect hairs. Now, as shown to the right of Fig. 4.2, molecule $C$ feels, compared with molecule $B$, a slight *upward* force due to the molecules to its right and left, which pulls it up and tends to restore a flat surface.

The water surface is therefore similar to the elastic membrane of an inflated balloon, which resists if you try to push a finger inside. This distinctive elasticity, just stemming from the loss of molecular force balance on a surface, is named *surface tension*. This analogy with an elastic membrane can be pursued further. If we make a little slit in a fully inflated balloon, it obviously bursts. What actually happens is that the two sides of the slit grow very rapidly apart, tearing the membrane. Clearly, cutting water with a regular knife like we cut rubber is impossible, but if we had a "molecular cutter", capable of severing the bonds between the molecules lying on the surface of a thin water film and of preventing them from rejoining, the two sides of the cut would break apart just as for the balloon[5], pulled by a force (telling us how much the membrane is stretched) that is clearly stronger the longer is the slit we made. Then, if we call $\ell$ the cut length and surface tension, usually indicated with $\gamma$ (gamma), the force *per unit length*, the total force is given by $F = \gamma \ell$ (see the left panel in Fig. 4.3).

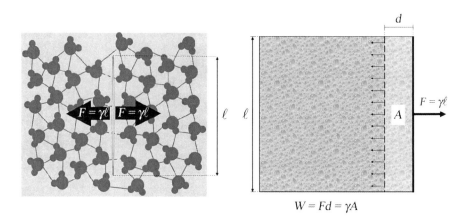

**Fig. 4.3.** Surface tension as a force per unit length (left), or energy per unit surface (right).

Sure, imagining such an "ideal cut" in a liquid is far from easy. We can, however, give another definition of surface tension that is rather different at

---

[5] We shall shortly see that this can be done using soap bubbles.

first sight, but actually fully equivalent. Suppose we try to stretch the surface of a liquid, as in the right panel of Fig. 4.3. To do this, we have to apply a force $F$ that is at least equal to the value $\gamma \ell$ that our elastic membrane can stand. Imagine that we do this, and that we extend the surface by $d$. Then the surface stretches, accumulating an amount of elastic energy that, just like a stretched piece of rubber, the film can give back by shortening. How much is this elastic energy? As always, to change the energy content of something we have to perform *work* on it, which is equal to the force times the displacement[6]. Thus this work is $Fd = \gamma \ell d$, which must be equal to the change in the surface elastic energy. However, $\gamma \ell d = \gamma A$, where $A$ is the increase in surface area, so that the surface tension is nothing but *elastic energy per unit area* associated to the surface.

This equivalent view of surface tension is much more interesting and helpful. In a nutshell, it means that increasing the surface area by an amount $A$ "costs" an energy $\gamma A$. Many effects related to surface tension are better grasped by exploiting this idea. For instance, our bugs do not sink because, to do so, their little feet would have to bend the water surface further, and this amounts to increasing the area with respect to a flat surface. In fact, the water surface is able to stand a weight per unit surface at least tenfold larger than that of a bug before yielding. An even simpler example concerns the shape of a falling water drop. Why is it spherical? Simply because, for a given volume, a sphere is the shape with minimal area, and therefore costing the least energy. Air bubbles rising in a champagne flute behave similarly (here water is outside and air inside, but this does not matter). For a less familiar example, note that water falling from a tap sooner or later starts dripping (the weaker the stream, the sooner this happens). Once again, this is because, at constant volume, water drops have a lesser surface than a cylindrical trickle of water. Or look at the dewdrops on a cobweb. Here too, dew at first condenses on the threads uniformly, and then splits into droplets to reduce its contact surface with air[7].

Not only the surface of water, but the surface of any liquid in contact with air is in tension. More than that, if we bring into contact two liquids that do not mix, such as water and oil, or even a solid with a fluid (either liquid or gas), we can define a kind of "interfacial" tension as the energy per unit area required to form a separation surface between these two media.

---

[6] This work does not depend on the time it takes. Doing the same amount of work in a shorter time requires only more *power*, which is just the work divided by the time requires to perform it.

[7] An intriguing observation is that these droplets are spaced *very regularly*. Explaining why is, however, too hard for the purposes of this book.

## 4.2 Surfactants, a split personality

The moment has come to play a slightly nasty, but effective trick on our bugs, begging in advance the pardon of any Association for Bug Protection, if it exists. Just pour some liquid soap or bubble bath into the lovely pond where they float, skate, and do many other things. A small amount; you really don't need a lot. Better still, soap yourself and have a nice bath (I sincerely apologize to the Association for Clear Lakes Preservation too). The result is disconcerting. Within a short time, our bugs start to flounder, and then, inexorably, to sink[8]. What happened? Apparently the remarkable elastic skin that keeps them afloat has disappeared. It is really so; the water's surface tension has plummeted. This puzzling effect can actually be unraveled by considering the technical name of the active ingredients of soaps and of detergents in general: *surfactants*, namely, molecules acting on the surface tension.

What is so special about these molecules – and about a wider class of chemical compounds named amphiphilic molecules, or simply amphiphiles – that lets them produce such a remarkable effect even when they are present at low concentration? As we have seen, as far as their feeling towards water is concerned, all substances can be split into two main classes, those fond of water (molecules, compounds, surfaces that are *hydrophilic*), and those detesting it (their *hydrophobic* equivalent). True, there are compounds too that are rather non-committal – they like water, but not too much. Nevertheless, the only consequence is that they are soluble in water, but only to a moderate extent (in oil, they behave similarly). Rather than being hesitant, amphiphilic molecules have a split personality, to the point that they can be physically divided into in two easily distinguished regions, the first madly in love with water, the second being an unrestrained lover too, but of oil (or, in general, of hydrophobic compounds). In a nutshell: like Janus, the double-faced Roman god who showed either his peaceful or his warlike face, these "molecular hermaphrodites" have in themselves *both* a hydrophilic and a hydrophobic nature. After all, in Greek, the word "amphiphile" means exactly "that loves both things."

We shall meet a lot of different amphiphilic molecules, but, for the moment, let us just get acquainted with a particularly important representative, at least for historical reasons, which will guide us around the world of these precious molecules. When I was a kid, I would ask myself where the name "Lauril", a washing powder very popular in Italy at that time, came from. The name seemed to have more to do with roast spices or university degrees than with tidy socks and collars. Later, working on similar stuff, I realized that this name comes from the main active component in that detergent, sodium lauryl sulfate. Actually, this is an old name (which does indeed have something to do with laurel), whereas its correct, but trickier, name is **sodium dodecyl sulfate**. Luckily, it is usually shortened to "SDS", and we shall do the same.

---

[8] Someone *really* did this experiment. Just check on YouTube.

How is SDS made, then? The hydrophobic part is a hydrocarbon chain similar to polyethylene, but very short, for it is made of just 12 carbon atoms. The hydrophilic part is instead a simple salt, sodium sulfate, which, like any salt, loves water. A simplified chemical representation of SDS is the following:

$$CH_3 - (CH_2)_{11} - O - \overset{\displaystyle O}{\underset{\displaystyle O}{\overset{\|}{\underset{\|}{S}}}} - O^- + Na^+$$

If this scares you, just take into account that this is one of the simplest surfactants (don't worry: luckily, we shall not need many formulas). When SDS dissolves in water, the sodium ion $Na^+$ is released into solution, whereas the negatively charged $(SO_4^-)$ sulfate group stays attached to the chain. The whole molecule is just a couple of nanometers long, and looks like a tadpole, with a hydrophilic "head" (the sulfate ion) attached to a hydrophobic "tail". A more realistic picture of SDS is shown in Fig. 4.4.

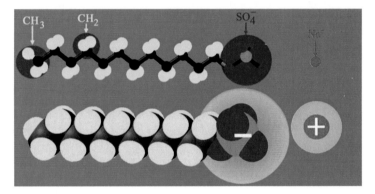

**Fig. 4.4.** SDS molecule.

In fact, this is the general structure of all amphiphilic molecules, which vary in the nature of hydrophilic head and in the kind, and *number*, of hydrophobic tails. Indeed, as Virgil did for Dante, SDS will lead us only on the first part of our journey. To proceed further, we shall need the help of a kind of "Beatrix" with *two* tails, who will guide us to the real Paradise of amphiphiles. These angelic two-tailed molecules are not so useful to wash hands, dishes, or clothes, but rather to make something very close to our hearts. More specifically, you and me.

What happens when we add a surfactant such as SDS to water? First, the amphiphiles try to settle in the single place where their double "emotional needs" are both satisfied – on the surface, between air and water, with their tails, which dislike getting wet, towards the air, and the heads, which are happy to do so, towards the water[9]. What builds up at the surface is then a *monolayer*, a very thin coating of surfactant with a thickness comparable to the length of an amphiphilic molecule. Because this monolayer is so thin, a ludicrously tiny amount of surfactant is enough to cover a huge surface. For instance, to cover an Olympic pool, it takes a bit more than *one gram* of SDS. An immediate result of monolayer formation is a substantial drop in the surface tension. At this stage, indeed, air is no longer in direct contact with water, but rather with the surfactant tails, and this is like being in contact with a hydrophobic oil, whose surface tension is much lower. The addition of SDS, for instance, roughly halves water surface tension, lowering it to a value comparable to dodecane, the hydrocarbon that forms the SDS tail. But then, as the Romans used to say, *in nomen omen*, or the destiny is in the name: "surfactants" necessarily had to do something to surface tension!

What we have just seen has a curious consequence. Physicists are well aware that surface tension has much to do with sea waves; not the big ones, whose rise and fall is due to gravity, but rather those ripples driven by wind (although large waves too can stem from these ripples). But seafarers guessed this long before the physicists. Pliny the Elder, in his *Naturalis Historiae*, tells us how Roman sailors, to calm down storms, used to pour olive oil into the sea (well in advance of the predicted storm). Now, this great foodstuff, which is much more than a simple oil, also contains molecules that are quite similar to amphiphiles[10], and the oil film on the sea surface noticeably reduces its surface tension. Eighteen centuries later, Benjamin Franklin, with some little experiments made on a clear pond in England (at that time, there was no Association for Clear Lakes Preservation to prevent him from doing that), had the merit[11] of bringing attention back to this effect, opening up the route to understand surface tension. It is rather surprising how right from the start, at a time when the existence of atoms and molecules was still more a philosophers' speculation than an accepted scientific fact, Franklin had a clear view of what was happening, in particular that the layer was just one molecule thick. But it is even more stunning that, once he realized this, he did not make the next step – that of reckoning the length of a molecule from the ratio between the amount of poured oil, which he knew, and the area of the spot that forms on the pond, which he could have measured (in his defense, I have to say that the spot is not easy to see, unless special tricks are used). That would have really been a historical step for science, made over a century earlier than the first experimental evidence of the existence of atoms!

---

[9] The same thing happens on the container walls, if it is made of a hydrophobic material such as plastic.

[10] To be precise, therefore, when we talk about oil, we shall not mean olive oil, which is not totally hydrophobic, but rather "mineral oil". So, basically, hydrocarbons.

[11] Besides those of inventing the lightning rod, designing a revolutionary stove, and being the first signee of the Declaration of Independence, and still finding the time for an appreciable number of love affairs. An energetic fellow, without a doubt. . .

It was only in the first two decades of the XX century that another American, Irving Langmuir, laid the foundations of the quantitative study of surface films, a still active research topic for which he got the Nobel prize in 1932. If the credit of being the true "founding father" of surface physics goes to Langmuir, here I wish instead to celebrate another scholar who made a seminal contribution to the experimental development of this science, and whom history, as often happens, has put aside. Maybe this is just because Agnes Pockels was a woman, and moreover not even an official scientist (which she could have become, if German universities had not been off limits for women at that time), but a simple housewife – albeit a housewife quite fond of science, to the point of using her own kitchen to perform surface tension experiments. Yet, just because she was a housewife, she also realized that these studies required very clean surfaces. This drove her to develop a method, to build experimental apparatus with her own hands, and to make observations that, in terms of ingenuity, are almost incredible.

Sure that what she had done was of interest, Agnes dared to write to as famous a figure as Lord Rayleigh, the physicist who, by the way, explained light scattering. Rayleigh, grasping the importance of her results, had in turn the merit of giving Agnes the deserved credit, helping her to publish in *Nature*, the leading scientific journal back in 1891 (when Langmuir was 10 years old)[12]. A toast to Agnes, please, from my lady readers!

You can amuse your children with a sign of the effect of surfactants on surface tension which can be seen using a small boat working by mean of a "soap engine". You only have to take a toothpick, dip one of its ends in liquid soap, and then place it on the surface of a basinful of water. From the soaped end, which behaves as the boat's stern, the surfactant molecules spread rapidly over the water, reducing its surface tension. Now, however, water "pulls" the toothpick surface *more weakly at the stern than at the bow* and, as a result, it darts forward[13]. A toothpick is not a great boat, but I am sure that keen parents can do much better. . .

As we said, surfactants form films on all hydrophobic surfaces. In particular, then, they form films on the surface of colloidal particles which are often made of stuff that does not like water. This is exactly one of the ways to create that fluffy down which prevents the particles approaching closely enough for attractive forces to make them coagulate. Surfactants are therefore widely used to stabilize colloidal suspensions even in the presence of salts that screen charge repulsion. Before going on with our story, let us dwell a bit more upon surfactant films, since these lie at the roots of one of our childhood wonders.

---

[12] As luck would have it, Agnes' younger brother Friedreich is much more renowned than she is among today's physicists, for having discovered an optical effect of particular interest in laser science. To my mind, however, his sister's overall contribution to science is much more significant. Incidentally, both Pockels, although from a German family, were born in Venice.

[13] Clearly, the boat does not move forever, but only until the toothpick is fully encircled by a soap film, so that the surrounding surface tension is uniform.

## 4.3 Soap bubbles: a paradise for kids and math nerds

Imagine you dip a straw in water, and then blow air through it. Could you get a water bubble that inflates like a balloon? Obviously not, because expanding the surface costs too much energy, so that our bubble would burst immediately. However, experience tells us that, provided we add a bit of soap, it is easy to make those gorgeous iridescent bubbles which left us openmouthed when we were kids. What is happening here? As we have seen, the surfactant forms a film both on the external and on the internal surface of the bubble. Lowering of the surface tension allows a "balloon" to form, where the elastic membrane is actually a thin water film covered on both sides by surfactant (rather than soap bubbles, then, we should really call them water bubbles)[14]. By the way, even if this works with any surfactant, our SDS is particularly suitable for making soap bubbles.

That this film is really *very* thin is witnessed by the myriad of colors on the bubble surface that used to amaze (and I hope still amaze) all of us. Why so many colors, when water is transparent? The light hitting the bubble is partially reflected by both the external and the internal surfaces of separation between air and water. When the film is thick, nothing special happens, and we just see some reflected light summing up both reflections, with no special color. This takes place for a newly made bubble, whose film thickness is around a few tens of microns. But the film changes fast, both because the weight pulls water to the bottom, so that the top part of the bubble becomes thinner, and because water evaporates fast from such a wide surface, rapidly thinning the film. When the film thickness becomes very small, comparable with the typical wavelength of the light (which, if you recall, for visible light is of the order of a fraction of a micron), the story changes, because of the optical phenomenon known as *interference*. In some cases, the two reflections help each other, yielding a total light intensity larger than their sum (in fact, if the two reflections are equal, *fourfold* larger), whereas in other cases they cancel each other out. Whether they help or fight each other, however, crucially depends on the specific value of the wavelength. So, for instance, if we illuminate the bubble with white light, the "red components"

---

[14] To tell the whole story, the surfactant actually plays a subtler role, for it also stabilizes the bubble. If the film is stretched in some region, becoming thinner, the film is diluted (the amphiphiles move farther apart from each other). But then the surface tension in that region grows, drawing water from the surroundings, and, as a result, the film tends to recover its initial thickness. In other words, the film "self-heals". This phenomenon, which is known as the *Marangoni effect*, has a lot to do with our soap-boat. It also explains why, when you are immersed in a bath tub covered by bubbles and skim the surface with a sponge you have rubbed with soap (*different* from the bath foam), a large fraction of the bubbles rapidly comes to a miserable end. Or, more subtly, why we often see whisky or cognac "tears" sliding down the glass wall. Giving a detailed explanation of all these effects is really beyond this book.

of the two reflections may kill each other, so that we end up with a reflected light that no longer contains the red, therefore looking blue-greenish[15].

The many colors of soap bubbles are therefore a kind of "fingerprint", implicitly telling us how thick the film is at each point. To read these fingerprints we do not need any C.S.I.: physicists, usually not so keen on everyday questions, can do the job. "Interferometric analysis" based on reflected colors has long become, in industry, a simple and reliable method to determine the thickness of a thin film made of a transparent material. In soap bubbles, moreover, water is not still, but generally has a whirling motion, mostly due to evaporation. This turbulent whirling changes the local film thickness just a bit, but enough to originate the color turmoil you surely have already appreciated.

We do not need to make bubbles to see these gorgeous colors. It is enough to dip a metal wire frame (shaped into a ring, for example) into soap water and then take it out gently. The result is an open soap film attached to the frame, showing a rainbow of magnificent colors changing with time for the very same reason we discussed for bubbles. A quick search on the Internet, and you will discover how varied and surprising the resulting chromatic patterns can be. However, I thought I would show you in Fig. 4.5 a rather special example, created by making a soap film "dance" to the beat of music. This technique, developed by Giuseppe Caglioti in my university, charmingly combines visual and auditory sensations. By making the frame supporting the film more elaborate, complex and fascinating shapes can be obtained. For instance, using the edges of a cube as a frame, we do not get surfaces corresponding to the sides, but instead we see flat films inside the cube, joining in a complicated but definite manner. Something even stranger happens if we dip two wire rings into the soap water and then hold them parallel at a given distance, allowing the soap film to form a tube between them (but only on the lateral surface; that is, not on the flat faces of the rings). We might expect the film's surface to be cylindrical, but it is not so, for the surface bends inwards in a very precise way.

What controls the shape these films take, the way they join, their possible curvature? Much of what we know is due to the painstaking work of a Belgian physicist, Joseph Plateau. From childhood, Plateau, who had also a good hand for drawing[16], was fascinated by the mechanism of vision, in particular by the way images persist on the retina. This led him to build some real ancestors of cinema such as the "phenakistoscope", which used animation mechanisms

---

[15] If the film gets much thinner than visible wavelengths (as happens just before the bubble bursts), something even more spectacular happens. The two reflections fully cancel each other for *all* wavelengths. As a result, the film no longer reflects anything, and becomes fully transparent (if there is a black screen behind the bubble, the film itself then looks black, and is in fact technically known as a "black film".

[16] Apparently, while his fellow villagers were trying to escape the battle of Waterloo, which was fought close to his birthplace, Plateau stayed in the surrounding woods to paint the gorgeous colors of butterflies.

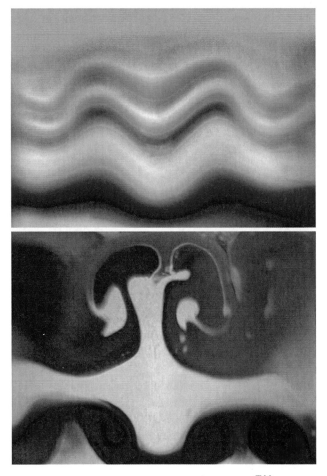

**Fig. 4.5.** Polychromatic dance of soap films. [*Musicolor$^{TM}$ images, courtesy of Giuseppe Caglioti*]

based on this effect. Plateau's main contribution to science, fruit of decades of observation[17], is his 1873 publication, *Statique expérimentale et théorique des liquides soumis aux seules forces moléculaires*, where the laws governing the organization of soap films are first stated. What is really surprising is that, when this work was published, Plateau had already been totally *blind* for 30 years. Struck by serious illness, whose progress he had the strength to

---

[17] Plateau used mixtures made of three parts of Marseilles soap solution with two parts of glycerol. The role of glycerol is to increase the viscosity of water to slow down film thinning due to gravity. This effective trick is still commonly used to make long-lasting soap films and bubbles.

describe in detail, just after starting to study surface tension phenomena[18], Plateau was helped by his wife Fanny, who everyday would read him papers and write out his ideas, by his sister Joséphine, also a good draftswoman, and by other friends and relatives, to carry on (I would say heroically) his work.

For soap films built on wire frames, Plateau's observations can be summarized as follows:

1. Each edge of the frame supports a single film.
2. Films that meet in threes always do so in a *line*, making angles of exactly 120° between themselves.
3. Conversely, *four* films must always meet in a point, making angles of slightly less than 110° (exactly those at the vertices of a tetrahedron, the regular solid with four triangular faces).

Recalling the meaning of surface tension, it is not hard to understand why it must be so. In particular, since the forces exerted by the films on the common edge have to balance, the second law is almost trivial (it should be easy to see why).

What is really curious is that soap films make it possible to solve experimentally some "minimization" problems in geometry that, though apparently easy, are in fact very challenging. Suppose for instance that you want to link up with roads four towns, placed on the corners of a square with a side of 100 km (to simplify, we shall also assume that there are no geographical obstacles to stop us building straight roads). A trivial way to do so, surely the most convenient for travellers, is shown in diagram 1 of Fig. 4.6, where all the towns are directly connected. This amounts to building four 100 km long sides, plus two diagonal roads of length $\sqrt{2} \times 100 \simeq 141.5$ km, adding up to 683 km of asphalt. But because our budget is limited, and moreover we do not like to overbuild the plain unnecessarily, what we really wish for is a road system having the *minimum* total length. We might then think of following diagram 2, which still allows any town to be reached from another with only 300 km of road, although people who have to move from $A$ to $D$ or vice versa would not particularly appreciate this. Surely we can do much better by exploiting crossings. In diagram 3, where we build only the crosswise roads in diagram 1, using the crossing to change road, we need only $2 \times 100 \times \sqrt{2} \simeq 283$ km to solve the problem, with a much fairer distribution of the travellers' time and energy. This seems to be the ideal solution, but it is not. Using only elementary geometry, it is easy to show that diagram 4 yields a total road length which is 4% shorter still[19].

In this specific case, proving rigorously that arrangement 4 is the shortest (although not unique) is not difficult, but the matter gets trickier as soon as the number of towns increases, or when they do not lie on a regular polygon.

---

[18] That his blindness was a consequence of a lengthy naked-eye observation of the Sun, as often reported, has no historical grounds.

[19] More precisely, the total length in this case is $100(1 + \sqrt{3}) \simeq 273$ km.

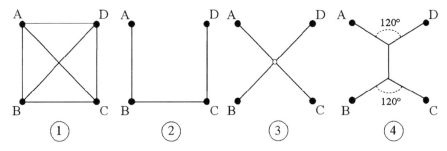

**Fig. 4.6.** The road problem.

Yet the exact solution can be simply obtained by using two Plexiglass sheets spaced by small rods standing for the town positions, dipping the sheets in soap water, and observing the structure of the soap films connecting the spacers. The reason is very simple. The total film surface, and therefore the total length of their sides, *has* to be minimal just to minimize the surface energy associated with surface tension.

Even more interesting for mathematicians is the study of *curved* soap films. To understand why, we have to come back to soap bubbles to introduce a concept which will be very useful in what follows. To inflate a balloon, you must *blow* air inside, so that the internal pressure counteracts the elastic tension of the membrane, which would like to retract. Something similar must then happen for a soap bubble (or even for a water drop in air and an air bubble in water). Namely, *pressure must be higher inside than outside.* How much higher? Starting from the definition of surface tension, this is not too hard to work out. But we shall use a trick that will allow us, if not to calculate its exact value, to see at least how this excess pressure depends on the radius $R$ of the bubble: so-called "dimensional analysis". We have just two "ingredients" for constructing the pressure excess, $\gamma$ and $R$. But $\gamma$ is a force per unit length, so the only way to obtain a pressure, which is a force per unit *surface*, is to divide $\gamma$ by a length that *has* to be $R$ (it is the only one we have at our disposal). Thus, the *Laplace pressure*, as this excess pressure is called, has to be proportional to $\gamma/R$. In fact, a rigorous calculation shows that it is just equal to $2\gamma/R$ for a liquid drop or a gas bubble, whereas it amounts to $4\gamma/R$ for a soap bubble, because we have *two* interfaces (obviously, here $\gamma$ is the surface tension in the presence of the soap films, which is much lower than for plain water). A curious feature is that *the smaller the bubble, the larger is its internal pressure.* If we could connect two soap bubbles of different size without bursting them, the smaller bubble would deflate and the larger inflate. Moreover, the more the smaller bubble size decreases, the *larger* its internal pressure gets, so that this process gets faster and faster[20]!

---

[20] This does not hold for a rubber balloon, since, as we have seen when discussing rubber elasticity, beyond a given radius its elasticity changes with size.

A curved soap film therefore implies a pressure difference between the two regions it separates. But what happens when we have two rings, held apart, and we make a soap film only on the side surface, leaving the ring planes *open*? Here the internal and external pressure must be the same, since the "outside" and the "inside" are actually connected. Yet, to follow the contour of the ring, the film *must* bend, and this seems to be contradictory. In fact, as I already mentioned, the film surface is not a straight cylinder, but curves *inwards*. In other words, if we examine just half of the surface, it looks like a saddle (see 4.7). From a mathematical point of view, these two opposite curvatures can be combined in such a way that, if we properly define a "total" surface curvature, this total vanishes. Finding surfaces with zero total curvature is one of the hardest problems in higher geometry. Once again, we see how a little soap water allows us to build surfaces of zero curvature with a contour bordered by the wire frame, provided that the soap film is not totally closed, so that the pressure is the same on the two sides. Soap films are therefore a resource of great interest for mathematicians, who, at least on this occasion, become experimenters. But not only for them. The analysis of these structures finds application in modern architecture, where soap film models are used to design *tensile structures*, such as giant domes for events and performances.

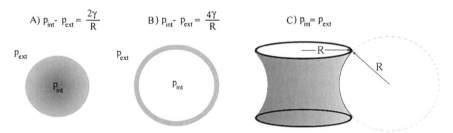

**Fig. 4.7.** Laplace pressure (internal minus external pressure) for a droplet (A) and for a soap bubble (B). A zero-curvature shape for an open soap film supported by two wire rings is sketched in (C).

Soap foams are of even greater interest for other applications. Note that the structure of a common foam, such as that used for shaving, *evolves* with time. At the beginning, it is made of many soap bubbles in contact, but when, owing to its weight, water is gradually drained to the bottom along the films, the films get thinner, and the foam becomes a jumbled structure made of interconnected irregular polyhedra. This first stage of evolution, which takes place rather rapidly, is followed by a second, much slower phase, when the films rearrange to satisfy Plateau laws, reaching a stable structure. A good friend of mine, Joel Stavans, is now professor at the Weizmann Institute, the most prestigious scientific institution in Israel, where he deals with quite complex topics at the boundary between physics and biology. Yet one of his most interesting investigations is still the work he did, together with James Glazier, when he was just a PhD student in Chicago. Joel and James took two spaced Plexiglass sheets, filled the gap with shaving foam, and then followed the development of the foam structure simply by... placing the sheets in a Xerox machine and taking copies. More than a true experiment, this may seem a typical diversion for young physicists during their spare time. Yet, even with this simple planar geometry, the time evolution of a foam is far from being trivial[21], so much that predicting what would happen required the genius of John von Neumann, one of the greatest mathematicians of the XX century and the real father of the computer, and is still of great interest in apparently unrelated areas such as crystal growth.

## 4.4 Micelles: when surfactants find peace

Let us now ask ourselves what happens if we *go on* adding surfactant. When they have to judge whether to dissolve or not, surfactants face an apparently insoluble dilemma. Should they please their hydrophilic heads, which are eager to do it, or listen to the hydrophobic tails, which shun contact with water?

In fact, *any* substance, even the most hydrophobic, must dissolve a bit in water. Once again, this is because of the crucial role played by entropy that we stressed when discussing polymer elasticity. Suppose we drop a wax droplet from a candle in a bathtub full of water, for instance. We have seen that wax is made of paraffins, long hydrocarbons that are among the least water-soluble substances. However, just look at the problem the way a single paraffin molecule would. Even if dissolving is expensive in terms of energy, for its relationship with water is very bad, the chance of moving into a sea of water, a practically unlimited space where it can zigzag at will, is too alluring in terms of entropy gain. Therefore, the first molecule will surely do it. As the wax molecules little by little dissolve, however, this situation changes. Now the entropy gain is much lower, since the free space in the bathtub is shared by all the molecules that have already dissolved. Quite soon, therefore, taking into account the energy cost, the game... isn't worth the candle! In practice, something like paraffin is soluble in water only in absurdly small amounts.

For surfactants, the matter is far easier, thanks to the hydrophilic heads, but even here, when the solute concentration increases, the cost of dissolving free molecules becomes large. Unlike a simple hydrophobic compound, however, a surfactant can find a really original solution to the problem, one that appeases its split personality. Once a specific concentration value, known as the *critical micellar concentration* (cmc), is reached, quite strange events take place. First, the surface tension, which initially decreased as the surfactant was added[22], reaches a minimum and does not vary any further, meaning that

---

[21] What can be immediately guessed is that only *hexagonal* bubbles are stable, just because the sides of a hexagon make angles of 120° between themselves.

[22] The surfactant has then divided itself between the solution and the surface.

the contact surface with water is *saturated* with surfactant. Second, when we shine a light through a solution beyond the cmc it looks slightly bluish, similar to diluted milk, meaning that particles of some kind, larger than surfactant molecules, have formed. Last but not least, if we add a little oil to the solution, the oil now *dissolves* (in other words, we see no floating oil drops). So, the incompatibility between water and oil has mysteriously disappeared.

What has happened? Simply, the amphiphiles have resorted to the principle "a trouble shared is a trouble halved", spontaneously forming aggregates called *micelles*. The number of molecules that make up a micelle, called the aggregation number, and the shape of the aggregates too, strongly depend on the kind of surfactant and on the solvent properties (for instance, whether salt is present or not). We shall later see what fixes size and shape in general. Anyway, in the simple case of SDS (in fresh or lightly salted water), micelles are just little balls with an aggregation number $N \simeq 80$. Unlike the random colloidal aggregates we formerly met, however, micelles have a neat structure. The hydrophobic tails stay inside, as much as they can, so that the heads, which lie on the micelle surface, screen them from contact with water. Thus everyone is happy. Each hydrophobic chain, avoiding the contact with the solvent, is hosted in a droplet made of other chains it likes, whereas the hydrophilic heads enjoy full immersion in water. For a surfactant molecule, the small price to pay is losing its freedom of moving around all alone, since this has now to be shared with a fair number of its fellows.

Usually, the cmc is very low. For SDS in fresh water, it is slightly more than two grams per liter, whereas for other surfactants we shall shortly meet it can be a hundredfold smaller. A very rough sketch of a SDS solution beyond the cmc is shown in Fig. 4.8. Here micelles are in equilibrium with a concentration of free surfactant equal to the cmc and with the surface layer. Moreover, the $Na^+$ ions, highlighted by white dots, are mostly amassed into little clouds around the micelles, playing the same role as counterions in charged colloids. Micelles should not, however, be regarded as static aggregates, as if a SDS molecule, once it has slipped into a micelle, stays there forever. On the contrary, the free surfactant molecules exchange continuously with those in micelles, so much so that the "time of residence" of a given molecule in an aggregate is just a few milliseconds. Thus, micelles are quite dynamic objects. Nonetheless, if we could take a snapshot of what is there in solution[23], we would see that, at any time, the aggregates are made up of more or less the same number of molecules.

Even if Fig. 4.8 yields a very rough picture of micelles, it is not hard to realize that, to fold into a sphere, the hydrophobic chains cannot be as straight as ramrods, but must twist about, as polymers in solution do. The difference is that, compared with the latter, surfactant tails are much shorter. Hence,

---

[23] In fact, this can be done. Although no microscope is so powerful as to "see" micelles in solution, their size and aggregation number can be precisely obtained from light scattering measurements.

below a given temperature (remember, at low temperature entropy counts for less), an extended zigzag configuration is more favorable because it costs less, so micelles cannot form anymore, the surfactant solubility drastically decreases, and the latter precipitates in the form of small crystallites. As we shall see, this effect is to be avoided if we want to make detergents allowing us to wash at low temperature.

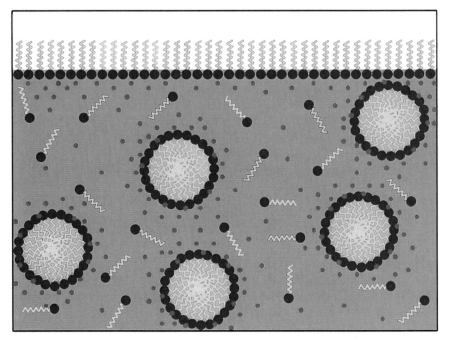

**Fig. 4.8.** A rough sketch of a SDS micellar solution.

## 4.5 As white as can be: the science of cleaning

There is a natural gesture we have made thousands of times in our life, for instance when we get up or before dining: washing our hands. We are so accustomed to using soap we probably take for granted that this marvelous stuff, which makes it possible to cleanse our skin of fats and other hydrophobic substances, has accompanied mankind since its dawning. Maybe it is really so, for even a Sumerian tablet from the XXIII century BC (when soap preparation was already known to the Egyptians) describes how to make it from oil of cassia, the Chinese cinnamon. Yet the history of soap is not so straightforward. The Romans, for instance, apparently did not use it, at least before the Late Empire period, and Biblical references are, at best, free interpretations by the translators. Only with the Arabs did the production of simple and aromatic soaps from olive oil become widespread, first in the Islamic world and much later in Europe.

Traditionally, saponification (which means just the making of soap) was done by treating vegetable oils or animal fats with caustic soda. The reaction yields sodium salts of the *fatty acids*, vitally important compounds that we shall meet again in the last chapter. Soaps are similar to SDS, albeit much less "aggressive", and are in every respect surfactants, with long hydrophobic tails and negatively charged head groups. Incidentally, since soap was not at all cheap, until recently laundry, home cleaning, and sometimes even personal care were done with lye, manufactured by boiling ashes (remember Cinderella?) to obtain dilute caustic soda solutions with detergent, degreasing, and disinfectant properties. While it simply scrapes off floor surfaces, washing soda, being strongly basic, has a softening effect on fabrics, which then release trapped dirt. In fact, it also induces partial saponification of organic grime, which improves its detergent power.

How then does soap work? What we have just learnt gives an important clue, by telling us that hydrophobic substances dissolve *only* in the presence of micelles, which are therefore essential for something to act as a detergent. The mechanism is very simple. Oils and fats, which dislike contact with water, are instead well suited to the environment in the micellar core, where they are easily solubilized. Similar to what happens for adsorption on colloid surfaces, hydrophobic compounds are carried along by the micelles. This property – making surfactants efficient carriers of water-insoluble substances – is crucial not only for all basic strategies for cleaning our skin, fabrics, and tableware, but also many mechanisms of internal transport in living organisms. For instance, cholesterol, a name that may remind you of stern advice from your doctor, but which is at any rate a vital compound for your body, can be carried around only thanks to the joint effort of lecithin and bile salts, which form micelles. When the micelles can no longer cope with this burden, the going gets really tough for arteries, where cholesterol settles and sticks. We should then literally keep the question of solubilization in micelles . . . near to our hearts.

## 4.6 A large and varied family

Soaps and SDS are just two representatives of a very large family of surfactants, or, more generally, of amphiphilic molecules, both natural and man-made. We do not use the same detergent to wash our hair and our car, nor do we clean our teeth with dishwashing powder. Besides detergency, moreover, surfactants have countless other uses, so much so that their total market value amounts to tens of billions of dollars. At this stage, to understand why these molecules are so valuable for the modern economy, it is worth meeting some other members of this large family. As we shall see, differences between amphiphile classes mainly stem either from the *type* of hydrophobic head, or from the *number* of hydrophobic tails. For the moment, we shall only consider those ones that are properly called "surfactants", coming back later to another wide class of amphiphiles, which are essential for the existence of life itself. In particular, setting physics almost completely aside, I shall mostly focus on the chemistry of the hydrophilic head, which is particularly relevant for most purposes. I beg the pardon of my chemist friends if what I am going to say is rather incomplete and and definitely... superficial. After all, with surface-active molecules, it has to be so.

### Anionic surfactants

The simplest examples of these surfactants are SDS and soaps made from fatty acids, where the tail is a hydrocarbon chain, whereas the head is made of a salt that ionizes in solution. The ion remaining bound to the chain is negatively charged (it is an *anion*), whereas the counterions released into solution are positive. Anionic surfactants, the first that were synthesized, have long been the most widespread kind of surfactant, in particular for their high efficiency as detergents, their foaming properties, and their low cost. Washing powders, degreasers, and floor detergents usually contain a large amount of anionic surfactants, which are also used at lower concentration in shaving foams, shampoos, bubble baths, and even toothpastes. Aside from detergency, anionic surfactants are widely used in industrial sectors ranging from pigment stabilization to cement making, where they increase frost resistance by promoting the formation of micro-bubbles preventing the cement from fracturing. They are also used as additives in some methods for enhanced oil recovery we shall discuss later. Last but not least, they also play an important role in biotechnology (for instance, many basic methods to investigate proteins are based on our familiar SDS).

Usually, anionic surfactants form charged globular micelles, similar, in this respect, to the colloidal particles we have already met, but much smaller (they have a size of a few nanometers). As we mentioned in Chapter 2, multivalent counterions – the sort that bear more than a single charge – can have deep effects on the behavior of charged colloids, and this is true for the detergent properties of anionic surfactant solutions too. In solutions of simple soaps, even a tiny amount of calcium $Ca^{++}$ or magnesium $Mg^{++}$ ions, which are abundant in hard or brackish waters, induces the formation of insoluble salts that, as anyone who has tried to wash with sea water knows, precipitate on the skin as a greasy film. But even in solutions of SDS or, to a lesser extent, of more complex anionic surfactants such as the **alkyl-benzene sulfonates**, which still form the main active ingredient of washing powder, divalent ions raise the "Krafft temperature" below which the surfactant precipitates. This is one of the reasons why, until a few decades ago, hand or machine washing required hot water.

Hence, chemists have made determined efforts to develop new surfactants, like the non-ionic types we shall shortly discuss, or suitable mixtures allowing for cold washing. In any case, this generally required them to add chemical compounds such as phosphates that specifically bind to divalent ions. Phosphates, however, have a rather unfortunate impact on the environment, because they feed and help the spread of seaweeds, which take oxygen from the other inhabitants of seas and lakes[24]. Around the end of the past century, we realized that there was already too much phosphorus coming from fertilizers[25] to let ourselves pour surfactants galore into our waters as well. The crucial step to get rid of phosphates has been to add to surfactants other compounds, the **zeolites**, which are of extreme interest for the chemists. Zeolites are micro-porous minerals obtained as colloidal particles by precipitating solutions of aluminum and silicon salts. These particles then aggregate, making a colloidal gel, which slowly rearranges into the form of zeolite crystals with a size of a few nanometers. A unique property of these "nanocrystals" is the extreme geometrical and chemical uniformity of the pores inside them. Their size ranges from a few Ångstroms to a few nanometers, depending on the zeolite, allowing them to trap specific ions or molecules. In other words, they are true "molecular sieves." Large amounts of zeolites, specifically designed to subtract calcium and magnesium ions and avoid surfactant precipitation, are added to washing powders, in a total amount that can reach more than half of the product weight (whereas surfactants do not exceed 20–25%).

### Cationic surfactants

Even though these surfactants have an ionic head group too, here the charge bound to the hydrophobic tail is *positive*, whereas the released counterions are negative. Cationic surfactants are harder and more expensive to make, perform definitely worse as detergents, and, moreover, are often rather unpleasant for health. Why then take the trouble to make these molecules, which actually take a slice of the market that is far from negligible? Simply because these surfactants are *essential* in very specific and noteworthy applications.

The reasons are many. First of all, as we shall see, cationics often form very different micelles that, instead of having a globular shape, look like very long "spaghetti". These aggregates have a remarkable effect on water viscosity. For example, less than 1% of a common cationic surfactant such as CTAB (**cetyl-trimethyl-ammonium bromide**, but you absolutely don't need to remember its name) can, in suitable conditions, turn water into a *solid gel*. As a consequence, these surfactants are of major interest as thickeners or,

---

[24] If a dying sea covered by greenish slime does not impress you too much, then note that the many jellyfish whose stings you try to escape, if you take your vacation on a beautiful Mediterranean beach, also owe their proliferation to phosphates.

[25] Also containing, on the other end, surfactants like the cationics we shall shortly meet.

more generally, to control the flow behavior of fluids. In oil recovery, the high pressure generated by the formation of these gels is used instead to create micro-fractures within the porous rocks where crude oil flows. There are also situations where the specific chemistry of their hydrophilic heads can be exploited to use cationic surfactants for particular purposes, above all as bactericides. Many commonly used disinfectants are based on a cationic surfactant, **benzalkonium chloride**, which is able to disrupt the membranes that enclose bacterial cells, leaving ours totally unhurt[26]. For the very same reason surfactants like CTAB are contained in eyedrops, nose sprays, and mouthwashes[27]. Another important industrial application is based on the strong affinity between the cationic heads and metals such as aluminum or steel. The head groups strongly attach to the metal surface, protecting the surface from acids in solution and reducing corrosion.

Yet the most successful commercial application of cationic surfactants is once again in the consumer market of beauty treatment. Most of you, after washing your hair, may have felt it to be unpleasantly dry, dull, sometimes "electric" (meaning that, refusing to obey your hairbrush or comb, it just does not want to stay down). This is because shampoos, beside carrying away grease, also remove that natural layer of fats that gives hair the sensation of softness, shine, and ease of combing. In the past this was remedied by dressing the hair with essential oils like the once notorious *Macassar*, a palm and coconut oil used in the Victorian age, or those jojoba and tea-tree oils that are fashionable at present because of the New Age and aromatherapy waves. The fast expansion of the market over the last century led instead to the development of less natural but cheaper products such as grease, an example of the emulsions we shall encounter, and lacquers, based on polymers. In the past few decades, however, hair conditioners to be used immediately after shampooing have invaded the market. Besides substances meant to regenerate, moisten, and protect hair against the blast of hairdryers, the basic ingredients of the conditioners are surfactants like CTAC (a close relative of CTAB, where the counterion is chlorine instead of bromine), where the polar head is a "quaternary ammonium" cation (informally called "quat") made of a nitrogen atom bound to four other chemical groups[28]. Quats strongly bind to keratin, the protein that makes up 97% of hair composition, because keratin is negatively charged. The hydrophobic tails of the surfactants remain exposed to air, making the hair soft and easily combed. Better still, when hair is damaged by aggressive dyeing or other treatments, the keratin surface

---

[26] Not all of them, actually. The same surfactant is also used in well-known contraceptive methods as a spermicide.

[27] Since, as we shall see in a few lines, cationics and anionics do not get along too well, using a mouthwash just after cleaning teeth with toothpaste (often containing anionic surfactants) is not a clever move, unless we first rinse our mouth carefully!

[28] Quats are also responsible for the bactericide action of benzalkonium chloride.

charge *increases*, so that the conditioner binds more effectively just to the damaged areas.

The fact that conditioners are made of cationic surfactants, whereas anionics are a common ingredient in shampoos, has been a hassle for the cosmetic industry, for these two classes of compounds are rather in conflict. Indeed, because of their oppositely charged heads, they would just make complexes that precipitate, with the result that hair would be neither clean, nor soft and brilliant. To obtain those convenient combinations of "shampoo + conditioner", so common in hotels, meant substituting the anionics with other surfactants we shall shortly meet. Obviously, if at home you use a regular shampoo, just remember to rinse your hair carefully before applying the conditioner.

Fabric softeners, often used for washing wool fibers, which again present a negative surface charge, work quite similarly. In this specific case, the key role is played by cationic surfactants rather different from those we have met so far. These amphiphiles have two hydrophobic tails, so that our double-headed Janus has also become double-tailed. This different molecular structure has a strong effect both on their solubility and on the kind of aggregates they form. Usually, indeed, they are only weakly soluble in water, so much so that they positively prefer oil, where they form "inverted" micelles with the heads inside and the tails outside. To make them soluble, special strategies are required, which we shall learn about shortly.

Many different cationics can therefore be added to the already large number of anionic surfactants. However, we are not done yet with charged amphiphiles, for there are also surfactants, known by the complicated name of **zwitterionics**, that combine both natures, since their heads have both a positive and a negative charge[29]. Most of them are positively charged when they are in acid solutions, whereas the negative charge dominates in basic conditions (in this case they are also called *amphoteric*). Even if they are much more expensive, they are preferred for sophisticated cosmetics because of their high tolerability and extremely low toxicity. Moreover, as we shall see, the heads of amphoteric surfactants, and in particular of **betaines** (the most commonly used in cosmetics) have some relation to amino acids, the little blocks proteins are made of.

## Non-ionic surfactants

We have seen that ionic surfactants create some problems both for washing, because of the effect of hard water on the Krafft temperature, and in personal care, where they are often too aggressive. Moreover, they are not particularly good for the environment. These drawbacks can be bypassed by using *non-ionic* surfactants where the hydrophilic head, rather than being charged, is made of a chemical group that usually forms with water just the same hydrogen bond that keeps the solvent molecules together. The class of non-ionic

---

[29] In German, actually, *zwitter* just means "hybrid" (but also "hermaphrodite").

surfactants is very wide too, but two specific kinds are particularly interesting both for commercial uses and for our purposes.

First, we have **ethoxylate alcohols**, which are close relatives of PEG, the simple hydrophilic polymer we met in the previous chapter. Here, while the hydrophobic tail is chemically similar to SDS (with a variable number of $CH_2$) groups), the hydrophilic part is a short chain too, but made of $CH_2 - CH_2 - O$ (ethoxylate) groups, whose oxygen can bind to water. Concisely, they are represented by $C_nE_m$, where $n$ is the number of carbon atoms of the hydrophobic chain, and $m$ the number of hydrophilic ethoxylated groups. Compared with ionic surfactants, where the repulsion between the charged heads makes it harder to gather them into micelles, the cmc of these surfactant is extremely low (typically a few *milli*grams per liter), so that a tiny amount of them is enough for effective detergent action. Moreover, their solubility in water, but also in oil, can be carefully tuned by changing the ratio between the number of hydrophilic and hydrophobic groups, so that these surfactants are suitable for many applications[30].

Other substances that easily form hydrogen bonds with water are sugars, which constitute a wide class of chemical compounds of primary biological importance (from saccharose, the common table sugar, to starch, a polymer of glucose, itself the main product of photosynthesis). Indeed, as we shall see, amphiphilic molecules with "sugary" heads are abundant in biological systems. Sugars, in particular sorbitol, together with polymers such as PEG allow surfactants such as **polysorbates**[31] to be made. These, because of their good (although not full) biocompatibility, are widely used as dispersants in cosmetics, drugs, and even foods.

Non-ionic surfactants are therefore much more suitable for personal care, have in general a much lower cmc, are not influenced by divalent ions, can be mixed with anionics without problems, and are often good foaming agents. Because of this, even though they were the latest to be put on the market, their market share is progressively increasing, so much that, in the past decade, they have overtaken the anionics, which were previously the top-selling surfactants.

## Polymeric surfactants

We have seen that polymers can be either hydrophilic or hydrophobic, and that copolymers with more than one kind of monomer can be made. We can then design a copolymer so as to obtain a chain behaving as an amphiphile, but a very long one, to form what is called a polymeric surfactant. The simplest way to do this is to synthesize a block copolymer such as

$$[A - A - A - \cdots - A] - [B - \cdots - B - B - B],$$

---

[30] Industrial ethoxylate alcohols, which are quite often mixtures of $C_nE_m$ with different values of $m$, are commercially knows as Brij[TM] – to be pronounced like the well-known French cheese.

[31] Commercially known as Tween[TM].

where A and B are, respectively, hydrophilic and hydrophobic monomers. But we can also follow more complex schemes such as

$$[A - A - A - \cdots - A] - [B - \cdots - B - B - B] - [A - A - A - \cdots - A],$$

where we have two hydrophilic chains separated by a hydrophobic block. Polymeric surfactants of the latter kind are very good stabilizers of colloidal suspensions. The polymer sticks to the particles with its hydrophobic middle part, leaving free in solution two hydrophilic chains that prevent the particle from coagulating. Obviously, the trick works the other way around too. If we have colloidal particles with a hydrophilic (for instance charged) surface, and we want to stabilize them in a non-polar solvent, we may simply choose A to be hydrophobic and B hydrophilic. This is quite useful, since stabilizing non-aqueous colloids is usually difficult. But the most interesting applications of polymeric surfactants are in the emulsion systems we shall later discuss. We can also imagine more complex schemes, where the polymer is made of a hydrophilic chain (or hydrophobic, if the solvent is an oil) to which we *side*-attach some shorter hydrophobic (or, respectively, hydrophilic) chains. These polymeric surfactant, called **graft polymers**, are particularly promising for future applications. This is the case of polymeric surfactants based on *inu-line*, a polymer of fructose naturally obtained from the roots of chicory, whose uses as an additive both in emulsion preparation and in cosmetics are rapidly growing.

Those are just a few examples of the many applications of the millions of tons of surfactants produced every year on the planet, used in almost all industrial sectors (from the consumer market, food included, to advanced technology), and eventually transferred to the environment. In fact, industrial products have always a complex formulation, where different kinds of surfactants, but also polymers and many other chemicals, contribute together to ensure success. A whole book would not be enough to tell the success story of surfactants in full. Just to get an idea of this, take a look at the following table, which summarizes the main ingredients of a typical washing powder (detailed compositions are obviously proprietary).

| Component | Weight % |
|---|---|
| Anionic and non-ionic surfactants | 10–25 |
| Zeolites | 30–55 |
| Bleachers | 13–25 |
| Fluidizers (sodium sulfate) | 5–30 |
| Corrosion inhibitors | 2–5 |
| Enzymes | 0.5–1 |
| Anti-foaming agents (colloidal silica and silicones) | 0.1–2 |
| Fluorescent compounds (leucophores) | 0.1–0.3 |

As you can see, surfactants and zeolites apart, there are many other components such as bleachers, additives to prevent excessive foaming and to protect the washing machine from corrosion, or proteins such as the enzymes that we shall meet. As a curiosity, note the presence of those "fluorescent compounds", aimed to increase the impression of whiteness compared with fabrics that have simply been washed, which are not so white. Some readers may remember the "window test" in advertisements, stating that a true white can be judged only in full sunlight. In fact, this is just because these fluorescent substances do not fluoresce in artificial light (unless it is from a halogen lamp). That is, they do not work. Obviously, this "whiter than white" is only an additional bonus, which then vanishes if you wash the fabric with just water and soap. . .

## 4.7 Questions of shape

The variety and complexity of surfactant chemical structures is enough to cause sleepless nights, if not for a chemist, surely for a physicist like me. For the same reason, understanding all the different possible shapes for the aggregates of these complex molecules may seem to be hopeless, for this could be very system-specific. Rather surprisingly, some simple considerations, which have to do more with geometry than chemistry, allow us to grasp, if not the full picture, at least some general guidelines to guess how a specific amphiphile aggregates. In short, it is sufficient to know how big its head is and how long its tail.

To be more precise about this, let us look at amphiphile aggregates with a "geometrical eye." Suppose that $N$ is the number of molecules that make up an aggregate, and that each molecule takes a volume $v$, so that the total aggregate volume is simply $V = Nv$. Within the aggregate, the tails can be rather warped but, even if they are fully stretched, they cannot be longer than their extended length $\ell$. Next, we can assume that each hydrophilic head takes a given area on the surface of the aggregate exposed to water, but here we should be more careful. Take for instance SDS. In fresh water, the charged heads strongly repel, but if we add salt this repulsive force decreases, so that the heads may able to stay closer together, each taking up a smaller area. Then, you may understand why we cannot immediately say how big a head is, for it is a physical more than a geometric quantity. Nonetheless, these effects can be simply accounted for by stating that, in specific conditions, each head will cover a given surface area $A$, although this does not coincide with its geometrical size and may be quite hard to evaluate. Starting from this very basic information, we shall now discover the large aggregate zoo that amphiphilic molecules can form. While you read the next paragraphs, keep an eye on Fig. 4.9, where all the structures we shall meet are sketched.

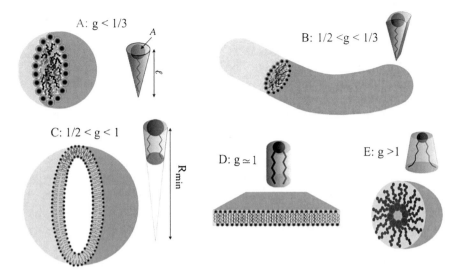

**Fig. 4.9.** Basic aggregates of amphiphilic molecules: spherical micelles (A), cylindrical micelles (B), vesicles (C), double layers (D), and inverted micelles (E). By each of them, the sketched shape of a molecule in the aggregate is shown.

### Spherical micelles

Surfactant molecules prefer to aggregate into a sphere. This is the shape ensuring the minimum surface contact with water for a given number of hydrophobic chains that have managed to get together and hide from contact with water. Yes, they would strongly prefer it, but the real question is whether a specific surfactant can *manage* to make spherical aggregates. Suppose that the sphere has a radius $R$. We can then evaluate the aggregation number by dividing the total volume of the sphere $(4/3)\pi R^3$ by the volume $v$ of a single molecule. However, $N$ can also be calculated by dividing the surface of the sphere $4\pi R^2$ by the area taken by a single head, and this different calculation should obviously give the *same* result. In other words, we must have

$$\frac{4\pi R^2}{A} = \frac{4\pi R^3}{3v} \implies R = \frac{3v}{A}$$

where the second equality, which is obtained by simplifying the first, shows how the radius is related to the molecule volume and to the area of a head. Now, however, it is time to think about the tails which, remember, can reach a maximum extension $\ell$. Therefore, *the radius* also cannot be larger than this value, otherwise we would leave a hole in the center! From the value for $R$ we have just found, it is easy to show that this holds only provided that the quantity

$$g = \frac{v}{A\ell},$$

which we shall call the *geometric parameter*, is equal to 1/3 or smaller. Recalling that the volume of a cone is just 1/3 of the base area times its height, this means that each molecule must fit into a little cone having a base equal to the head size and a height equal to the micelle radius. In simple terms, to be able to make a sphere the amphiphilic molecules must have "a big head and a slim body" (namely, they should be long and not too fat). This is usually true for anionic surfactants like SDS in solutions when there is little salt present, so that the electrostatic repulsion increases the effective head size, or for non-ionic surfactants, whose sugar head groups are really bulky. But what happens when we greatly increase the salt concentration of a SDS solution, or if we consider surfactants such as CTAB, where the cation forming the head is considerably smaller than an anion? In this case, the surfactant body is too bulky compared with the head to fit into a sphere. Thus, we have to look into different solutions to the problem.

## Cylindrical micelles and "living polymers"

If the head becomes too small to allow the formation of spherical micelles, the amphiphiles can aggregate in the form of long cylindrical "sausages", closed at the ends by two hemispherical caps to avoid contact between the hydrophobic chains and water. Even if this shape increases the surface area, cylinders offer some additional opportunities. Call $R$ the radius of the cylinder, $L$ its length, and try to repeat the simple evaluation we have just done for a sphere. Once again, we can write the aggregation number in two different ways:

(a) Dividing the cylinder volume, given by the base area $\pi R^2$ times its length $L$, by the volume $v$ of a molecule.

(b) Dividing the area of the lateral surface of the cylinder, $2\pi L$, by the area $A$ of a head group.

Equating these two results as before, we easily find that the radius is now equal to *twice* $v/A$. Therefore, to have $R$ smaller than $\ell$, it suffices that $g \leq 1/2$. When the geometric parameter is smaller than or equal to 1/2, but larger than 1/3 (otherwise, spherical micelles are preferred), "sausage-shaped" micelles will then form. As I already mentioned, this is the case for most cationic surfactants with a single tail, such as CTAB.

Note that, whereas the molecular volume and the area of a head exactly fix the radius, our calculation does not tell us anything about the cylinder *length*. Apparently, our sausages may have any possible size. So what does fix the maximum length of the micelles? Once again, this is a matter of competition between energy and entropy and, curiously, it is just the molecules we have neglected, those lying on the caps, that lay down the law. The latter are by no means comfortable (they would prefer to lie on the cylinder too), so that building the two micellar caps is expensive in terms of energy. It looks as though the longer are the sausages, the better things should go. However, convincing so many molecules to wander around together costs a lot of entropy, so that, beyond a certain length, our sausages break down into smaller cylinders. In any case, micelles of widely different shape, from very short cylinders to long sausages, are always simultaneously present in a solution of these surfactants. Usually, moreover, these cylinders, in particular the longer ones, are not rigid but flexible. Making a comparison with typical Italian products (which you may or may not know – if not, it is worth trying them), they look more like a northern-style fresh sausage than a matured spicy one from Calabria.

For some special surfactants, building the hemispherical caps is really expensive, so that they form spectacular sausages, with a length that may be *thousands* of times larger than their radius, which never exceeds a few nanometers. Moreover, since they are so long, these aggregates are also very flexible. At a distance, even if they are not simple monomer chains, they really look like polymers and, on increasing the surfactant concentration, they also form very tangled meshes. Yet there is a basic difference from polymers. Micelles are spontaneous aggregates, which form and dismantle continuously. Therefore, unlike a polymer chain, whose length is fixed once forever, these long aggregates continually fall apart, and each fragment may join another aggregate, in a flurry of divorces and new encounters. That is, somehow these are *living* polymers.

Surfactants displaying this behavior, such as a close relative of CTAB bearing the atrocious name of **cetyl-pyridinium salicylate**, are of great industrial interest. As for the polymers, these long aggregates greatly increase the viscosity of a solution and are therefore very useful as thickeners. However, they are not pre-assembled, but form spontaneously on adding even a tiny amount of surfactant to the solvent. In some cases, a concentration lower than 1% is enough to turn water into an elastic solid that, when shaken, trembles like jelly! Moreover, the formation and dissolution of living polymers can often be driven by small changes of the temperature or salt content of the solution. This feature provides a degree of control which is hard to obtain for polymer solutions.

### Vesicles (liposomes)

What happens, however, if the surfactant molecule has *two* tails? Now the hydrophobic body becomes quite plump and, head size and tail length being equal, the volume $v$ taken by the molecule is so large that even the condition $g \leq 1/2$ is not satisfied any more. Here, even cylinders do not offer a way out, and this is rather troublesome for the tails which, one way or another, *must* shelter from water. But a solution does exist. Since we cannot form a solid aggregate, whatever its shape, we may think of leaving a hole in the center. As two single people, no longer able to withstand the cost of living alone, may decide to pair off, maybe reluctantly, two amphiphiles can put together their chains, which are happy to do so, forming a complex that now has two hydrophilic heads on opposite sides. Clearly, the story is not finished yet, since these two chains are still exposed laterally. But a lot of couples of this kind can form together a double layer of surfactant that folds around and joins to make what is called a *vesicle*.

Now we have a very peculiar aggregate: water outside, but also water *inside*, separated by a double layer of amphiphiles. This truly is a revolutionary solution, introducing a separation of the environment (the "outside" and the "inside", with the amphiphiles playing the role of border police) without which, as we shall see in the last chapter, I would not be here to tell this story, nor you to listen to it. Actually, even more than surfactants, the molecules that are really skilled in making vesicles are a class of biological amphiphiles that we shall meet later. For the moment, let us just observe that vesicles are still small, but much larger than a micelle. Indeed, the requirement of keeping an optimal area per head group both for the external and for the internal layer fixes a minimum size for the vesicle. With some geometrical considerations, one can shown that, for a given value of the parameter $g$, this minimum radius is about $\ell/(1-g)$.

Usually vesicles do not form spontaneously, but rather require a little "spur to action", for instance by applying ultrasound. The vesicles formed using this method have a size or the order of tens of nanometers, but with more refined techniques, for instance by progressively hydrating a surfactant film lying on a surface, giant vesicles with radii of tens of microns can be obtained. With more elaborate methods, even subtler aggregates can be made, such as "multilamellar" vesicles made of a double layer enclosing a second double layer, itself enclosing a third double layer, and so on and so forth, till we get a kind of onion-like structure.

Vesicles are extremely interesting for pharmaceutical purposes, where they are better known as **liposomes** ("fat bodies"). The possibility of enclosing drugs in liposomes, enabling them to flow undisturbed through blood vessels until eventually released in the target tissue, is indeed quite alluring. Till a few decades ago, this looked just like science fiction, but from the 1990s it has become the standard way some drugs in tumor chemotherapy are given, with the great advantage of a substantial reduction of the dosage (since no drug gets lost elsewhere) and, consequently, of the heavy side effects of these treatments. There are different strategies that can be exploited to reach this goal. We may for instance acidify the interior of a liposome, so that the drug becomes charged and is not able, in this condition, to enter the cells. Little by little, as the liposome flows in the blood, its acidity reduces and the drugs is now able to penetrate the cells. A crucial step to obtain these results has been careful "masking" of the liposomes, which the immune system does not look upon favorably, before they reach their target. To do so, the vesicle is covered with a layer of PEG which, as we said, is a polymer readily tolerated by the human body[32]. However, since a few years ago, there have also been novel "DNA vaccines", often working the other way around. Here, in

---

[32] These vesicles that worm their way into the body are known as *stealth liposomes*, like the famous radar-invisible bombers.

contrast, macrophages, the killer cells of our immune system, are *tempted* to eat up liposomes containing a specific DNA molecule that the patient's body must learn to recognize and fight. But the most sci-fi applications can be found in the field of biotechnology, where some advanced techniques of "genetic transfer" exploit liposomes. Although they are still less effective than methods based on viruses, into which genes inserted in the so-called bacterial "plasmids" are transferred, liposomes contain no alien proteins that may trigger the immune response, can carry a large amount of gene material, and, above all, cannot replicate like viruses, or genetically recombine with bacteria to generate pathogens. For these reasons, gene liposomal transfer is particularly interesting for *in vivo* applications, so much that it has already been tested in gene therapy of organs such as spleen, lungs, and liver.

### "Inverted" micelles

The larger the geometric parameter $g$ gets, the less inclined the surfactant molecules are to form strongly curved surfaces. For instance, we have seen that SDS forms very small micelles, whereas the vesicles formed by double-tailed surfactants have a radius at least tenfold larger and, necessarily, a hole in the middle. When $g$ is *exactly* equal to one, the volume of the molecule is equal to the product of the head area times the chain length, which is like saying that the surfactant can fit in a cylinder. There is no way to make an array of cylinders bend, and therefore the only structure that amphiphiles of this kind can form is a virtually unlimited flat double layer.

However, nothing forbids a surfactant from having a body bulkier than the head; some double-tailed amphiphiles can have a geometric parameter *larger* than one. Whereas SDS may physically remind us (apart from the feet) of Eega Beeva, the famous Disney time-traveller, a molecule of this kind is more like Wimpy, Popeye's famous friend with the small head and big belly. A shape of this kind rather suggests that these molecules may get together to form aggregates with the *reverse* curvature, that is, with *the heads inside and the tails outside*. This is what actually happens, obviously provided that what is outside is not water, but a non-polar liquid. These surfactants, which are named *lipophilic* (meaning that they mostly like fats) are indeed barely soluble in water but very soluble in oils, where they form *inverted micelles* like those shown in the last panel of Fig. 4.8. Among them, a particularly interesting example is a double-tailed surfactant called **Aerosol OT**, or AOT (its full name is too complicated and of no interest for us), which we shall meet soon.

So far, we have mostly dealt with the uses of aqueous surfactant solutions, and in particular of detergency, namely, of strategies to dissolve hydrophobic compounds in water. However, there are almost as many, perhaps less familiar, situations where it is useful either to add a surfactant to a non-polar liquid, or to dissolve hydrophilic substances in oil. Just to give some examples, many industrial lubricants, like those for reducing corrosion (for instance in your car's motor) are surfactants, or else require the addition of a small amount of surfactant to be solubilized. We shall also see that the addition of a little water to fuels such as diesel oil, again possible thanks to surfactants, yields a noticeable reduction in polluting exhaust fumes. Last but not least, there is a familiar liquid (at least to an Italian like me), olive oil, which contains quite a high amount of natural amphiphiles that, besides solubilizing a small amount of water (which is what actually "fries"), enable it to keep in solution valuable antioxidants such as polyphenols. These short notes on lipophilic surfactants bring us to the topic that, after a brief intermezzo, will take up all the rest of this chapter.

## 4.8 A mischievous break: watch the label!

The connection between surfactants and environmental issues is relevant and intricate. Just recall the huge figures for the production and consumption of surfactants I mentioned before (half of which, remember, refers to *household* uses), and you will realize that the problem should not be underestimated. I already hinted at the problem of sea eutrophication (where excess nutrients cause algae to flourish) due to the phosphates that used to be added to washing powder, and, surely, it is not pleasant for plants to find underfoot (or, better, "underroot") those surfactants we use daily at home (please, *never* pour your dishwater on the roots of a magnificent camellia, even if it is blooming in the garden of a next-door neighbor you cannot stand). Nor do I advise you to swallow surfactants such as the nonylphenol ethoxylates (close relatives of the $C_mE_n$), recently banished from the market but formerly widely used, which, besides suspected of being carcinogenic, may affect hormone regulation so severely as to induce a change of sex in fishes (which in itself may not be terrible, except that no one asked them in advance what they thought about it). I am afraid I have not got the expertise to discuss this topic in detail, which would otherwise take up a good part of this book. However, I trust that it is safe to state that most problems come from using *too much* of this stuff. As we have seen, indeed, good cleaning requires just a small amount of surfactant, surely much less than we commonly use to do the dishes or shampoo our hair.

Nonetheless, I wish at least to take the side of an old friend of ours, mostly to put you on the alert towards a certain environmentalism by hearsay that sometimes, aiming its darts at some demonized substance, make us lose sight of much more serious problems. Recently, some alarm has been spread about our familiar SDS, best known on blogs and websites as SLS (because of the "lauryl"), accused of the most pernicious effects on health. Some of you may have even happened to see signs on "health stores" stating that all products on sale are absolutely "SLS free". Now, as we have said, sodium dodecyl sulfate is a strong surfactant, and thus rather aggressive for use on the skin (or in the mouth, in the case of toothpastes), so much so that in shampoos it is usually replaced by the better tolerated non-ionics[33]. Yet, apart from a little skin irritation, I am not aware of any more serious consequences.

Nonetheless, since I pay special attention to shampoos (see the earlier footnote), I once asked for advice about particularly gentle products, to be promptly reassured by the sales clerk that the one she was suggesting contained no trace of SDS. However, once I had recovered my precious glasses for early long-sightedness (which, as I usually do, I had left at home), and carefully read the composition on the label, I was rather taken aback. There was no SDS, in fact, but that shampoo contained *ammonium* lauryl sulfate, which is just like SDS in chemical structure (in particular as concerns the presence of the sulfate group, which is the skin irritant), except that the counterion is an ammonium ion. Now, the ammonium ion is surely no safer than sodium, which is abundant in the sea water we swim in with pleasure. Yet that's it. The diabolical SDS is banished, and the customer is satisfied. Therefore, now that you know something more about surfactants, watch the label!

## 4.9 Small but mighty emulsions

Detergency is based on dissolving hydrophobic stuff in the hydrophilic core of a micelle. But this power is limited. As we have seen, a specific surfactant wants to make aggregates with a precise curvature, dictated by the size of its head and the length of its tail. In practice, this means that substances like fats can only make a niche for themselves among the hydrophobic chains, and cannot be stored beyond a rather low amount without swelling the micelle too much. However, there are many applications where we would like to dissolve a substantial amount of oil in water, maybe 10 or 20%. If we could only rely on the spontaneous inclination of surfactants to make aggregates, there would be little we could do about this, for these aggregates are subjected to the strict laws of geometry we have mentioned. Yet, if surfactants do not comply with our desires *spontaneously*, we may try to *force* them to do so. Practically, this means that we have to shake a mixture of water, fats, and surfactant vigorously, rather like we do when we prepare mayonnaise. The energy we transfer to the solution by mixing is just what is required to break the oil into droplets.

If we only had oil and water, these droplets would rapidly fuse into a single blob, a condition that is much more favorable than the initial state because the boundary surface between droplets and water costs a lot of energy (as we know, this is just the product of the surface area we created by mixing times the interfacial tension between water and oil). Yet surfactants, by their own

---

[33] Not that the latter are totally innocent. Some time ago, the production manager of the Italian branch of a large multinational skincare company frankly confided to me that he used to wash his hair with simple soap, to avoid any risk. The fact that he was almost bald, however, did not encourage me to follow his example.

nature of go-between, will cover the droplets, placing themselves with their heads toward water and tails toward oil, and this means things change quite a bit. First, the surfactant lowers the interfacial tension, and therefore the energy cost of the droplets. But there is a subtler reason. In order to coalesce, two droplets that occasionally meet must first open up the surfactant films covering them, so that they merge into a single film surrounding the larger drop that forms. As for soap bubbles, however, the films covering the droplet have the prodigious feature of "self-healing". In other words, the surfactant molecules come to the aid of the zones where the film tends to tear. Because of this, the process leading to drop coalescence is severely slowed down.

A system of this kind, made of oil droplets in water, covered by a surfactant layer, is called an *emulsion*, and shows some relevant differences compared with a micellar solution. First of all, the droplets are much larger than micellar aggregates, and have a size which typically ranges from some microns up (they are therefore giants compared with micelles). The more energetically we shake the mixture, the smaller the droplet size we get. In addition, the size is just an average value, for the actual diameter varies from drop to drop. But, above all, the main difference is that, unlike micelles, an emulsion is not a stable structure, for water and oil unavoidably tend to separate, for reasons we shall shortly see. Nonetheless, with care, it is possible to keep an emulsion stable for a very long time, or anyway long enough for the purposes of their application.

What we have just described is properly named an *oil-in-water* (or O/W) emulsion. We can, however, do the opposite, stabilizing *water* droplets in oil, thus obtaining what is called a *water-in-oil* (W/O) emulsion. Clearly, the surfactants favoring the formation of W/O emulsions are different from those forming O/W emulsions and, for those dealing with emulsions, the capability of forming one or the other kind of emulsion is the most important feature of a given surfactant. This natural propensity is usually summarized into a single number, known by the mysterious acronym HLB(*Hydrophilic–Lipophilic Balance*), which is applied at every turn by my chemist friends, whereas it usually baffles physicists. Anyway, in practice, surfactants that are much more soluble in water than in oil more easily form O/W emulsions, whereas W/O emulsions require more lipophilic surfactants, such as those forming inverted micelles.

A large number of common products, such as beauty creams, ointments, lotions, and suntan creams, are surfactant-stabilized emulsions. In cosmetics, O/W emulsions based on anionic and non-ionic surfactants are generally used to release oils or waxy substances that protect, moisturize, or simply soften the skin. In pharmacology, emulsions of diverse kinds (mostly based on more bio-compatible non-ionic surfactants, but also on cationics, because of their antibacterial properties) are used to control drug dosage accurately, or simply to make them more palatable. In farming, insecticides, fungicides, and pesticides are often spread as emulsions, so that large watering systems can be used while limiting the amount of active principle supplied. When they are not made of solid particles but of dispersed liquid droplets, pigments and paints are emulsions too[34].

However, the area where emulsions, both natural and man-made, surely play a chief role is the food industry. There is even an emulsion without which not only the human race, but the whole class of mammals we belong to would not exist. It is exactly that magic fluid, milk, that has cropped up ever since the first chapter, which is an aqueous suspension of hydrophobic nutrients such as fats. Obviously, what stabilizes the fat droplets here is not surfactants (I guess no mom would like to see her baby burping up soap bubbles after being breast-fed), but rather proteins, which show many points in common with amphiphiles, as we shall see in the final chapter[35]. When, to highlight the diffusion of blue light by a particle suspension, I suggested using skimmed milk, I did so because this means an absence of just those larger droplets that would give a cloudy, whitish appearance even to a dilute milk solution. Besides milk, some other examples of emulsified foods are mayonnaise, vinaigrette (where the emulsifying agents are mustard proteins), ice-creams, and many sauces, which owe their stability to the presence of proteins, natural amphiphiles, or, in many industrial products, the addition of edible surfactants such as polysorbate.

Taking into account all these applications, you may realize that extending the lifetime of an emulsion as much as possible is a primary technological goal. No one likes to discover that a suntan cream bought last year has turned into a useless separated mixture of water and oily stuff. But why does an emulsion die? As we mentioned, the most relevant process is drop coalescence, taking place when two droplets stay in contact long enough to allow the surfactant films to merge together. Clearly, for this to happen, it is not enough that two droplets meet fleetingly during their random Brownian zigzag (which is very slow in any case, when they are large). Instead, gravity brings them together, by making them settle to the bottom for W/O emulsions, in which the droplets are denser than oil, or rise to the surface, in the opposite case of O/W emulsions.

The successful strategy for stable products seems therefore to consist in making very fine emulsions, where the droplets are so small that they sink or rise very slowly – and maybe requesting the consumer to shake vigorously before use. In fact, this works rather well, just as well as adding polymers that increase the solvent viscosity, or using polymeric surfactants as stabilizers, which strengthen the surface film and noticeably increase the emulsion lifetime. Nonetheless, an emulsion heads for a slow death even *in the absence* of coalescence processes, namely, without the requirement that the droplets

---

[34] In spite of their name, photographic "emulsions" are not properly such, but colloidal dispersions of silver halide (the material sensitive to light) in a jelly, deposited on a supporting film made of cellulose acetate.

[35] Even the caoutchouc latex is an emulsion stabilized not only by natural surfactants, but also by proteins.

meet. The mechanism causing this progressive ageing of emulsions has to do with the Laplace pressure, together with the fact that the solubility of a hydrophobic substance in water may be small, but is never exactly zero[36]. The smaller are the droplets, the larger is their Laplace pressure and, as originally grasped by Lord Kelvin (who, by the way, was the first to predict the "entropic death" of the Universe), this causes the oil contained in small droplets to be *slightly more soluble* in the surrounding water. Hence, little by little, the oil gets out from the smaller bubbles, which shrink, and passes through the water to get into the bigger ones, which inflate. Unless the droplets are very small, this effect is quite limited[37], and therefore the process is very slow. However, as for soap bubbles, the process is self-feeding, because when a droplet get smaller its pressure increases too, accelerating the shrinking process. So this effect, known as *Ostwald ripening*, makes the bigger droplets grow at the expense of the smaller, which progressively vanish. This competition clearly goes on until, in principle, a single drop survives. A curious example of Ostwald ripening is the so-called "ouzo effect", taking place when the Greek liquor of this same name, the French anisette, or the Italian sambuca, are rapidly diluted in water. As some of you may have noticed, the solution immediately turns cloudy, scattering around a lot of light. This is because the essential oil of anise, which is the base ingredient of all these spirits, is not soluble in water, and separates from it in the form of emulsion droplets with an initial size around a micron, which then slowly grow for just the reason we have discussed.

But is it possible to get droplets that are stable *forever*? We have seen that the major problem is the surfactant spontaneous curvature, which does not allow a micelle to swell without limit. Yet this limitation can be overcome using a simple trick, which consists in adding to the solution some alcohol with a rather high molecular weight, for instance pentanol. These molecules look like surfactants, because they have a short hydrophobic tail, but their head is too small to enable them to form micelles, so that they are almost insoluble in water. However, in the presence of surfactants, they intersperse within the micelle, which becomes a mixed surfactant/alcohol aggregate. The small heads of alcohols, however, *reduce* the average spontaneous curvature of these aggregates, which are then much happier to swell. This is the way to make what is called a *microemulsion*, made of droplets that can be as large as ten nanometers (a big gain compared with a micelle, in terms of volume). That microemulsions are stable solutions is suggested by the fact that they form spontaneously, with no need to supply mechanical energy by stirring

---

[36] Here we discuss the case of an O/W emulsion, but the same is true for water droplets in oil.

[37] What fixes the droplet size below which it becomes relevant is the ratio $\gamma v/kT$, where $v$ is the volume of an oil molecule. This quantity, which is a length, is typically of the order of a few nanometers. Therefore, the effect is negligible for micrometer-sized droplets, but is the main cause of the death of those "nanoemulsions" we shall shortly discuss.

vigorously. In fact, they can be stable for many years. It is even easier to obtain water-in-oil microemulsions by using a surfactant such as AOT. In this case, the aggregate size can be varied across quite a wide range, because the two tails of the surfactant open up or close so as to tune the curvature suitably. Actually, the size of an inverted AOT micelle can be increased almost tenfold simply by changing the ratio between the concentration of the surfactant and added water.

The high stability and ease of preparation of microemulsions is the reason that, in the past few years, they have progressively replaced traditional emulsions not only in many cosmetic applications, where their ultra-fine dispersion state yields a much more effective release of active compounds, and in the formulation of pesticides, but also in many hi-tech areas. For instance, anti-rejection drugs such as cyclosporins, allowing transplant patients to survive for longer, are given in the form of injectable microemulsions. Using surfactants that are fully bio-compatible, moreover, it is possible to deliver microemulsions that easily mix with blood using a very low dosage of active components, which are released only in the target tissues. For these reasons, microemulsions have become vehicles for carrying many innovative drugs. Moreover, microemulsions are widely used in chemical processes, where they behave as "liquid microreactors", or in biotechnology, where they allow proteins to be trapped or enzymatic reactions to take place in a controlled micro-environment. In a completely different area, the presence in diesel oil of a small amount of (micro)emulsified water noticeably reduces the emission of nitrogen oxides and of carbon monoxide, because water vaporization decreases the combustion temperature, reducing the formation of particulates. To point out an application of particular interest for my home country, a research group in Florence led by Piero Baglioni, a well-known soft matter chemist and a good friend of mine, have used microemulsions for many years in advanced art restoration methods, with brilliant results.

One of the reasons microemulsions are so versatile is their rich and varied behavior. The very same system can turn from an O/W to a W/O emulsion if we change the oil to water ratio, often passing through an intermediate state, called a bi-continuous phase, where oil and water are intimately mixed in comparable amounts without separating, just because of the presence of surfactant interfaces. Advances in soft matter science have been crucial to understand the full range of opportunities offered by water + oil + surfactant systems. These advances are currently leading to the development of new systems, called nanoemulsions, which mix surfactants with different HLB values (the Hydrophilic–Lipophilic Balance, if you remember) to obtain droplets stable over long periods, with a size that can be as large as a tenth of a micron[38], overcoming one of the major limits of microemulsions.

---

[38] In a curious inversion, *nano*emulsions are therefore much larger than *micro*emulsions, which are made of nanometric droplets.

An application of emulsions, but of microemulsions too, where colloids, polymers, and surfactants display their most intimate relationship, is colloid synthesis. In the first chapter we talked about "polymeric" colloidal particles, made for instance of polystyrene or PMMA, that can be prepared with a perfect spherical shape and pretty uniform size. We shall soon see that particles of this kind are extremely useful, if not crucial, to investigate some basic aspects of the organization of matter. But how are they made? Just take polystyrene, for instance, whose monomer is not water soluble at all. Hence, there is no way of synthesizing these particles directly in water, which is a such a bad solvent for the polymer. On the other hand, were the synthesis to be performed in a good solvent, we would surely obtain not rigid balls, but fractal coils. The trick, then, is to add a surfactant, even one as simple as SDS, and to use micelles as "reactors" in which the polymerization takes place. In the presence of SDS, monomers dissolve in the micelle, so that, if an "initiator" is added and the solution is heated up, polymer chains begin forming within the aggregate core. The micelles then swell, progressively becoming emulsion droplets whose cores become *solid* when the chain molecular weight gets sufficiently large. Gradually, a compact lump of plastic forms, which retains a spherical shape and behaves as a seed for the growing particle. Part of the SDS that has been used remains trapped, giving the particle its surface charge, whereas the remaining surfactant can later be removed, or left on the particle surface as a stabilizer.

## 4.10 Black gold

For the end of this chapter, I wish to dedicate a few lines to a fluid of vital importance, whose composition, extraction, and use sum up many of the aspects of soft materials we have met so far: crude oil, the black gold so crucial not only for the development, but also for the geopolitics of today's world. Let us start by dispelling some woolly thinking. First, there is more than one kind of oil, ranging from the valuable, low viscosity Arab petroleum (which usually is not black, but straw-yellow), to the heavy and very viscous black oils extracted in other oilfields (for instance in Venezuela), to bitumen, which is not even liquid but a solid hidden in the sands and shales of Canada. An oilfield is moreover something *much* more complicated than a cavity in the underground filled with this precious liquid. Instead, crude oil is finely dispersed within micro-porous rocks, where it gathered over geological ages, together with gaseous hydrocarbons such as methane and, almost always, water. How it formed and, above all, how it got there rather than, for instance, our backyard is still a mystery (if we knew that, it would be easier to find it, or to guess how much of it is still there).

Just because of the way crude oil is dispersed, it is far from easy to get it out. The lucky Texan who just digs a hole in his own plot of land to see a black gold fountain gush forth is no more than a movie character. Actually, at the beginning oil does gush out spontaneously, pushed by the internal pressure underground, but this takes place only for a short time, and then the upthrust drains away. Had we just exploited the internal pressure, most oil resources would still be below the ground, trapped forever. To take a little more out, the most common and less expensive method is to give it some help by pumping pressurized water or vapor into the well to push it out. Yet even this technique works only to a limited extent. Whereas it is easy to push a liquid with low viscosity, like water, with a more viscous one, the opposite does not work. For precise physical reasons, the less viscous liquid finds its way through, forming "fingers" that leave the more viscous one behind. As a result, after a while you pump water in, and you get water out.

Till the 1970s, when crude oil was still cheap, this was not a serious problem. When an oil well began spitting out water, it was abandoned without much hesitation. Yet, starting from the first oil crisis, things have changed a lot. Oil companies have begun wondering (where "wondering" actually means investing a lot of money in research) whether it is worth adding surfactants to the residual oil, so as to extract it *together* with water as an emulsion to be split at their leisure above ground. This is an example of an *enhanced oil recovery* (EOR) strategy. Easy to say, but expensive to do, for the cost of the surfactant, even in small amounts, would have added too much to the price of the barrel, so much that, once the turmoil was over, the big oil groups soon stopped throwing money into what seemed to be... a bottomless pit.

As we know, the new millennium has seen new and substantial geopolitical problems arise that are spurring the oil industry to reconsider this decision[39]. Today, however, owing to the rapid development of the science of complex fluids, EOR strategies are quite different. The research is mostly focused on the development of methods for *water shutoff*, techniques aimed at blocking just those channels through which water would like to get out, leaving oil behind. A possible approach could be to feed into the reservoir water-soluble colloidal particles made of a kind of polymer gel and, in suitable conditions, let them swell so that they plug those pores the water flow through. However, considering what we have said about our scanty understanding of colloidal filtration processes, there is still a long way to go. Even so, the exploitation of colloids, polymers, and surfactants in the oil industry is widespread, ranging from the preparation of drilling muds to the design of highly water-resistant cements, and to the development of additives protecting metals from corrosion (a serious problem, considering the temperature, pressure and salinity conditions of an oil well).

Anyway, besides issues of extraction (and transport), for soft matter science crude oil is an interesting fluid *per se*. One of the most relevant problems is that quite often, even if we add nothing at all, crude oil comes out as a

---

[39] You may then understand why researchers in this field such as myself are not totally unhappy when the barrel price goes sky-high. When we think of it as private citizens, however, things are different.

very viscous emulsion, which is quite hard to separate. And this has to be done in the field, both because it is hard to sell, instead of gasoline, a sort of molasses full of water, and because it is not nice to discharge into a river or the sea fluids that are not exactly environment-friendly (luckily, the legal limits for oil content in waste waters are very low). What is even worse, similar emulsions form spontaneously as a consequence of crude oil spills, for instance from tankers. Getting rid of this sticky sludge is much harder than removing a hydrocarbon layer (as the cormorants know to their cost).

But how does it happen that petroleum forms emulsions spontaneously, when it does not contain surfactants? This is something we have started to grasp only recently. Even if they are mostly made of hydrocarbons and water, the fluids gushing out of a well always contain a large quantity of solid particles, mainly clays, which are mostly hydrophobic, but not *completely*. As a consequence of their slightly equivocal nature, these particles tend to stick at the interface between oil droplets and water (it is a kind of universal law of surface science that filth always goes to the interface), where they feel most comfortable. Clearly, they do not lower the interfacial tension appreciably, but nevertheless they form a barrier, a kind of wall that is a serious obstacle to coalescence. For two droplets to merge, the particles must first be pushed aside, and this can be quite hard, in particular owing to the presence at the interface of certain big organic molecules, the asphaltenes. These molecules, which are just the ones that give petroleum its black color, and are clearly the main component of asphalt, make the adsorbed particle layer very viscous. In general, emulsions of this kind, stabilized not by a surfactant but by colloidal particles, are named *Pickering emulsions*. Currently, they are being widely investigated, because they may prove very useful. For instance, using polymer colloids to make Pickering emulsions, and then fusing these particles by special methods, David Weitz of Harvard University has obtained rigid but permeable shells useful to encapsulate drugs or other substances, which he has dubbed "colloidosomes."

# 5

# Nanoarchitecture

*Kepler, Bernal, and your greengrocer: in the end, it's worth being tidy – Opals, colloidal crystals, and tomorrow's computers – Standing to attention, but in disorderly ranks: the gaily colored troops of liquid crystals – From glasses to gels: a brief journey across solids that aren't – A question of shape: when it takes an MRI scan to grasp what a glass really is – Sand, dust, and grains: when suspensions lose their suspenders – Sandcastles and quicksand: from silos to hourglasses, from nuts to jams.*

I guess most of you have been through that stressful, dispiriting step that often goes along with leaving for a long-awaited vacation – the moment when we discover that the family luggage, in spite of all our efforts, will not fit in the trunk of the car. The aim of this chapter is to make sure that the next time this happens, it will not be just an excuse for mutual domestic reproaches, but an occasion to reflect upon the fact that packing is a fine art, which, even in the simplest case, may baffle the best of scientists.

We have discussed some of the countless uses of substances that should be regarded as supramolecular systems, complex fluids, or soft materials, because of their structure or of the way they respond to mechanical stresses. In doing so, however, we have slightly neglected an aspect that I tried to stress in the preface: as well as their practical interest, these systems provide a great opportunity to tackle challenging problems concerning the way matter is organized internally. We shall now try to remedy this deficiency, by showing that many of these basic questions can be addressed by looking at the way this stuff can be "packed". Hence, we shall not so much be discussing new materials, but rather the way the systems we already came across are organized. Making an exception to what I promised earlier, I have given these complex structures the collective name of "nanoarchitecture" because, in this specific case, the prefix *nano* is often appropriate and effective. We shall, however, briefly encounter two other kinds of materials, liquid crystals and granular fluids, which, even if they are not strictly soft matter (the first are often small

R. Piazza, *Soft Matter*, DOI 10.1007/978-94-007-0585-2_5,

molecules, the second lack a basic ingredient, the solvent), are close relatives of the latter.

In fact, both the topics we shall broach and the way we shall do it should please our "theomads". Admittedly, until now I have slightly neglected them, and they may be a bit unsatisfied with what I have written so far. This chapter has therefore been written for them, but also for those of you who believe that scientific research is not just about making things work, but also understanding *why* they work. Above all, however, it is dedicated to those who believe that doing science, besides contributing to the development of technology *and* culture, is also a way to keep alive the natural curiosity that all of us had as children. Among this kind of people, there are many scientists whom, just because they share this view, I regard much more as friends than colleagues. In what follows, I shall take the opportunity of introducing them to you.

## 5.1 Kepler, Bernal, and your greengrocer

Have you observed a greengrocer while he sets out a crate of oranges, trying to fill it as full as possible? If you ever have, you will agree that he would start filling it more or less as shown to the left of Fig. 5.1, carefully putting in each fruit according to an ordered layout where each orange is surrounded by six others forming a regular hexagon[1]. To lay down the second layer, the greengrocer will surely make use of the "sockets" left in the first, to get a second plane of hexagons, but staggered with respect to the first. If the crate is deep enough, there is now room for a third layer. But at this point, the greengrocer is faced by a dilemma, for there are actually two ways to do this. He could place the oranges exactly on top of the first ones, repeating the initial scheme. Going on this way, he would obtain a sequence of alternate layers of type "A" and "B". However, looking back at the first layer, you may notice that there are *two* distinct kinds of sockets, respectively pictured in black and white. To add new layers, we have so far just used the black ones, so that in the sequence ABABAB...there actually "holes" corresponding to the white sockets. Yet, if your greengrocer is sufficiently creative, he could make a different choice, placing the third layer according to scheme "C", which does not overlap with the first one. At this stage, he could repeat the whole procedure, obtaining the packing ABCABCABC..., which is *different* from the former. Going on adding layers, but gradually reducing the number of oranges, we obtain a kind of pyramid, which is actually the usual way cannonballs or oranges on display are stacked.

As you can see, even packing spheres is not completely trivial. Anyway, a good greengrocer knows by experience that, in terms of the number of oranges

---

[1] Obviously, if it is one of those modern crates with preset packaging, this does not require any mental effort, but here we suppose that the crate is just an empty and sufficiently deep container.

he can pack in the crate, the two methods are fully equivalent. It can indeed be shown that in both cases the oranges take up about 3/4, or more precisely 74.05%, of the crate volume. No one will convince him that he can do better.

A physicist too may take for granted what our day-to-day experience teaches us. But with mathematicians it is quite another story. The first who faced the problem of packing a collection of spheres (which we shall regard as *hard*, namely, not squishy at all, something reasonably true for oranges) was Kepler, the scientist to whom we owe to the laws governing the orbits of the planets. In 1611, Kepler took cannonball stacks (oranges were in short supply in Prague, where he was at that time) as a model to explain why snowflakes, when magnified, always show an clear hexagonal symmetry. Since, although crazy about geometry, he was also a physicist, Kepler held it to be self-evident that this way to arrange spheres is the most closely packed. But life is not so simple as that. For almost four centuries his surmise, now known as the *Kepler conjecture*, has spoiled the nights of many first-class mathematicians.

To understand why, let us first look at the way oranges are distributed in space in the two arrangements we have just considered. This is much easier for the ABC stacking, where, as I have tried to show in Fig. 5.2, we can divide up the internal space in the crate (which now we regard as very large, to avoid wall effects) into a lot of little cubes placed side by side. In each of them, we find an orange on each corner, and one at the middle of each side[2] (if you cannot

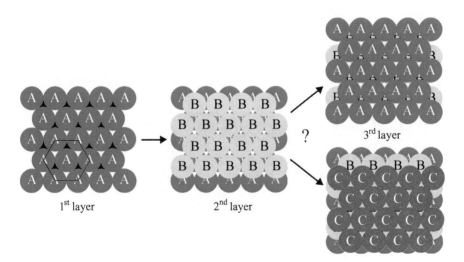

**Fig. 5.1.** Two neat ways to fill a crate of oranges.

---

[2] Clearly, each sphere lying on a side is shared by two adjacent cubes, whereas those on the vertices (the cube corners) belong to eight cubes. Hence, each cube actually contains the equivalent of four spheres. Since the diagonal across each face of the cube is twice the orange diameter, one then finds that the volume fraction taken by the spheres is just equal to $\pi/\sqrt{18} = 0.74048\ldots$

picture this easily, just buy some oranges, which are also good for your health). What we are looking at is a crystal structure with a *face-centered cubic unit cell* (or a FCC crystal). The elementary cell is much harder to picture for the ABA stacking, which is called *hexagonal close packing* (HCP) and consists, as we have seen, of a series of alternating staggered hexagonal planes. Anyway, as for the FCC packing, each sphere is surrounded here too by as many as 12 other spheres.

However, in math it is not wise to put bounds on the imagination. In fact, there could be more complex cells, possibly containing tens of spheres, allowing us to build a denser crystal. It took the greatest math genius of all times, Friedrich Gauss, to show that FCC and HCP are indeed the closest packed *crystal* arrangements, meaning arrangements with a repetitive order. But this is not nearly enough, for we cannot say anything yet about packings that are *not* ordered. Some doubt may actually arise from considering that, if we take four spheres and put one at each corner (vertex) of a pyramid with a triangular base (what is called a *tetrahedron*), we get a better packed heap, because here the spheres take up almost 78% of the volume. Tetrahedra, however, do not fill space regularly; that is, they do not allow us to make a crystal, and therefore this packing value cannot be reached. All the same, we could think of adding some spheres to "fill holes" in an irregular arrangement of tetrahedra, still retaining a packing larger than 74%. It was only in 1998 that Thomas Hales, of Michigan University, having reduced the problem to the analysis of "just" 5000 or so possible random arrangements that could in principle pack better than FCC, claimed that the Kepler conjecture is correct. It took a full board of mathematicians four years more to announce that Hales' proof (which was 250 pages long) was correct "at least to 99%"[3]. Do not try to explain this to your greengrocer, if you care about your reputation.

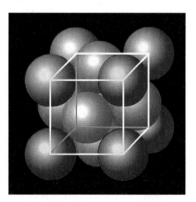

**Fig. 5.2.** Face-centered cubic structure (FCC), with its unit cell outlined.

---

[3] Currently, Hales trusts that a shorter proof, not requiring all cases to be checked one by one, will be obtained by working hard (as part of a large group) for about... twenty years.

Taking it for granted that hard and well-ordered spheres can be packed to fill about 74% of the total volume, we may now ask ourselves what is the largest volume fraction they can reach if their arrangement is fully *chaotic*. In simple words, how many oranges will our greengrocer fit in the crate, if he throws them in at random? This is actually a rather vague question, for it is far from easy to say what "at random" really means. Besides, speaking about *maximum* random packing is to some extent self-contradictory, since this is a *very distinctive* situation, satisfying a precise condition. And true chance does not follow rules (something that lottery systems players still do not seem to realize). Therefore, it is not surprising that, in contrast to the Kepler conjecture, mathematicians have tended to avoid this question.

It took an Irish experimental physicist, John Desmond Bernal, who was also quite well known among biologists, to give a rather amazing answer. Besides an outstanding scientist, Bernal was also a many-sided character. A brilliant mind, a man of mixed Italian–Portuguese–Spanish–Jewish blood, an active and ardent communist[4] but nonetheless strenuously committed to his country during the Battle of England and Operation Overlord, a man with a rich and complex love life, he was totally unlike the cliche of a scientist as a book (or lab) worm. "Sage" Bernal, as he was usually known, was largely responsible for the birth of crystallography (the investigation of crystal structures), and this was thanks to his strong interest in packing problems.

Unconcerned with the inaction of mathematicians, Bernal began doing what all experimentalists are supposed to do: experimenting, by filling all kinds of containers with spherical objects and counting how many of them he could fit in. Clearly, there is a huge number of ways to place balls "at random" in a given volume, and in principle we may expect to obtain very varied results. But what Bernal found was that, provided some precautions long known to corn merchants[5] were followed, the values he got were surprisingly similar. The volume fraction was always, or at least in the vast majority of cases, about 62–64%. Since then, a lot of experiments, made on the most diverse systems, have fully confirmed Bernal's finding, so that today no one doubts that a maximum value of about 0.64 for the *random close packing* (RCP) of spheres can be assumed, at least empirically. However, why the RCP value is so well defined and, above all, what real physical meaning it has, are still debated questions, which go to the heart of our (lack of) understanding of what order and disorder really mean.

But what have the Kepler conjecture and Bernal's experiments to do with solids, liquids, and, last but not least, soft matter? To explain this, I have to tell you another story, where the value of colloids as "model systems" emerges in full.

---

[4] Which, at least scientifically, was not of great help to him, for he eventually subscribed to the ludicrous anti-Darwinian ideas of Trofim Lysenko, the puffed up farmer turned by Stalin into a hero of Soviet science.

[5] Basically, tapping the container.

## 5.2 Colloidal crystals: ordered by entropy

Understanding liquids is far harder than solids, and it is easy to see why, at least in a general sense. Solids and liquids, making up what is called "condensed matter," stay together because of some kind of attractive force acting between atoms and molecules, whose strength depends on distance: the farther apart two molecules are, the less they feel each other. In solids, atoms are roughly trapped in fixed positions, according to a specific crystalline order, and therefore their mutual distances are known. For instance, to find how elastic a solid is, or how much heat it absorbs, we just have to work out how much the atoms "vibrate" around these positions, knowing the strength of the interatomic forces. Just because *we know* where each atom sits, this is feasible, although not necessarily easy. For liquids things are quite different, because atoms and molecules are free to move around, so they actually look for the best places to minimize the total energy (or, to be precise, what we have called "free" energy). In other words, the molecular positions *are set by forces between them.* We seem to be trapped in a vicious circle: to calculate the force between two molecules we have to know how far apart they are, but their mutual distance is, in turn, fixed by the force itself.

On the other hand, it is not even easy to grasp *why* liquids should exist. Why isn't Nature content with *a single kind* of disordered matter, the gaseous state? Why does water suddenly turn into a very different substance, vapor, when it boils in a pot? It was Johannes Diderik van der Waals, the same scientist who introduced dispersion forces, who first understood that, for the liquid state to occur, attractive forces between the molecules are *required*; just because of them a gas condenses into a liquid when it is cooled. However, van der Waals also understood that this is true only below specific values of temperature and pressure: beyond them, gas and liquid can no longer be told apart, and a *single* indistinct fluid state exists. For instance, if heated to $400°C$, $H_2O$ in the sort of pressure conditions found in the depths of the oceans is neither "water" in the usual sense, nor vapor. Rather, it is a totally different fluid with very peculiar chemical properties, for it dissolves oils, fats, and even plastics with no effort[6]. The difference between a gas and and a liquid, which look so unlike each other (a liquid may be thousands of times denser than a gas, and behave mechanically almost like a solid), is then really very subtle. Therefore, it is not surprising that the physics of liquids has developed much later than solid state physics.

One of the pioneers of the investigation of liquids, back in the 1930s, was the great American chemist and physicist John Gamble Kirkwood. At that

---

[6] These conditions are probably common for the very hot springs in the ocean depth, called *hydrothermal vents*. As a consequence, these springs shelter a strange, almost alien ecosystem, mostly made of "extremophile" bacteria that do not exploit sunlight to extract energy as all other creatures do (maybe in a roundabout way, by eating something like a cow that eats grass, which uses the Sun's energy directly). Instead, they make use of methane or sulfur compounds.

time, trying to deal with problems of this kind was more a game of chance than a science, with the scientific prestige of the player at stake. Yet, maybe because of his middle name, Kirkwood was undaunted. To make his task easier, he began by trying to understand what happens for a gas made of spherical particles where no force is present, except when the spheres bump into each other and bounce off like billiard balls. This is exactly what we mean by a "system of hard spheres", differing from what is called an "ideal" gas only because molecules are not dot-like, but rather have a certain volume.

Since there is no attractive force, hard spheres do not know the difference between gas and liquid. Hence, Kirkwood hoped to understand more easily how the structure changes as the fluid gets denser, namely, as the fraction of the total volume taken up by the spheres grows. To do this, he developed some mathematical techniques, still used to study liquids, which immediately proved very useful to investigate how the molecules are distributed in the fluid. Yet he soon found a very strange result: beyond a given concentration, the problem not only becomes very challenging, but it does not even seem to have any solution. Thus, with a stroke of genius, back in 1939 Kirkwood wrote:

> ... One would conclude therefore that a limiting density exists above which a liquid type of distribution and a liquid structure cannot exist. Above this density, only structures with crystalline long range order would then be possible.

However, this was quite a premature conclusion for those times, and the problem of what happens for a system of hard spheres at high concentration fell into oblivion for many years, partly because Kirkwood's methods were so hard that few dared to embark on them, not wishing to stake their reputation on such questionable results. Around the middle of the 1950s, however, computer science shifted the problem's ground, for physicists began practising numerical simulation, which means asking computers to use the laws of physics to see how a virtual but equivalent system behaves. Thus, in a short paper in 1957, Bernie Alder and Tom Wainwright, two of the founding fathers of computer simulation, announced to the world an amazing result: beyond a given concentration, a hard-sphere fluid *spontaneously* orders into a crystal. A scientific row quickly erupted, for many theoreticians could not believe that this was possible. Their line of reasoning was more or less as follows. If attractive forces are crucial to keep molecules together in a liquid, how can a *solid*, where the molecules are much more constrained, possibly come about in the absence of any force to keep the molecules together? After all, it is not easy to claim they were wrong, for the idea that billiard balls can arrange themselves spontaneously is anything but intuitive. But computers, which were meanwhile growing more and more powerful, soon confirmed Alder and Wainwright's result. What is more, they showed with precision that this happens when the room taken by the spheres gets larger than about half the total volume.

The basic reason that many theoreticians could not accept this result is probably related to some confusion over the real meaning of entropy, rather fuzzily identified, as we have already mentioned, with "disorder". A crystal is definitely more ordered than a liquid, so it *seems* to have a smaller entropy. Hence, they reasoned, only attractions can force the spheres to swim against the tide and gather together. Alas, the opposite is true! Even though the crystal looks like an ordered state, it has a *larger* entropy than the fluid, and by forming crystals the spheres do nothing but increase their entropy, as thermodynamics commands. To understand this, we just have to use *our* idea of entropy as "freedom of motion", take into account what our greengrocer knows, and put two and two together. What does it really mean to say that spheres pack better in regular order than at random? Simply that when we arrange them regularly, each sphere has more room to move about. This is easy to grasp by considering spheres at a volume fraction of 0.64. In the disordered state, this is the maximum density, so all spheres are jammed tightly in place, whereas if we order them neatly as we did for oranges, we are still left with an additional 10% room for other spheres. In simple terms, our balls would prefer to cruise around the whole volume as they do in a fluid and, provided there are not too many of them around, this works fine. But when the neighborhood gets too crowded they hamper each other too much. By agreeing to confine themselves in a crystal cell they do lose their freedom to roam anywhere, but they get more space to move about *individually*. In this case, as you can see, order means freedom. How is it that billiard balls understand this, whereas my fellow countrymen hardly seem to do so?

Real atoms and molecules are, however, much more complicated than hard spheres, and the way they interact is not as simple as billiard ball collisions. The result obtained by Alder and Wainwright seemed likely to remain just an oddity, confined to the virtual words of computers, or at most an amusement for our theomads. But this is where colloids come into play, because hard spheres, though not generously provided by Nature, can actually be made. Let us see how. First, we had better use a solvent that is not aqueous, to get rid of annoying electrostatic forces. Suspending particles in a non-polar solvent is not easy, but if we make them "hairy" by sticking surfactants or short polymer chains to their surface, the suspension becomes stable. To tell the whole truth, dispersion forces are still there, but one can show that they can be strongly reduced by using particles and solvent with a similar index of refraction[7]. This is the route followed by Peter Pusey and Bill van Megen[8], who used PMMA spheres dispersed in organic solvents to obtain the best approximation of hard spheres existing outside a computer memory. Their results are visually summarized in Fig. 5.3, showing a series of suspensions prepared at a particle volume fraction, denoted by the Greek letter $\Phi$ (Phi), which increases from 48% to almost 64%. The top row, showing the samples just after they have been vigorously mixed, shows that the suspensions have initially uniform colors, though much more vivid than the pale blue we saw in diluted milk, which is an effect of their high concentration. However, look at what has happened the day after (central row). While for $\Phi$ lower than about 0.5 (50% in volume) the suspensions still have a uniform appearance, for a slightly higher volume fraction iridescent sparks suddenly appear. On further increasing $\Phi$, the sparks progressively invade the cell, until they fill it completely for $\Phi \simeq 0.55$. These colored spots show that *colloidal crystals*, little grains have grown, made of colloidal particles clinging together to form an orderly crystal lattice.

Where does this color cascade stem from? When light with a wavelength comparable to the distance between the "layers" (think of the oranges) passes through a crystal, it is *diffracted*, that is, diverted from its original direction. Diffraction is an effect strongly related to scattering; in fact, it is nothing but scattering from an *ordered* structure, where part of the incident light (or more generally, the radiation), instead of being diffused all around, is scattered only at specific angles that depend on the specific symmetry of the crystal. In fact, by analyzing these diffraction lines, known as *Bragg peaks*, we can obtain the crystal structure; we can tell, for instance, if it is FCC, HCP, or something more complicated. Because of this, diffraction is the prime technique on which crystallography is based. In atomic or molecular crystals, the typical distance between layers, or planes, of particles is a few tenths of a nanometer, so to observe diffraction we have to use X-rays, which have a wavelength around this size. But in colloidal fluids the characteristic distance is of the order of the

---

[7] Dispersion forces actually have an electric (or rather, electromagnetic) origin too. The physics behind them is very complicated, but for our purposes it is sufficient to know that they depend on the refractive index difference between particles and solvent, and that, under conditions of index-matching, they are very weak (except when the particles are suspended in water, which is a peculiar solvent, as we have said). Working close to index-matching, moreover, is the only way to see something inside; otherwise, because of scattering, the suspension would be as white as milk.

[8] Peter is a true British gentleman, with a natural modesty that often makes him candidly admit that he has not understood something, a quality that is far from common among scientists of his caliber (and because of this, he tends to grasp most things before and better than many others). Bill, Dutch by birth, but Australian by adoption and spirit, is a true original, with a vein of eccentricity that makes him look like a kind of Crocodile Dundee of science. The improbable mix of these excellent friends has begotten splendid results for the physics of matter.

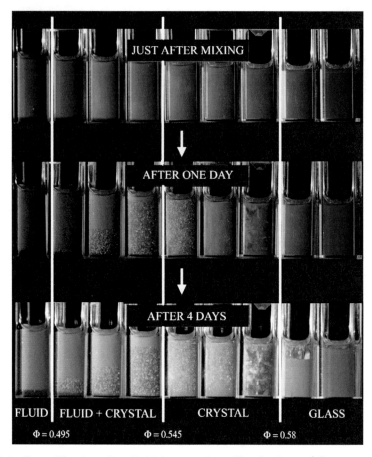

**Fig. 5.3.** Crystallization of a colloidal suspension of hard spheres. [*Courtesy of Peter Pusey and William van Megen*]

particle size, so it is *visible light* that is diffracted. The different colors are due to single crystallites, oriented so as to show different planes of particles, and therefore diffracting different wavelengths. This does not mean that between the spots there are no crystals, simply that no Bragg peak is directed so as to reach the camera lens. Since there are two basic kinds of hard-sphere crystals, what type are these colloidal structures? Usually, it is a balanced mixture of FCC and HCP, with the sequences AB and ABC following one another randomly and "democratically", so as to be equally represented.

That colloids may form crystal structures had already been known for more than a decade, but the 1986 experiment by Pusey and van Megen, which agreed with what simulations predicted, finally showed that hard spheres do not exist only in the virtual world[9]. As for many other discoveries and inventions, however, Peter and Bill had been long preceded by Nature. Even now, perhaps one of my readers is wearing on her finger one of the loveliest examples of colloidal crystal: an *opal*. These iridescent gems form by slow precipitation of silica from water as pellets with a size ranging between 150 and 300 nm. Water evaporation, taking place over geological timescales, then leads to a gradual concentration of the silica spheres and to the growth of a compact colloidal crystal. It is crucial for the evaporation to be so slow, otherwise the strain due to water surface tension would rapidly break up the crystal. Indeed, most opals are found in Cretaceous geological sediments more than 100 million years old (over 97% of opal extraction occurs in Australia). Another basic requirements is that the spheres do not differ in size by more than about 10%, otherwise what we get is a disordered solid structure, which we shall deal with shortly. Actually, many natural opals, known as *potch*, have no commercial value, for they do not show colors, but just a whitish appearance due to regular (non-Bragg) scattering from a disordered structure. Moreover, as jewelers know, opals are rather fragile gems and they are very sensitive to humidity. Because of this, they are often prepared as a thin layer sandwiched between a glass or quartz sheet and a template made of a black stone, such as obsidian, that enhances their colors. Interestingly, in 2008 the NASA space probe Mars Reconnaissance Orbiter probably detected the presence of opals on Mars, witnessing the existence of liquid water on the planet much later than we formerly believed.

How "hard" are colloidal crystals when compared with ordinary crystals? Once again, we can make good use of dimensional analysis. The dimensions of both the elastic modulus and the breaking load of a material are a force divided by a surface area. Since energy has the same dimensions as work, and work is force times displacement, i.e. times a length, it is easy to see that elastic modulus and breaking load also have the dimensions of an *energy per unit volume*. In atomic crystals, the typical energy is given by the strength of the bonds holding the atoms in position, whereas the only characteristic energy for a colloidal crystal is the thermal energy $kT$, which is usually quite a lot smaller. What is more important, however, is that the volume we have to divide by has to be comparable to a unit cell, in other words to the cube of the typical spacing of the crystal lattice. For a closely packed structure, the latter is not much larger than the size of the particles forming the crystal, and colloidal particles are thousands of times bigger than atoms. Thus, both the elastic modulus and the "strength" of a colloidal crystal are billions of times smaller than those of their atomic equivalents, so much so that, when the cells

---

[9] In my own small way, I have also made a modest contribution to this story. Some years later, Vittorio Degiorgio, Tommaso Bellini and I showed that not only crystallization, but the whole thermodynamic behavior (what is called the "equation of state") of a hard-sphere colloidal suspension fully conforms to theoretical predictions.

in Fig. 5.3 are shaken, the Bragg peaks disappear, showing that the crystals "melt"[10]. Nonetheless, hard-sphere crystals are not so very soft and, to melt them, the cells must be shaken vigorously (note that, even though the solvent molecules are free to move, the whole structure stays together). Some colloidal crystals can be much softer still. Since crystal formation takes place even in the absence of attractive forces, what happens if we make the particles even *more* repulsive, for instance by charging them? Here, since the surrounding ion clouds try not to overlap, it is as though each particle takes up a much larger "effective" volume. For strongly charged colloids and no added salt, this leads to colloidal crystal growth even when $\Phi$ is very low, in some cases lower than 1%. These crystals, which sometimes have a different structure[11], are really feeble, so weak that they can be *poured* just like water. When we turn the container, what actually happens is that the crystal melts under its own weight and flows, reverting to crystal form (more rapidly than for hard spheres) once the fluid has come to rest. You may see some "homemade" (where "home" stands here for "my lab") examples of these crystals and of their gorgeous colors in Fig. 5.4.

Colloidal crystals have rapidly become a real test bench for models of the growth and structure of crystals, and are therefore of considerable interest for condensed matter physics. But do they also have any practical application? In the long term, yes. There is still a long way to go, but the rewards could be great. Colloidal crystals may play a vital role in the development of tomorrow's computers, exploiting optical circuits where it is *light*, and not electric current, that carries information. In these circuits, controlling whether a light beam passes or not could be regulated just by tuning the diffraction properties of an ordered structure. Actually, simple optical circuits are already built within optical fibers, currently the fastest and most effective way to transfer information. But the chance of making optical switches, selectors, or even transistors not just in one but in three dimensions is really alluring, for these components could be used as logic circuits or memory elements and could open up unlimited frontiers to computer science. The first step along this way is being able to make *photonic crystals*, three-dimensional structures that control the passage of light of different wavelengths just by Bragg diffraction. One of the approaches being followed is to use suitable colloidal crystals as templates, substituting a solid matrix for the solvent without changing the position of the particles. The particles themselves are eventually removed to leave "holes" that amplify diffraction effects by maximizing the difference in refractive index with the solid matrix. Hence, we get a kind of "ordered Gruyère", suitably selecting which wavelengths are transmitted and which diverted. As I said,

---

[10] Because the crystals then reform rather slowly, and since we can use visible light to observe them, the birth and growth of crystals is much easier to study for colloidal than for atomic systems.

[11] It is the body-centered cubic structure, or BCC, where there are no spheres on the sides, but just one at the center of the cube, which has a maximum packing of about 68%.

**Fig. 5.4.** A colloidal crystal show. The three images in panel A refer to the *same* sample, observed at different angles with respect to the direction of the incident light (the same is true for my lab's logo, shown in C). Panel B displays a suspension of charged colloidal particles at settling equilibrium. The sediment is made of a lower colloidal crystal phase, over which there is a liquid phase that does not show Bragg peaks. Panel D shows a natural opal.

the road to achieve practical results is still very long, apparently endless. Yet developments that were unforeseeable a few decades ago, such as MP3 players or E-paper, has taught us that, with technology, we had better avoid playing the. . . crystal-gazer.

## 5.3 Glasses and gels: when hate and love yield similar results

Coming back to Fig. 5.3, other changes can be seen to take place in the samples when $\Phi$ further increases. Once the crystallites have spread throughout the suspension their spots fade away, to be replaced by a totally different tex-

ture that resembles suspended "sheets" with a blue-green shade, indicating a
change in the colloidal crystal morphology. Most surprising is what happens
next, when $\Phi$ goes beyond about 58%. The suspension reverts to a uniform
color, first green and then blue, as if no crystallites were present any more.
Yet the suspension is quite rigid, even more so than colloidal crystals. With
their experiment, Pusey and van Megen had shown the birth of a hard-sphere
*glass*, a structure that is "mechanically" solid, but fully disordered.

Where does this transformation into a vitreous state stem from? A hint
comes from what happens at slightly lower concentration. The particles *would
like* to arrange themselves into a crystal (after all, they could rearrange into
crystals at densities up to $\Phi \simeq 0.74$) but, to do so, they must move around
and reorganize. This becomes harder and harder to do because, crammed in as
they are, they hinder each other's movement. In fact, those sheet-like crystals
forming just before the glass transition do not grow in the core of the suspension, but close to the walls, the only place where this is still feasible. When
$\Phi$ increases slightly more, even this surface-growth mechanism is suppressed,
for each particle gets trapped in a cage formed by its neighbors, as shown in
Fig. 5.5B. The only way out of this cage requires all surrounding particles to
move in a coordinated way, leaving an escape route. This suggests that glass
is not a truly stable phase of matter, but rather an *arrested state* that may
take very long time to escape from.

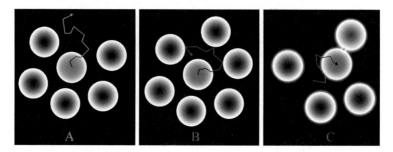

**Fig. 5.5.** Particle diffusion in a fluid (A), in a "repulsive" glass (B), and in an
"attractive" glass (C).

We have already encountered glass phases, in particular when talking
about polymers. For these (and even more for "molecular glasses", such as
common window glasses) things are much more complicated, because they
are not simple hard spheres, but the basic mechanism leading to glass formation is very similar. In fact, many aspects of the behavior of molecular
and polymeric glasses[12], for instance the viscosity or the way these systems
vibrate, have a precise equivalent for a system of billiard balls. Just because
of their simplicity, hard-sphere glasses have become a model system to inves-

---

[12] Or at least of a wide class of vitreous systems, called "fragile" glasses.

tigate the origin and the structure of common glasses. However, many basic aspects have yet to be clarified. For instance, is it really unavoidable for the system to arrest? In fact, in computer simulations some little crystals usually form before this "dynamic arrest" takes place, rapidly leading to the crystallization of the whole system. The very same type of samples as used by Pusey and van Megen seemed to crystallize if prepared and mixed in space, on board the International Space Station. Once brought back to the ground and remixed, however, they again became stuck in a glassy state. Apparently, therefore, the tiny sedimentation effects due to gravity have a huge effect on crystal formation. For that matter, even the samples used by Pusey and van Megen begin to crystallize after a few days, starting from the free surface at the top (see the last row in color plate 6). Very probably, therefore, a system of hard spheres will sooner or later crystallize. Yet the experiment we have described shows that, around $\Phi = 0.58$, something special does happen.

In the past few years, physicists have realized that arrested states may originate from two distinct situations. To understand this, we have to consider a model slightly more complicated than hard spheres, namely what is called a system of *sticky hard spheres*. The basic idea is that we also introduce *attractive* forces between the particles, but of a very simple kind: a sort of "glue" spread over the particle surface that, when two spheres touch, binds them with a given strength that can be controlled. This can be achieved rather easily, for instance by adding to the suspensions surfactants or polymers (which means other particles, but much smaller, such as micelles or little coils) that do induce exactly this kind of force between the colloidal particles. We shall not inquire too deeply into these interactions, known as "depletion forces". Let me just point out that these forces arise once again from the demand by the extra particles to maximize their entropy. They are small, but there are also a lot of them, all very annoyed by the presence of the big particles, which deny them freedom of motion. Thus, they join forces and try to pack the big guys all together, so that they take up less room, True, this way the latter *lose* entropy, but there are few of them, so they count almost for nothing. After all, entropy is basically a question of numbers, or, as Miguel Cervantes used to say, *Muchos pocos hacen un mucho*, a lot of little guys have a big effect. All things considered, the big particles, in this special solvent made of smaller particles, will tend to attract and stick to each other. A system of this kind shows a much richer behavior than hard spheres, because the strength of the glue can be tuned between very low values, of the order of the thermal energy, and much larger values, comparable to the strength of dispersion forces. For example, in specific conditions the suspension may separate into two phases, similar to the gas and liquid phases for a molecular fluid[13].

In the presence of the surface glue, as shown in Fig. 5.5C, a particle can become trapped even if there is not a fully closed cage around it. In principle, each particle could get out of the surrounding cluster, but in practice it gets stuck because, before getting free, it touches another particle and sticks to it. The dynamic arrest of the system can therefore take place at much lower particle volume fraction than is required for hard spheres, even for a concentration as low as 0.1% if the glue is really strong. Clearly, the disordered structures that form are much more "open" than hard-sphere glasses. In fact, they are quite similar to the fractal aggregates stemming from colloidal aggregation. This observation leads us to regard these arrested systems (called *attractive* glasses to set them apart from hard-sphere glasses, conversely called *repulsive*) as a prototype of what we have called colloidal *gels*. The aggregates arising from van der Waals (dispersion) forces on adding salt to charged colloids would be an extreme example of structures of this type, obtained when the adhesion strength becomes very large.

Hence, this point of view allows us to find a common framework for all those disordered solids, such as glasses, gels, and colloidal aggregates, which have long been considered separately. Obviously, we are still talking about systems with a much simpler structure than a polymer mesh, but this track seems to be the right one. In particular, what characterizes a vulcanized rubber is not only that the bonds holding the chains together are strong, but also that the monomers that bind to each other are a small fraction of the total. We are getting some idea about this situation too. Francesco Sciortino, for instance, a brilliant theorist in the University of Rome "La Sapienza", has for some time been studying the behavior of spheres that stick to each other only at specific points, as if the glue, instead of being uniformly spread on the particle surface, were concentrated only in localized spots. On varying the number and strength of these spots, Sciortino's spheres show a very rich behavior. In some cases they behave like monomers, forming long chains that, in their turn, link up at a few points to make an elastic mesh. The properties of these nets mirror many aspects of rubber. What is more, the model describing them is in every respect equivalent to those developed in polymer science. As you can see, by toying with these simple tiny Lego bricks, physicists can eventually manage to act the "little chemist."

---

[13] The sticky hard-sphere model has proved to be extremely useful to understand some basic aspects of intricate problems, ranging from protein crystallization to the structure of ricotta cheese. We shall meet them again when discussing biological systems.

## 5.4 The world is not (just) a ball

We have spent quite a bit of time in describing how spheres can pack, and what consequences this has for the way they spontaneously organize in solution. Although this is surely the favored course for our theomads, we should not forget that world is not just made of rigid or sticky balls. How, then, do objects with a different shape pack? Obviously the shapes we could consider are many, but we shall mainly focus on two specific geometries, sticks and "eggs" (or, as we should call them, *ellipsoids*). Both these kinds of particles can teach us a lot about the way matter can be organized. About ten years ago I was in Utrecht, in the Netherlands, discussing some recent results with Albert Philipse, a brilliant colloid chemist who is able to make particles with the most diverse shapes. Entering Albert's office, I noticed something rather odd, to say the least. His room was full of jars containing pills, nails, toothpicks, and copper-wire bits; in addition, there were vials filled with stuff that, as Albert pointed out, was rod-like too, but rods of microscopic size, which were actually colloidal suspensions of hematite and other mineral particles. What Philipse was doing was studying how objects with this shape pack randomly, trying in particular to find how the maximum packing fraction depends on the ratio between the length $L$ and the diameter $D$ of the rods, a kind of parallel to the RCP value for spheres.

What experiments show is that the RCP value decreases very rapidly on increasing the ratio $L/D$ (see the left panel in Fig. 5.6). For rods ten times as long as they are wide, the maximum packing fraction shrinks to about 35%, whereas randomly putting in very long sticks, with $L/D \simeq 100$, only about 5% of the container can be filled. If you are used to dealing with spaghetti (as an Italian, I surely am), you should not be too surprised. Unless we put them in neatly, very few strands of spaghetti can fit in a pot, at least before they are cooked. The same happens for any kind of rod or stick, for each particle has to work very hard to find space among the others[14]. Thus, the rods have to come to an agreement that, as in the case of spheres, ensures each of them a greater freedom of motion. Here, however, the compromise does not consist in organizing into a crystalline phase, for the centers of the rods can be placed at random, with no special requirements; what matters is only that, as in a box of spaghetti, the *directions* of the rod axes are as close as possible. Thus, we are facing a new kind of Middle-earth between liquids and solids: the rods can flow freely as in a liquid, provided they remain suitably *oriented*. These materials in between solids and liquids, showing order in their orientation, but not necessarily in their positions, are named *liquid crystals* and, once again, originate from the quest for larger freedom of motion, i.e., entropy.

---

[14] After all, if you think about it, when a rod rotates it "sweeps" a space equal to the volume of a sphere of diameter $L$, which is much larger than the volume of the rod itself. Thus, to be able to move without hindrance, it would need a huge amount of room.

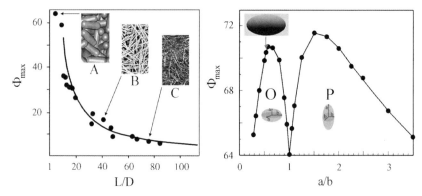

**Fig. 5.6.** Left: mSaximum packing fraction for rod-like particles as a function of the ratio between length and diameter (the insets show some of the particles actually used by Philipse). Right: maximum packing fraction for prolate ellipsoids (those longer than they are wide) and oblate (short, wide) ellipsoids. The arrow shows the position of M&M's on the graph.

I am sure you have already heard of liquid crystals; indeed, among computers, TV sets, and digital watches, you can surely find many examples of them around the home. Practical liquid crystals are not, however, made of colloidal particles, but of small molecules. In this case, what leads them to align spontaneously is not so much (or not only) their shape, but rather the fact that the *forces* acting between these molecules tend to act along a given direction. In some aspects, however, the effect is rather similar, and proper "colloidal" liquid crystals *do* exist. Many interesting questions about liquid crystals have been unraveled by studying suspensions of certain rod-like viruses that we shall encounter at the end of this book. Some biological macromolecules, for instance collagen, cholesterol, or the proteins that make silk (which we shall talk about in the next chapter), are organized as liquid crystals too. Finally, just because of their elongated shape, many kinds of surfactant molecules can form liquid crystals, but at much higher concentration than those typical of micellar solutions.

In fact, there are many liquid crystal structures, covering the whole Middle-earth between liquids and solids, which are collectively called *mesophases*. Some of them are shown in Fig. 5.7. In the simplest case, where the molecules (or the colloidal particles) are oriented on average along a given direction, but their positions are random, as in a liquid, we speak of *nematic* liquid crystals[15]. But as well as having a preferred orientation, the molecules may also be organized in regularly stacked planes, although they are still randomly

---

[15] The word has the same Greek root as "nematode", and refers to the worm-like lines one sees under a polarizing microscope.

positioned within those planes. These so-called *smectic* liquid crystals[16] can be of several kinds, depending on the way the orientation axis is directed with respect to the planes. Things can be even more complicated when the average orientation axis does not keep a fixed direction, but turns progressively from plane to plane, so that the molecules arrange themselves along a kind of helix. These liquid crystals, called *cholesteric* because they resemble those formed by cholesterol molecules, are of great technological interest, because the "pitch" of the screw can be controlled by applying a voltage. What makes liquid crystals so interesting is just the chance of controlling the orientation rapidly and in specific spots. Since their optical properties are rather peculiar too, this allows those familiar displays to be made. The world of liquid crystals is so rich as to form a research field of its own, one that would require an entire book to itself, whereas here I cannot go beyond a brief outline. Anyway, if you have realized how several aspects of their behavior are related to soft matter, I have reached my goal already.

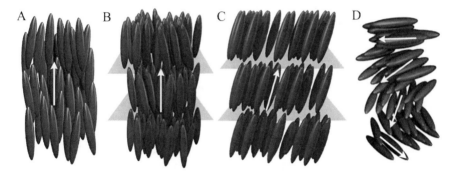

**Fig. 5.7.** Structure of nematic (A), simple (B) or "tilted" (C) smectic, and cholesteric (D) liquid crystals. Source: *Wikimedia Commons*, revised by the author.

Rods are quite interesting, but they do not give a direct answer to the basic problem we wish to investigate, namely, can particles with another shape pack better than spheres? Let us then consider objects with a shape rather similar to spheres; say, something like a sphere squeezed or stretched out along an axis. To picture these "little eggs", which are called *ellipsoids*, just think of rotating an ellipse (a shape I hope you have met before, maybe when learning how planetary orbits are shaped) around one of its axes. If you turn it around its longest axis, you generate what is called a *prolate* ellipsoid, which is a stretched sphere, whereas turning it around the shorter axis you get an *oblate* ellipsoid, namely, a sphere flattened at its poles, rather like the Earth. An ellipsoid is therefore something like a "symmetric egg" with two axes, which

---

[16] From a Latin word meaning cleansing, for these substances often have a soapy consistency, whereas nematics are viscous liquids.

we shall indicate with $a$, that have the same length, and a third axis, which we shall call $b$, that can be shorter or longer depending on whether we have, respectively, an oblate or a prolate ellipsoid.

Do ellipsoids pack better or worse than spheres? This problem has been tackled by Paul Chaikin, one of the most ingenious condensed matter physicists currently on the market, but also one of the wittiest, funniest, and most original fellows I have ever met, particularly when he sports a long, curled mustache making him look like a scientific double of Salvador Dalí. Mustache apart, his scientific originality mainly consists in trying to deal with topics that, although apparently simple, actually conceal deep and delicate physical questions. As we shall see, ellipsoid packing is a problem of this kind.

The rooms where soft matter scientists work are full of surprises. We have already noticed this on peeping into Albert Philipse's office, but Paul's room in New York University is even more amazing. Among trinkets, trifles, and toys fit for a kindergarten, what stands out is a big yellow bin, completely filled with... chocolate candies. Not generic candies, however, but specifically those (and here I necessarily have to do some advertising) commercially known as M&M's.

What is the purpose of a candy bin in the room of one of the brightest sparks in physics, besides being a dispenser of presents for guests (something I enjoyed quite often)? To make this out, we should take a look at the numerical simulation of ellipsoid packing by Paul and colleagues. The right panel in Fig. 5.6 shows that the maximum packing fraction $\Phi_{max}$ has a very peculiar trend. As soon as the ratio $a/b$ becomes slightly larger or smaller than one (a one to one ratio corresponds to a spherical shape), $\Phi_{max}$ grows well beyond the value $\Phi_{max} \simeq 0.64$ we found for spheres, rapidly reaching a value larger than 0.7, and then decreases again when the axis ratio is larger than 1.5 or smaller than 0.5. What is really curious (and experimentally quite useful) is that the actual shape of the M&M's, which are oblate ellipsoids, is very close to optimal packing[17]. So, ellipsoids that are not too long or too fat definitely pack better than spheres. Why? Perhaps because they are not randomly oriented, but rather line up as in a liquid crystal, filling the container better? No, this does not work. As we shall see, ellipsoids that are not markedly different from spheres do pack at random. Moreover, if we stretch or compress them too much, so that they look like cylinders or coins, things get worse. Indeed, while neatly stacked coins pack very well, their *random* packing fraction is much lower than for spheres. The real point is that an ellipsoid may rotate so as to *fill the holes* existing in a random packing configuration (this obviously does not work for a sphere, no matter how it turns).

---

[17] When news of these results broke not only in scientific journals, but also on CNN, Paul called the management of the company that makes M&M's, hoping to get something in return for this free advertising (which, for a scientist, means *financial* support). What he got was the bin I have mentioned...

Perhaps there is a more intuitive way to understand why ellipsoids can fit together more tightly than spheres. To see this, we need to introduce a concept with the rather pleasant name of the *kissing number*, which is the number of particles that, in a given arrangement, touch a selected one. For example, in FCC and HCP, the kissing number $K$ is 12, which is also the maximum value that can be obtained for an arrangement of spheres. For a looser packing, this number will be quite a lot lower, and in general will be different from sphere to sphere, yet it still makes sense to inquire about the minimum number of contacts we can have in RCP conditions. This question is hard to solve, but a reasonable conjecture due to Shlomo Alexander and Sam Edwards states that $K$ is, on average, equal to twice the number of "constraints", that is, of the conditions that we have to set for fixing a particle in place[18]. For instance, to hold a sphere in position we must prevent it from moving along the three space directions, so we have $K = 2 \times 3 = 6$. In fact, from Bernal's experiments it is already known that the average number of contacts per sphere in RCP conditions is slightly larger than 6. For ellipsoids, however, this is not enough, for we must also prevent them from rotating. It is not difficult to see that, both for prolate and for oblate ellipsoids, one has to fix the position of two axes, so the total number of constraints is 5. To pack decently, ellipsoids must then have on average *ten* contacts, and therefore their RCP structure must be denser than for spheres. Does the conjecture by Alexander and Edwards work? To test it, and actually to do much more than this, Paul Chaikin had the brilliant idea of packing M&M's into a glass globe with a size slightly larger than a human head (see the original "big candy head" in Fig. 5.8) and, with the complicity of a physician friend, slipping it on the sly into a MRI machine (obviously at daybreak, to avoid taking up real patients' time). Doing this, he was able to reconstruct the position of every particle, verify that all axes are placed at random, and reconstruct all the values for $K$, which actually turns out to be slightly *lower* than 10 (we do not yet know why, but probably this has to do with entropy again).

Anyway, both for prolate and oblate ellipsoids, $\Phi_{max}$ is still lower than the value for the hard-sphere crystal. Can we do better? The suggestion by Alexander and Edwards led Chaikin to consider randomly packing less symmetric ellipsoids, where *all three axes* are different, which require a minimum of 12 contacts. He discovered that, by choosing the three axes in the ratio $1.25 : 1 : 0.8$, one gets a $\Phi_{max}$ value equal to 0.747, this time *larger* than for the FCC and HCP structures. Unluckily, this random packing still cannot outdo the crystal structure that *this kind* of particles can form, where $\Phi_{max}$ is at least 0.757 (discovering what the real maximum value is will probably take Thomas Hales and co-workers the next million years).

Contrasting the density of a disordered structure with that of a crystal allows us to address a basic problem in condensed matter physics. We have already seen that a hard-sphere glass is just an "arrested" structure since, as

---

[18] This happens for those structures that civil engineers call *isostatic*.

**Fig. 5.8.** The "Big M&M's Ball". [*Courtesy of Paul Chaikin*]

soon as they can, these balls prefer to order so that they maximize their entropy. For ellipsoids things are no different, although the random and ordered packing fractions are so close that the glass may be very stable. Yet, if we were to find a shape with a disordered packed structure *denser* than the crystal, the reverse would be true. For particles of this kind, glass would cease to be just an accident of Nature, instead assuming the status of a stable phase of matter, a very interesting result for us physicists (probably laymen could not care less, but we are used to it). So far, Paul has failed in this, although he is still trying hard with promising shapes such as tetrahedra. Knowing him, I am sure he will not give in easily.

A short digression. Last year, as I was about to return to a conference room, I got stopped by Daan Frenkel, probably one of the world's greatest experts in computer simulation, who told me I *absolutely* had take a look at a picture before going back in. The illustration, reproduced in Fig. 5.9, showed four so-called *briquettes*, weighing about 50 grams, as commonly used in coal stoves. These briquettes, however, are not just coal "eggs". Rather, they are nice ellipsoids with three different axes of 55, 42, and 34 mm respectively, so the the major and minor axes are respectively 1.3 and 0.8 times the middle one. The ratio between the major and the middle axes is therefore only 4% larger than the value for Chaikin's "ideal" ellipsoid, while for the shorter axes it is practically perfect! Thus, when thrown at random into a bag, these briquettes pack in beautifully, saving a lot of room. No one knows who found this solution, or how it was found, but we know that exactly this shape has been commonly used for more than a century, which makes me wonder whether physicists' "discoveries" are sometimes just. . . Columbus eggs[19].

**Fig. 5.9.** Briquettes for coal stoves (source: *Wikimedia Commons*).

## 5.5 Sandcastles and shifting sands

Maybe the closing sentence of the last section is a bit too harsh. Packaging problems, besides being very stimulating, are actually of huge interest both in scientific and in industrial applications. A relevant example for chemistry and biology is chromatographic techniques, among the most useful methods for identifying and separating compounds ranging from simple molecules to proteins. These techniques use columns filled with spherical and usually porous grains, able to capture the "prey" that has to be identified, or more simply to select molecules on the basis of their transit time[20]. But think also of how

---

[19] If you don't know, or don't remember, the story of the Columbus egg, just check on the web.

[20] Curiously, quite often the smaller compounds are the ones that take longer to get though, for they wander through the grain pores, which the bigger particles ignore.

many materials, from rice to cereals, from coffee grains to *muesli*, from sand to washing powders, are in grain form. Thus, it is very important for the industry to understand not only how they can be packed, stored, or conversely kept sufficiently soft to avoid the grains sticking together (a serious problem for "hygroscopic" materials like table salt, which attract moisture), but also how they can be transported through pipes. Hence, in the past decades many physicists and engineers have devoted themselves to the investigation of these systems, named *granular materials*, which have some remarkable surprises in store.

What is so different between a granular material and a very concentrated colloidal suspension? One might say that granular materials are nothing but suspensions that have "lost their suspending agent", which is certainly true if we compare them with a particle dispersion in a liquid. But there is actually a solvent there, air, which is always in between the grains *exactly* as in aerosols. What makes the difference, however, is the size of the grains. Comparing a typical aerosol particle with a grain of sand, let alone a coffee bean, is like comparing Ulysses with the giant Polyphemus. Thus, what is really different is that the thermal energy provided by the solvent (which, remember, is always $kT$, whatever the particle size) is trifling compared with the effect of weight. Dragged to the bottom by their own burden, unable to reorganize spontaneously because the thermal energy they possess does not enable them to make even a tiny jump upward, the particles become "jammed" to form bridges, arches, and other blocked configurations that do not allow them to settle into a true equilibrium structure. In fact, we have already seen that oranges, thrown at random in a crate, form a disordered structure and not a crystal, which would be their preferred state, whereas tiny colloidal spheres can often make it. Because $kT$ is so small compared with any other energy in play, granular materials can therefore be considered as "athermic" systems, ones where temperature counts for nothing[21].

Just because they are always "out of balance", granular materials show a very peculiar mechanical and elastic behavior. We already mentioned the dilatant property of sand – in other words, to deform when pressed, it must *increase* its volume (which explains why, when you walk on the shoreline, it dries around your footprints). Another distinctive aspect is that, unlike a liquid, whose pressure increases with depth and is the same in all directions (which should be evident if you ever dive), the distribution of the stresses due to weight in a granular material is much more complicated. For instance, in a grain silo, which we can roughly picture as a cylinder on top of a cone, the pressure at the bottom does not grow any more once the height of the

---

[21] The potential energy due to weight can be billions of times larger than the thermal energy. Even if the usual concept of temperature does not apply to a granular material, entropy remains a fundamental quantity, which still has to do with the number of different "configurations" the system can take. This allows us to define a kind of "effective temperature" that basically states how the entropy depends on the material volume, that is, on its "compactness".

grain pile becomes larger than a few times the container diameter. Where has the thrust of all the additional grain gone? On the walls, which therefore must withstand most of the stress. This is something that anyone designing a silo knows quite well. Moreover, just think what happens if you try to make a mound by letting sand fall on a table from your hand. As the sand falls, a taller and taller cone forms. Yet the mound never gets too steep because, when the angle at the tip of the cone becomes too small, little "avalanches" get more and more frequent, lowering the cone and enlarging its base. These "catastrophic" events are so abrupt and unpredictable that they draw the attention of many physicists and mathematicians for their links with similar events such as real snowslides, but also with apparently unrelated effects such as earthquakes, forest fires, and even stock market collapses.

More surprises come from considering how granular fluids flow, for sand and rice do flow too, but *very differently* from water. Think again of the sand mound. If you tilt the table, sooner or later the mound suddenly slides down. Yet not all the sand flows, only a surface layer that again looks like a little avalanche. Moreover, if you drill a hole at the bottom of a tank filled with water, the liquid at first rushes out and then, as the container empties, flows more and more slowly. Not so with the sand. The outflow velocity is always *the same*, whatever its level in the tank (ask yourself what has this to do with hourglasses)[22]. Even more different from the behavior of a simple liquid is what happens if you make sand flow through a long vertical tube. Instead of a regular flux, you get something that looks more like a traffic jam in rush hour, with sudden hold-ups for no reason[23], which move *backward* with respect to the flow direction and then suddenly fade away, once again with no apparent cause. The relation between granular flow and traffic is so close that some physicists deal with both problems as they were almost the same.

Since granular fluids have no spontaneous thermal agitation, the easiest way to set them in motion and allow them to explore different configurations is to "fluidize" them. This can be done, for instance, by blowing air upwards from the bottom of the container, a method widely used in industrial production to prevent granular matter from becoming too compacted, or by shaking them. The effects of shaking on granular fluids have aroused the interest of many researchers, for they often give rise to curious and unexpected effects. The best known of these was originally pointed out in a paper published two decades ago in *Physical Review Letters*, one of the most prestigious physics journals, under the eyecatching title of "Why the Brazil nuts are on top," and was then

---

[22] Clearly, in the case of sand, the shape of the tank bottom must be suitably designed, otherwise the outflow may stop because of lumps blocking the exit hole, caused by the build-up of "arched" structures between the grains. In addition, if the opening is not carefully planned, the flow involves only part of the sand, whereas the rest remains in the container.

[23] At least no apparent reason. When I get trapped in these situations, I still find it hard to convince my son that we do not necessarily have someone in his eighties ahead, whose way of driving is bringing the traffic to a standstill.

followed by hundreds of related studies. Contrary to what we might expect, when a mixture of grains of different size is shaken or vibrated, the largest ones *come to the surface* (this is exactly what happens with Brazils in a bag of mixed nuts). In other words, the mixture separates spontaneously, and different kinds of grain sort themselves into different places. If you do not believe me, do a simple trial by mixing refined with coarser raw cane sugar. First, place the cane sugar in a container and slowly pour refined sugar *on top* of it. Then shake the container gently, to mix the two kinds of grains (this will not come out too well), and then go on tapping the bottom of the container. Figure 5.10 shows what you should eventually observe. The simplest explanation of why large grains end up at the top is that the little ones more easily manage to get to the bottom by slipping through the holes left by the big ones. However, the story is not yet completely clear, for in some conditions the reverse takes place.

The investigation of granular flow is interesting on the large scale of geological or astronomical phenomena too. Particle separation based on size is obviously crucial in sedimentation processes, but think also of the way desert dunes move. Also of great interest for planetary geology are the soil fluidization processes caused by earthquakes or by the strong "shock waves" produced by meteor impacts. Finally, Saturn's rings form a spectacular granular flow lab, made of a multitude of ice and dust grains with a size ranging from less than a centimeter to a few meters. The incredible complexity of these rings, accurately observed by the Cassini space probe, which contain a myriad of subdivisions and are moreover perturbed by Enceladus, one of Saturn's moons which crosses their orbit, is still far beyond our powers of explanation.

**Fig. 5.10.** From left to right: white refined and brown cane sugar as prepared at the start, after a first quick mixing, and after patiently tapping the container on the table.

# 6

# Dreamtime

They say we have been here for 40000 years
but it is much longer.
We have been here since time began.
We have come directly out of the Dreamtime
of our creative ancestors.
We have kept the earth as it was on the first day.
Our culture is focused on recording the origins of life.

The Aboriginal Australia
Art & Culture Centre

*All together now: a rave party for soft matter – When double-faced Janus becomes double-tailed: the biological membranes – Organic origami: helices, sheets and acrobatics of the amino acids – Chemists, architects, truck drivers, aerial fitters, janitors, guards, sometimes even killers: a thousand jobs for proteins – Solar cells, molecular turbines and tightrope walkers: wonders of the biological nanomachines – Lunch break, with a high protein content – The cell: techniques of yoga breathing – A Chieftain with an excellent memory, and the magic of his faithful Shaman – Factories of dreams – Time dust or stardust?*

If you have managed to follow me up to this point (maybe plodding a bit), you will not regret it, for we have reached the climax of our story, where we shall really see for ourselves the stuff that dreams are made on (and we ourselves with them). From any point of view, living systems constitute a triumph of soft matter, where all the characters of Middle-earth cooperate to develop an incredible variety of strategies and expedients for surviving and populating any corner of this wonderful vale of tears. Here we shall find again polymers, amphiphiles, and particles of colloidal size that collaborate in developing structures, managing processes, and trading with their surroundings, within such a jammed and hectic environment that biologists have coined the expression "macromolecular crowding" to describe it.

Actually, a living organism is made almost entirely, or at least in all its crucial parts, from soft matter, but a kind of soft matter that is somehow "activated". Living beings have learnt to do three main things very well, namely, (a) *separating* themselves from the rest of the world, (b) sucking up *energy* from the surroundings, and (c) *investing* that energy by converting it

R. Piazza, *Soft Matter*, DOI 10.1007/978-94-007-0585-2_6,
© Springer Science+Business Media B.V. 2011

into a flexible form of cash available when, where, and how they need it (maybe for the future of their offspring). It is mostly this last credit in their personal record that allows them to *rule* soft matter, molding it for the purposes of the program written inside them and multiplying its potential *ad lib*. Science as such does not pretend to give full answers to the great existential questions, such as what life really is. It limits itself modestly, but stubbornly, to trying to explain the little "hows", rather than the big whys. Nevertheless, piece by piece, glimpses of the overall picture can already be seen. What can be said for sure is that life basically looks like a struggle against entropy, this time really meaning disorder, an epic attempt to claim that the unavoidable and icy end of time foretold by Kelvin is still remote, if it must happen at all. It is rather ironic that the weapons with which this battle is fought are based on those materials whose organization and running are so intimately connected with entropy. We might say that, faced with life, entropy is cheated and beaten.

This is not, however, a book on biology, and I have neither the time nor the competence to tell you the countless ways in which soft matter organizes in biological systems. In fact, I do apologize to the chemists and biologists right from the start for my inappropriate intrusion, and for the very many simplifications and (I hope few) blunders that will unavoidably go with it. If this can justify my efforts, I have the single, humbler goal of showing you that many of the concepts we have developed in earlier chapters may be guidelines that allow us to grasp something of these wonderful, but complex, biological machines.

## 6.1 *Concludo, ergo sum*

No, I have not decided to finish here. For those of you (probably many) who are not familiar with my ancestors' tongue, or those who have neglected it for too long (almost all the others, I guess, myself included), I should explain that, besides "to conclude", the Latin verb *concludere* can also mean "to hold inside", "to confine", even "to fence in". Thus, the title of this section, which could be read "I confine (myself), therefore I am", mimics René Descartes' famous phrase *cogito ergo sum* (which, in the case of a rose, a fox, or a baobab, might be a bit conceited), but takes the liberty of amending it. Indeed, as we are going to see, anything pretending to be a "self" must set precise limits to itself, and this comes well before "thinking" does.

Living matter has many powers making it very different from inanimate objects, the most distinctive being the ability to self-replicate. I imagine that, whatever your philosophical, religious, or political beliefs, you will agree on the following: prior to reproducing, any organism must be able to distinguish what is *inside* from what is *outside*. In other words, the first step in making an "individual" needs to be to construct a fence that divides it from the rest of the world. How can a simple organism, made of a single cell, incorporate just what it needs, leaving anything alien outside? The strategy adopted by all living cells is to build an external membrane, or **plasma membrane**, hosting the so-called **cytoplasm** inside, with a supporting framework that is nothing but a vesicle made of double-tailed amphiphiles.

There is an incredible variety of biological amphiphiles with different chemical properties, allowing them to perform a wide range of functions. But they follow just a few basic structures, because life keeps and consolidates those solutions that prove to be successful in the course of evolution. Broadly speaking, we can split them into two large classes, **phospholipids** and **glycolipids**, plus some other compounds, whose main representative is **cholesterol**, which are not properly amphiphiles because they are mostly hydrophobic and therefore practically insoluble in water. These three kinds of substances, together with proteins which are regular guests in the membrane, as we shall see, make up almost all the molecules in the plasma membrane. Hence, from the beginning, life has made use of one of the most distinctive properties of amphiphilic molecules, namely that they form aggregates like vesicles whose structure clearly defines an "inside" and an "outside". The plasma membrane cannot, however, be a rigid box, but should behave like a flexible bag, allowing the cells to take on different shapes and to change shape when needed. This is already true for an amoeba, but much more so for the cells in multicellular beings like ourselves. Take a red blood cell, with its donut shape that has to keep changing to move through the capillaries. Then picture how different this is from a neuron, which passes on nerve impulses by a "transmission cable", the axon, which can run from the big toe to the spinal cord with a length of over three feet.

For a cell, moreover, changing shape is crucial for sticking to a surface, multiplying, or simply growing. Usually, artificial vesicles are not very flexible because, within the surfactant double-layer, the chains are not disordered but rather well aligned, similar to what happens in a liquid crystal (after all *they are* "two-dimensional" liquid crystals). Many phospholipids have a perfect way to get around this limitation. One of the two hydrophobic legs is slightly shorter and "crooked", preventing the chains from packing regularly in the membrane. This lame leg always contains two double-bonded carbons and, from what we learnt for polymers, we know that this prevents the molecules from taking the zigzag configuration crucial for lining up neatly. As a consequence, the "skin" of phospholipid vesicles can be envisaged as a thin liquid film (the membrane thickness is just a few nanometers), fully disordered and very flexible. An important example of these crooked molecules is lecithin (more properly called **phosphatidylcholine**, shown in Fig. 6.1. A cell membrane, however, is something much richer, where the other lipids contribute to form regions of higher or lower flexibility, the latter useful to host membrane proteins, as we shall see. Moreover, the presence of lipids of different kinds allows "asymmetric" membranes to be made, where the composition of the inner layer facing the cytoplasm differs from that of the outside layer.

There is, however, another group of functions that makes flexibility even more important. A cell cannot be content to keep in what it cares for, it also

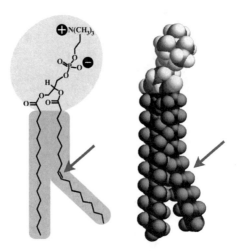

**Fig. 6.1.** A molecule of phosphatidylcholine. Its chemical structure, where the presence of charge of both signs is highlighted, is shown to the left, whereas the true three-dimensional shape is shown to the right.

has to take in what it needs (at least food!), and keep out what should be discarded. The plasma membrane must therefore *trade* with the surroundings. The wonderful molecular machines that are membrane proteins can easily take care of small things (salts, simple molecules, even energy), but *swallowing up* another cell (as an amoeba regularly does) is another story, and must be handled very differently. These processes, called **endocytosis** and **exocytosis**, by which a cell respectively brings matter in or takes it out, go well beyond food intake or waste disposal. In fact, they are an essential weapon for the army in charge of defending higher organisms, the immune system, where specialized cells like the macrophages engulf (and digest) dangerous invaders, as well as a basic mechanism for signal transport by neurons. Moreover, a cell is often forced to perform endocytosis and exocytosis willy-nilly, when a virus manages to penetrate the cytoplasm, multiply, and leave again to attack the neighboring cells.

As we shall see, proper molecular motors, forming a broad filament network that control the membrane like a puppet on strings, drive and run these large shape changes. However, many endocytosis processes, whether spontaneous or assisted by special proteins, require tiny "dimples" to be made in the membrane, with a size of a few tens of nanometers and therefore a large curvature. From what we have seen earlier, a large vesicle such as a cell, whose size may range from a few hundreds of nanometers in bacteria like the *Mycoplasma* to the 12 cm diameter of an ostrich egg (not to mention those of dinosaurs), does not have serious curvature problems, for the lipids can comfortably assemble while keeping the right area per hydrophilic head. Clearly, there is a cost to pay even to make a gently curved membrane (which, as we pointed out, behaves like a rubber sheet), but the cost can be afforded quite easily by those "muscular" structures within the cell that we shall shortly meet. Making tiny, sharply curved dents is quite another story. But Nature has found a valid solution, learning how to *control* the local curvature. In the membrane there are usually lipids such as lecithin, with a rather bulky hydrophilic head and therefore a geometric parameter $g < 1$ (like "cones" with the tips turned toward the inside of the lipid double layer) and, to a lesser extent, other amphiphiles with a smaller head and $g > 1$, for instance cephalin (or **phosphatidylethanolamine**), which we may picture as being an "inverted cone", with the tip pointing toward the cytoplasm or to the outside of the cell. As we said, lipids diffuse fairly freely about the membrane, and therefore these two kinds of amphiphiles can be transported and rearranged to form structures with different curvature, allowing dimples and dents to be formed. During endocytosis processes, these indentations grow deeper and deeper, until they form proper vesicles that detach from the plasma membrane and enter the cytoplasm: thus, the cell membrane "breeds". Where do these vesicles end up? To see this, we must take a look at the general structure of a cell, to learn that in addition to the plasma membrane, inside the cytoplasm there is a zoo of other membranes, vesicles, and lipid double-layers, hosting most of the vital processes of the cell. Lipid membranes are in many regards the "cradles" of life, without which life itself would not be possible.

Let us first distinguish between two very different kinds of biological cells, dividing living organisms into two broad classes. As you know, the most precious molecule for a cell is **DNA**, which is something like the hard disk containing all the information and programs that make a living being work. In the simplest organisms, which are known as **prokaryotes** (from the Greek "before the nucleus"), DNA is free to wander around the cytoplasm. Till the last decade of the twentieth century, prokaryotes were equated to what we call bacteria; in other words, all organisms without a cell nucleus were regarded as bacteria. But this was a rather biased view, which collected together all these organisms on the basis of what *we* (and all animals, plants, and fungi, plus other single-celled organisms) have, and they do not. Today, however, we know that things are very different. There is another entire kingdom of prokaryotes, the ***Archea***, which contains many organisms that were formerly classified as "extremophile" bacteria, for they live in truly extreme habitats or need to use very unusual energy sources to work (for their *metabolism*, as we shall say). Bacteria and *Archea* are actually quite similar in terms of shape and general structure of the cell, mostly because neither of them contains internal organelles. Moreover, since they cannot control the shape of their cells by means of those "pull rods" present in higher organisms, they are both enclosed by rigid cell walls, which behave like a kind of shelter for the membrane[1]. Curiously, however, as far as genetics and the basic cellular

---

[1] The membrane lipids of the *Archea*, however, are very different from those of all other organisms, and their cell wall is made of polymers that are rather different from murein, the main component of bacterial cell walls. On the other hand, even

metabolic processes are concerned, *Archea* are much more similar to eukaryotes (the group that contains all other living species), so much so as to make us suspect they are our real ancestors.

In the eukaryote ("with a good nucleus") cell, DNA is carefully enclosed in the **nucleus**, and set apart from the rest of the cytoplasm by *two* membranes. This is an important distinction from the prokaryotes, but the real and impressive difference is in the size. Although the latter can vary a lot, typical eukaryote cells have a diameter around $10 - 20$ μm, at least tenfold larger than those of prokaryotes. This means that their volume is more than a thousand times bigger, so that comparing a bacterial cell with a cell in our body is like comparing one person with a whole village. Because of this, the organizing complexity of eukaryote cells is really huge compared with that of bacteria, and in this organization a leading role is played by membranes.

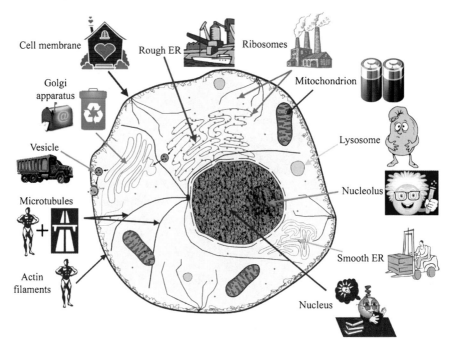

**Fig. 6.2.** (Animal) eukaryote cell, with some of its main organelles.

---

in bacteria the walls may have a range of different structures – thick walls usually reinforced by other polymers for the so-called *gram-positive* bacteria, much thinner walls enclosed in turn by a second external membrane for *gram-negative* bacteria. Because of this difference, the infections deriving from these two classes of bacteria often have to be fought with distinct types of antibiotics.

A sketch view of an eukaryote cell (specifically an animal cell, since plants show some important differences) is pictured in Fig. 6.2, where I have mainly tried to highlight the role played by the different "organelles" in the cytoplasm. Many of these are separated from the rest of the cell by membranes and lipid double-layers, or are even just made of them. We have already mentioned the nucleus, mostly filled by DNA in the form of what we shall call **chromatin**. Although it is the "headquarters" where most information transfer and replication processes occur, the nucleus takes up no more than 10% of the cell volume. It does not contain smaller organelles wrapped up in membranes, for even the **nucleolus**, whose function is mostly that of "manipulating" an important roommate of DNA we shall meet, the **RNA**, is not physically separated from the rest of the nucleus.

Let us now get out of the nucleus through one of those "pores" that you may make out in the figure, which are the gates through which RNA is exchanged with the cytoplasm, to try to identify in the rest of the cell the centers of the main vital functions. Clearly, a primary function is producing energy and turning it into a kind of "currency" that can be spent on all the processes essential for life. Unlike plants, which draw energy directly from sunlight, animals have to produce it starting from chemical compounds – "food", in other words. The most effective way to do this is to exploit complex reactions involving oxygen, which together constitute what is called *cell respiration*. What "breathes" is not, however, the cell itself, but rather those organelles called **mitochondria**, whose shape and size make them resemble little cells. Actually, a mitochondrion has many points in common with a cell. It is not only surrounded by its own lipid membrane and contains many other double layers, forming the wiggly structures sketched in the figure; it is also equipped with its own DNA, totally *different* from the DNA contained in the nucleus[2]. In fact, biologists today are pretty sure that mitochondria were originally fully independent prokaryote organisms, some of the first that learned to "breathe" oxygen. Possibly the eukaryote cells later managed to incorporate them into the plasma membrane, "taming" and turning them into proper internal power plants. In a typical cell there are a lot of these tiny power stations, to the extent that they can take up a quarter of the cell volume. This kind of "colonial imperialism" seems to be one of the successful strategies of eukaryotes, for plants may have done the same with **chloroplasts**, which are just the solar energy factories used by them to do what is called **photosynthesis**[3].

---

[2] In sexual reproduction, mitochondrial DNA is inherited *only* from the mother through the mitochondria contained in the egg cell. By studying its composition, we can therefore reconstruct the family tree on the mother's side.

[3] The "discovery" of photosynthesis by organisms like the blue algae had devastating environmental effects, just because it produces oxygen. Because of its extraordinary reactivity, this element, which today we see as the symbol of life itself, was actually a terrible poison for the first organisms, which did not respire oxygen (almost absent in the primeval atmosphere). Hence, incorporating mitochondria, and thus transforming a deadly danger into a brilliant way to exploit

Before food can be used to make energy by respiration, it must obviously be digested, by which we mean not the coarse pre-digestion taking place in higher organisms, but a proper transformation into the basic chemicals needed by the cell. These processes chiefly take place in **lysosomes**, lipid vesicles with a diameter of about $0.2 - 0.4$ $\mu$m which mostly contain enzymes, a fundamental kind of protein we shall soon meet. However, the places where lipids truly run wild in a fantasy of membranes and double layers are those complex structures called the **endoplasmic reticulum** (ER, in the figure) and **Golgi apparatus**. The role of these structures, which are much more "insubstantial" than the other organelles and visible under the microscope only if we use some clever dyeing techniques developed by Camillo Golgi, became clear only recently. You may regard the endoplasmic reticulum as a kind of stockroom, containing a lot of "tanks" in which protein complexes are amassed, and subsequently assembled into structures playing a primary role, for instance the lysosomes. A specific subgroup, the "rough reticulum", is then a kind of "port", like London's dry dock, where a myriad of **ribosomes**, those tiny and apparently negligible black dots that are actually *molecular factories*, are moored. From the reticulum, proteins are transferred to the Golgi apparatus, which is the cell's mailroom. Here proteins are marked with their final address, assembled in vesicles, and dispatched. Aside from these ordinary mailing tasks, the apparatus deals with "overseas trade" too, by assembling vesicles containing those proteins that have to be expelled from the cell by exocytosis. Thus, the Golgi apparatus is also a kind of dump where wastes are hoarded before being sent to recycling or trash separation and disposal. In fact, between the cell membrane and the Golgi apparatus, but also between the latter and the endoplasmic reticulum, there is constant heavy traffic, involving complex operations of vesicle detachment and fusion mediated by proteins such as **chagrin**, covering the vesicles with elegant "decorated" structures, or like the **SNARE proteins**, signalling specific berths on the membrane.

As we said, if the cytoplasm were just a bag of fluid, all complex transport operations performed by the cell, besides those required to change its shape, could not take place: clearly, something needs to do the hard mechanical work. This is the job of the **cytoskeleton**, made both by filaments of **actin**, an incredible protein, and by much more robust cables, the **microtubules**, which are champions of body building, but also valuable highways for cellular traffic. We shall shortly encounter these systems, which are wonderful examples of molecular motors. To do this, however, we first need to get acquainted with those macromolecules that, together with DNA and RNA (those we shall call **nucleic acids**), are the biological molecules *par excellence*: the proteins.

---

energy, was the result of a desperate attempt by the first eukaryotes to survive this "environmental revolution".

## 6.2 Proteins, a matter of molecular origami

When dealing with polymers, we have seen that it is often useful to make macromolecules using more than one kind of monomer. Copolymers greatly broaden the range of applications, particularly when they have an amphiphilic nature. In this matter, Nature really goes the extra mile, for proteins are copolymers made not of two or three, but of *twenty* different monomers, the **amino acids**. In fact, as you can see in Fig. 6.3, all amino acids have a similar chemical structure; three of the four bonds of a central carbon atom are always made up in the same way, with one bond to a hydrogen atom and two, on opposite sides, to groups called the *carboxylic* (COOH) and *amino* (NH$_2$) terminals. However, it is the fourth group, called the *residue* and indicated by R, which makes the difference. Residues can range from a single hydrogen atom such as in glycine, the simplest amino acid, to rather complex organic groups like that in tryptophane.

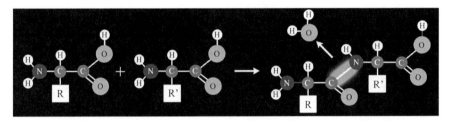

**Fig. 6.3.** General chemical structure of an amino acid, and formation of the peptide bond.

Amino acids are, in themselves, rather special chemical compounds because, while the carboxylic terminal is a weak acid, the amino group would rather behave like a base. As a result of this intrinsic double-nature, when amino acids are dissolved in a strongly basic solvent (for instance, a caustic soda solution) they behave like acids, releasing a H$^+$ group and becoming negatively charged, whereas in a strong acid solution they act like positively charged bases. But the most important point is that, much as an acid and a base tend to "neutralize" each other, two amino acids can bind, joining a carboxylic to an amino terminal by what is called a *peptide bond*. Hence, this condensation reaction yields a "dimer" which contains *two* residues which, as shown in the figure, may be different (shown here as R and R'). Note that the reaction also produces a water molecule. Thus, if there is a lot of water around, the reaction does not proceed spontaneously, but must be "aided" by providing energy (we shall come back to this when we talk about the "primeval soup").

We can then regard these two residues, R and R', as being like a word made of two letters that we can choose from the twenty available (the different amino acids, a similar number to the letters of the alphabet). Thus, with only *two* amino acids, we can make $20 \times 20 = 400$ different "words". But the dimer *still* has a carboxylic and an amino terminal. Hence, the reaction can go on, connecting up other amino acids to yield a polymer called a **polypeptide**, which constitutes the so-called "primary structure" of a protein, just as the letters of the alphabet make up the multitude of words of a language. The number of different polypeptides is almost too large to count, because in this special "tongue" the words are very long, with each protein usually made of *hundreds* of amino acids. Actually, just six or seven amino acids are enough to generate more combinations than the average lottery, even taking into account that the words are identical whether read from left to right or vice versa.

The following table shows the twenty amino acids that are directly "coded" by DNA for protein synthesis, according to a protocol we shall meet shortly[4]. Besides specific chemical differences, amino acid residues can be divided into a few large classes. First of all, some of them can ionize in solution, and therefore these proteins are *polyelectrolytes* which, as we have seen, may behave very differently from neutral polymers. Moreover, which amino acids ionize, and how many do so, depends on whether the solvent is basic or acidic, and therefore the total charge of a protein depends on solvent conditions. The table shows, for instance, which amino acids are charged in physiological conditions and what sign their charge has. But the most relevant thing, playing a crucial role in the way a polypeptide organizes to make a protein, is that some residues (obviously the charged ones, for example) are strongly *hydrophilic*, whereas others are decidedly *hydrophobic*. From experiments, it is possible to define a "hydrophobicity index", HI, shown in the table, whose value is larger the more hydrophobic an amino acid is, whereas it takes negative values for polar hydrophilic residues. (Ignore the final column for now; it concerns the transcription of amino acids in the genetic code we shall deal with at the end of the chapter.)

We have discussed at some length how the split personality arising from the simultaneous presence of chemical groups with opposite properties leads amphiphiles to form aggregates, so as to screen the hydrophobic parts from contact with water. Here, however, things are much more complicated, for there are no *separate* regions allowing us to tell a hydrophobic "tail" from a hydrophilic "head": instead, amino acids of both kinds can line up along the sequence in a very complex and (apparently) random fashion. Just look for instance at the amino acid sequence of **lysozyme**, a protein with bactericide properties (it attacks murein, the main component of the bacterial cell wall) abundantly present in tears, saliva, and egg-white, shown in Fig. 6.4. As you see, hydrophobic amino acids, shown as yellow dots, can be found all along the chain, often close to charged groups. Here, rather than split personality, we would be better to talk about molecular "introspection", for, to solve the

---

[4] Actually, there are at least two other amino acids that play a biological role, the first not directly coded by DNA, the other used only by a few *Archea*.

| Amino acid | Symbol | Q | HI | (RNA) Codon |
|---|---|---|---|---|
| Alanine | Ala/A | 0 | 1.8 | GCU, GCC, GCA, GCG |
| Arginine | Arg/R | + | -4.5 | CGU, CGC, CGA, CGG, AGA, AGG |
| Asparagine | Asn/N | 0 | -3.5 | AAU, AAC |
| Aspartic acid | Asp/D | - | -3.5 | GAU, GAC |
| Cysteine | CysC | 0 | 2.5 | UGU, UGC |
| Glycine | Gly/G | 0 | -0.4 | GGU, GGC, GGA, GGG |
| Glutamic acid | Glu/E | - | -3.5 | GAA, GAG |
| Glutamine | Gln/Q | 0 | -3.5 | CAA, CAG |
| Histidine | His/H | + | -3.2 | CAU, CAC |
| Isoleucine | Ile/I | 0 | 4.5 | AUU, AUC, AUA |
| Leucine | Leu/L | 0 | 3.8 | UUA, UUG, CUU, CUC, CUA, CUG |
| Lysine | Lys/K | + | -3.9 | AAA, AAG |
| Methionine | Met/M | 0 | 1.9 | AUG |
| Phenylalanine | Fen/F | 0 | 2.8 | UUU, UUC |
| Proline | Pro/P | 0 | -1.6 | CCU, CCC, CCA, CCG |
| Serine | Ser/S | 0 | -0.8 | UCU, UCC, UCA, UCG, AGU, AGC |
| Threonine | Thr/T | 0 | -0.7 | ACU, ACC, ACA, ACG |
| Tryptophane | Trp/W | 0 | -0.9 | UGG |
| Tyrosine | Tyr/Y | 0 | -1.3 | UAU, UAC |
| Valine | Val/V | 0 | 4.2 | GUU, GUC, GUA, GUG |

**Table 6.1.** The 20 basic amino acids, with their symbols, the charge sign (Q) in physiological conditions, the hydrophobicity index (HI), and the different nucleotide "triplets" coding for them.

problem, the protein has probably to focus all its attention on its complicated internal world.

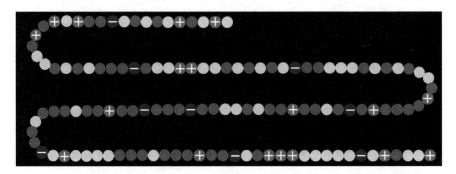

**Fig. 6.4.** The 130 amino acid chain of human lysozyme. Hydrophobic and polar amino acids are respectively shown in yellow and pink, whereas the sign of the charged ones is indicated.

What strategies do proteins use, then, to "fold", keeping the hydrophobic regions close to each other and screened from the solvent as much as possible?

Protein folding is infinitely more complex than the coiling of a simple polymer chain. However, billion years of evolution have selected a very general assembly line, which in some sense resembles the method we use to make a prefab, where we start not from single bricks but from fixed pre-assembled elements. To understand these basic elements, which make up what is called the *secondary structure* of a protein, we must take into account another crucial aspect concerning, rather than the residues, the "skeleton" of the polypeptide chain. When the peptide bond forms, the terminals, which are by themselves strongly hydrophilic, lose the inclination to ionize. Yet, looking at Fig. 6.3, we see how a single hydrogen remains attached to the nitrogen (N) atoms, whereas what is left of the carboxylic group is just an isolated oxygen (O) atom. Talking about water, we have seen that oxygen, because of its polar nature, likes to bind (if rather weakly) to the hydrogens of other molecules, giving rise to that "sharing" game that is responsible for all the special features of water. Thus, oxygens are really thirsty for hydrogen. On the other hand, nitrogen atoms are very generous, agreeing with good grace to share their little treasure with an oxygen atom through a hydrogen bond. From Fig. 6.3 it is easy to grasp that this is not possible with the closest oxygen, for it points to the other side; but it is feasible with oxygens belonging to *other* amino acids. Hence, bonds can form between different parts of the chain, making the chain much more compact than for a simple polymer.

In proteins, however, these bonds do not form at random, but in a very regular fashion. There are two basic ways that best satisfy this bonding desire, and these build up the two main substructures of the proteins. The first of

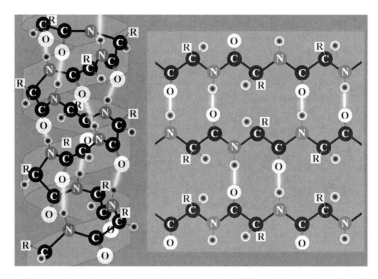

**Fig. 6.5.** Schematic view of an alpha helix (to the left) and beta sheet (to the right), showing the hydrogen bonds between oxygen and nitrogen atoms.

these structures, shown to the left of Fig. 6.5, is called the **alpha helix**. Here the chains wind up into a helical structure, so that each oxygen atom sits directly above the nitrogen atom belonging to the amino acid four steps further back in the sequence. As you can see, the helix has a very regular "pitch" and, because of this, alpha helices are quite compact and robust. Whether the helix is hydrophobic or hydrophilic overall depends on the residues. In any case, the polar nature of the oxygens is satisfied, so they make no objection to staying away from water. Yet there is an even simpler way to make a lot of hydrogen bonds, which consists in folding the chain *on a plane*, so that it looks something like the meanders of a river, and thus allowing oxygen and nitrogen atoms that are far apart to couple. These planes, firmly connected by hydrogen bonds, are dubbed **beta sheets**, and provide an inviting opportunity to hide hydrophobic residues. Indeed, if the residues are placed *to the same side* of the plane, and two of these planes are placed face to face, they form a kind of "sandwich" filled with hydrophobic stuff that is now screened from the outside. For this reason, beta sheets are one of the best ways to pack into the heart of a protein regions that are mostly hydrophobic[5]. The introduction of these two basic structures, which are present in almost all proteins (we shall meet some important exceptions), is due to William Astbury, a British pioneer in X-ray studies of proteins and DNA, and, above all, to Linus Pauling, the great American scientist who, with his theory of chemical bonding, was the first to establish a bridge between chemistry and atomic physics, besides being one of only four scholars to be awarded *two* Nobel prizes (for chemistry and for peace, for he was strenuously opposed to atomic weapons)[6].

Of course, not all amino acids in a protein are included in either alpha helices or beta sheets. Some of them organize into simple "filaments", called *loops*, that connect the helices and sheets. Loops are mostly localized close to the contact surface between the protein and the surrounding aqueous solvent, allowing the oxygens to satisfy their need for hydrogen bonds, and it

---

[5] When, for reasons we do not fully understand yet, some protein folds the wrong way, so that beta sheets are exposed to the solvent, this *really* produces a mess. Because of their hydrophobic surfaces, these deviant proteins aggregate very easily, forming "amyloid" plaques, which accumulate into tissues, inflaming them. There are many kinds of amyloidosis, diseases that so far leave us impotent, but probably the best known and most tragic is Alzheimer's disease. As you see, colloidal aggregation (because this is the issue here, even if these are much more complex phenomena than the simple ones we have met so far) is not just an academic problem.

[6] Pauling also strongly supported the use of vitamin C as a precautionary measure against tumors, himself eating at least three grams of it a day. His example was followed by another winner of the Nobel prize for medicine, the Italian Rita Levi Montalcini. Pauling lived till 93, and Montalcini has just reached her 101st birthday. You won't be surprised to hear that I myself, in my own small way, take plenty of it (in any event, it does no harm, for it is a water-soluble vitamin that does not accumulate in the tissues when in excess).

is therefore not surprising that they are mostly made of hydrophilic amino acids. Moreover, loops enable helices and sheets to combine, forming specific "structural motifs" that biochemists have learnt to recognize. Because of this, rather as we can classify architectural buildings from their characteristic motifs and building strategies, a classification of motifs has been developed in which structurally similar proteins are grouped together. This allows us to guess the evolutionary path of these fundamental biological macromolecules. Structural motifs and their relationships make up the *tertiary structure* of a protein, which is the highest organization level of a single polypeptide chain. Two specific amino acids play a distinctive role in the formation of the tertiary structure. On the one hand, we have **proline**, whose amino group differs from those of the other amino acids. Because of this, it can only accept, but not *give* hydrogens, and this strongly perturbs the formation of helices and sheets. **Cysteine**, in contrast, can noticeably strengthen the protein structure, because it contains a sulfur atom which can form sulfur "bridges" with another cysteine, similar to those we found in vulcanized rubber.

The need to screen hydrophobic groups is therefore the driving force in protein folding, but it is mainly the hydrogen bonds, mostly those located in the protein "heart", that have the final say in how this is done. All these complex stages lead to a really amazing result. While a simple polymer can take on a huge number of distinct but equivalent conformations, amounting to all the possible self-avoiding paths with a number of steps equal to the number of monomers, a protein has *a single*, or at most a few tertiary structures that are stable, in other words corresponding to a minimum total energy of the chain. This is the real secret of proteins, making each of them a "unique piece" suited for a specific role. Each protein, to work properly, needs to fold in that precise manner. At best, incorrect folding produces an inactive protein, but in the worst cases, defective pieces of this kind, such as the so-called prions, may lead to terrible pathologies, of which Creutzfeldt–Jakob disease (the human version of "mad cow" disease) is only the best known example.

How proteins can fold in such a precise way is a puzzle that has only partly been solved. Actually, the real problem is *how little* time it takes them to fold. If a protein chain had to explore all possible configurations at random, trying each of them for just a millionth of a second, it would take more than the age of the Universe to find the right one. Yet the whole process takes place in a few *thousandths* of a second, and sometimes much less! In addition, many "wrong" configurations have an energy differing very little from the minimum value, and it is strange that the protein is not "cheated" into remaining trapped in one of these misleading conformations[7]. Biologists have already found many important clues that may help in explaining this apparent paradox, showing for instance that the folding process must often be "supported". This role is

---

[7] After all, when we boil an egg and then let it cool (or when we treat it badly, as we did in the first chapter), the egg white certainly does not revert to the original state, meaning that the proteins it contains have suffered an irreversible change.

played by other proteins, the so-called **chaperonins**, whose job as *chaperones* is to escort the young chains to their debutante ball. Many proteins, moreover, after they have been synthesized, are modified within the endoplasmic reticulum or the Golgi apparatus: small sugar polymers become attached to them through a process known as *glycosylation*, which is crucial to obtain the correct folding. Nonetheless, some proteins, even when treated so badly as to be reduced to total idleness, rapidly revert to the native conformation if returned to the right conditions.

Quite probably, the secret of this astonishing skill lies in the preassembling technique, so that even if the protein has stopped functioning because it has partly "uncoiled", many substructures remain intact and protein reconstruction is then far easier. Yet many aspects of this basic process are still unclear, so it is not surprising that quite a few theoreticians, used to tackling obscure problems in fundamental physics, have become *aficionados* of the study of protein folding. Two examples of ternary structures, for proteins directly involved in DNA synthesis or breaking, are shown in Fig. 6.6. The way they are pictured, particularly dear to biochemists, just shows those structural elements we have discussed, without lingering over the complex chemical details of these macromolecules.

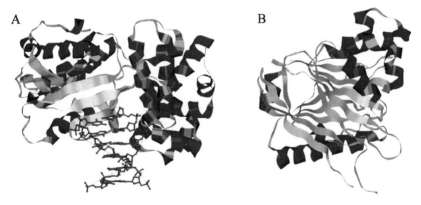

**Fig. 6.6.** Two examples of "helices and ribbons" pictures of proteins (a "ribbon" with an arrow is the usual way a beta sheet is pictured). (A) DNA-polymerase, the enzyme catalyzing the polymerization of DNA (a fragment of the latter is shown in red). (B) Deoxyribonuclease, catalyzing the cleavage of DNA into single nucleotides. (*Source: Protein Data Bank (PDB), structures 7ICG and 3DNI*)

Looking at the figure, it is natural to ask how it is possible to derive the structure of such complex molecules. In fact, for these proteins, we know the position of each single atom down to about a tenth of nanometer. The most common strategy is to study protein *crystals*, once again making use of diffraction (here of X-rays, because the size of these structures is in the nanometer range) and analyzing the position and intensity of the Bragg peaks.

Like simple molecules, proteins do crystallize, although with *much* more effort (we shall come back to this soon). The only difference (not at all a trivial difference) is that, whereas the unit cell of a crystal made by simple molecules contains just a few atoms, here the basic repeat unit is made of *thousands* of atoms. Hence, their diffraction patterns are dotted by a multitude of spots, and it is fascinating (if not incredible) how bio-crystallographers, step by step, manage to reconstruct the whole structure from these patterns[8].

The elegant reconstructions in Fig. 6.6 are, no doubt, very useful for protein biochemists, who are able to grasp from them much more information than I can tell you, or indeed understand. However, our look at colloidal systems suggests that, for the behavior of a protein in solution, what matters is not so much its complex internal structure, but rather the *contact surface* with the solvent. Thus, it is useful to take a look at a protein where, instead of helices and ribbons, we show the real space occupied by the atoms. In Fig. 6.7 you see what a simple protein such as lysozyme would look like. The first observation we can make is that, unlike a simple polymer, its three-dimensional shape is very compact. Because the internal structure is kept together by hydrogen bonds, those "holes" typical of the fractal geometry of polymer coils are totally absent, and the lysozyme molecule looks more like a globular surfactant micelle, impenetrable to the solvent. Yet there are important differences. First, we can observe that on the surface there are both negative and positive charges (many more of the latter, since in physiological conditions lysozyme behaves like a base), besides uncharged polar groups. Remember also that the charges are always $H^+$ or $OH^-$ groups, since charged amino acids are either acids or bases. The most important thing, however, is that the attempt to hide the hydrophobic regions has been only partly successful. On the surface there are large hydrophobic patches that surely do not like contact with water.

A lysozyme solution looks, then, like a suspension of almost spherical globules, with a positive net charge but also some hydrophobic patches on its surface. Thus, if the charge is screened, for example by adding salt, these particles tend to stick to each other to reduce the exposed hydrophobic area. Therefore, in these conditions, lysozyme particles behave rather like those sticky spheres we met in the last chapter. As we have seen, if the "glue" is strong enough, the particles easily form clusters that hinder them from ordering into a crystal, and these clusters rapidly precipitate from the solution. This tendency to aggregate without crystallizing is the bio-crystallographer's nightmare, for protein crystal-making becomes more a culinary art, where recipes and methods are just handed down, than a science. In fact, the difficulty of getting good crystals is the real bottleneck in the study of protein structure, for we know the primary sequence of many proteins, but far less about their tertiary

---

[8] In fact, it is possible to reconstruct the structure of proteins even when they are in solution, without crystallizing them, using *nuclear magnetic resonance* techniques (related to the MRI scans you may have had taken), but unfortunately this can only be done if the proteins are not too large.

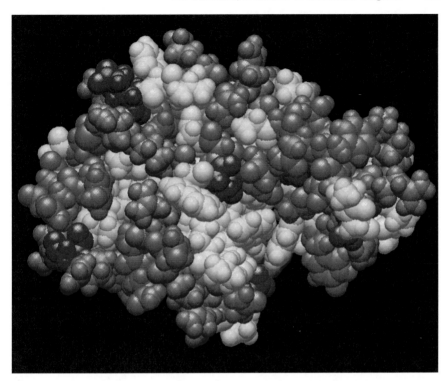

**Fig. 6.7.** Three-dimensional structure of the human lysozyme, with space-filling atomic groups. Hydrophobic, polar, positive, and negative amino acids are respectively shown in yellow, pink, red, and blue. (*Source: Protein Data Bank (PDB), structure 1IY4*)

structure. In the past few years, soft matter physicists have shown that the analogy between sticky spheres and a solution of globular proteins close to crystallization is deep and sound. Yet there are important differences, both because the behavior of hydrogen bonds with the solvent (which also stabilize the solution) is rather complex, and because the patchy nature of the protein surface makes them more akin to those spheres with localized spots (and thus "directional" bonds) that we mentioned earlier. Thus, there is still much to be understood about the colloidal aspects of protein behavior. Luckily, the opportunity to use new, powerful "synchrotron" X-ray sources has considerably lessened the practical need for large crystals (where by "large" we mean a few millimeters!).

Lysozyme is a very simple example of a protein. In fact, as we shall see, many proteins are made by *more* polypeptide chains suitably joined to make what is called the **quaternary structure** of a protein. For the moment, however, we leave this superficial glance at polypeptides to ask ourselves what all these structures are useful for. The answer is simple: *for everything*. All basic biological functions are made, mediated, or supported by proteins. Indeed, the key role of the information contained in DNA is just to make proteins, or rather to have them made by RNA, by means of "genes", which are nothing but pieces of DNA "coding" for a specific protein. Both within the complex city that is a cell and in the surrounding countryside, proteins have all kinds of jobs, playing the role of chemical and material engineers, truck drivers and janitors, surgeons and janitors, aerial fitters and sentries, architects and masons, bankers and reporters, policemen and even secret agents with license to kill. We may well ask ourselves which is more sound: the common view, where DNA is regarded as "the" molecule of life and proteins only as the servants that carry out its orders, or the opposite view, where the proteins are lords, having found in nucleic acids a useful way to transfer information. The world of proteins is therefore an extremely rich one, a world of infinite surprises, which cannot be covered in the few pages we have left. Let us try, though, to outline the curriculum vitae of some of these stars.

## 6.3 Little chemists

A couple of times earlier, we mentioned catalysts, molecules that act as mediators in those chemical reactions that, although energetically favored, proceed very slowly or not at all because they require a large activation energy (in our simple image, energy to get over a hill that comes between reagents and products). Suppose for instance that a compound AB may split into two molecules A and B, but that this hardly happens because the activation energy is far larger than the thermal energy. However, a catalyzer X can join to AB almost for free, forming the "complex" ABX, which can then split into A+B, leaving X free again. Thus, we have killed two birds with one stone, getting around the hill and getting X ready for use once again. In other words, catalysts are never used up in the reaction. For example, hydrogen peroxide ($H_2O_2$) tends to turn spontaneously into water plus oxygen, with a reaction releasing as much as $41kT$ per molecule. But the activation energy is quite high, to the point that even if you leave a bottle open, in two weeks no more than 1% of $H_2O_2$ will have split (besides, hydrogen peroxide for home use contains stabilizers), but if you add just a little of a catalyst such as manganese dioxide, a component of alkaline batteries, the reaction becomes much faster, so much that the hydrogen peroxide starts to "boil" vigorously (so please do not try this at home).

Because of this distinctive property, catalysts are among the most sought-after compounds in the chemical industry. However, for living systems this is not enough, for they also need accurate control of the chemical reactions taking place within them, fixing with absolute precision which reactions should or should not take place, and above all, *when and how fast* they can occur. Unfortunately, simple catalysts are not particularly "specific", often mediating several chemical reactions. Moreover, once the catalyst is there, the reaction proceeds with little chance of controlling its speed. Finally, most biochemical reactions are so complicated that no simple molecule can catalyze them. Where simple catalysts fail, however, **enzymes** triumph, because each enzyme catalyzes just one or at most a few of the chemical reactions taking place in the cell. Enzymes, by far the largest protein class, are vastly more effective than simple catalysts. For example, **catalase**[9], an enzyme that catalyzes the decomposition of hydrogen peroxide (a dangerous poison for cells), increases the spontaneous reaction speed thousands of billions of times. That is to say, enzymes are workers that do just one job, but do it supremely well.

For example, take a look at a simple but important enzyme, **carbonic anhydrase**, shown in Fig. 6.8. This protein is mainly present in red blood cells, and has the delicate task of controlling the blood's acidity with extreme precision (if it fails, we are done for!) by turning water and carbon dioxide ($CO_2$) into carbonic acid ($H_2CO_3$). A crucial ingredient for this reaction is a zinc atom, that black dot in the figure which is inside the enzyme's *active site*, made up in this case of three histidine residues bound to zinc. The latter, as you may notice, is just visible, buried as it is in the protein heart, and is therefore not particularly exposed, restraining its action. Nonetheless, it is accessible enough to the water molecules. A fourth histidine within the active site can take a hydrogen from the water, leaving an $OH^-$ group to the zinc atom which then catalyses the exchange of atoms according to the reaction.

How can enzymes be so effective and selective? When their action was first investigated, they were thought to operate with a kind of "key and lock" mechanism in which the shape of the active site allowed only those molecules participating in the reaction (the *substrates*) to insert correctly. But this "geometrical" view is a bit naive, for it does not explain where the lowering of the activation energy comes from. In fact, the substrate *interacts* with the active site, and it is really the bonding between site and substrate that slightly changes the state of the latter, yielding a marked reduction of the activation energy. By bonding, the substrate borrows some energy from the enzyme, which is later returned to the protein by the reaction products. The fact that enzymes do work shows how crucial the hydrogen bonds are between the amino acids. Remember that the conformation of a simple polymer constantly fluctuates because of Brownian motion, something that prevents it keeping a well-defined active site.

There is a third, even more important aspect that makes enzymes so deeply different from simple catalysts. For living systems it is not enough to catalyze reactions that would occur spontaneously, were it not for the activation energy. They also need to work "against nature", by investing energy in building *more energetic* compounds. An important feature of several enzymes is just to be able to operate the other way around, giving out of their own pocket the excess energy required by the products. For example, while in tissues, where

---

[9] All enzyme have names ending in *-ase*, even if sometimes customary names are retained, as is the case for lysozyme (which in fact is a *hydrolase*).

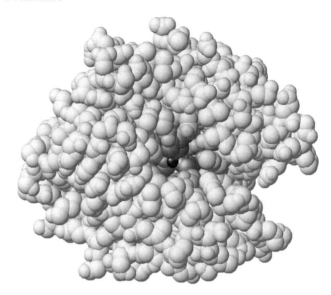

**Fig. 6.8.** Structure of carbonic anhydrase. (*Source: PDB, structure 1ca2*)

the concentration of carbon dioxide is high, carbonic anhydrase works the way we have described, within the lungs, where this gas is released as an end product of the respiration, it is carbonic acid that is to be converted into $CO_2$.

Where do enzymes take this energy from? Let us take a short break to discover the mechanisms by which the energy obtained from food or, for plants, directly drawn from sunlight, is stored, shared, transferred, and exchanged by living beings. There is an almost universal "currency" used for these trades, the rather simple molecule, **adenosine triphosphate** (ATP). Without entering into too many chemical details, ATP is made of a base plus sugar group called adenosine, which is one of the basic constituents of DNA, to which three phosphate groups similar to those in the phospholipid head are bound. Within the cell, an ATP molecule can *hydrolyze*, namely, bind to water by losing a phosphate group. Thus, it becomes **adenosine *di*phosphate** (ADP), releasing a lot of energy, amounting to about $20kT$ [10]. However, the reaction does not occur spontaneously, because of its large activation energy. This is just what enables living organisms to use ATP like an exchange currency, controlling where and when the energy is released thanks to enzymes called **ATP-ases** that catalyze the hydrolysis of ATP into ADP (see Fig. 6.9).

An even larger amount of energy can be obtained by persuading ATP to break up completely, losing another phosphate group and becoming a molecule known as cyclic AMP which, as a bonus, has a crucial function as a "messen-

---

[10] This value refers to a reaction taking place within the cell. In standard lab conditions the released energy is just $12kT$, but the internal conditions of the cell are anything but "standard".

ger" carrying the orders to be given for performing many important functions within the cell. Among them is the regulation of the level of **glycogen**, a glucose polymer[11] mostly made in the liver and muscles, constituting a ready-to-use energy supply (a longer term stock is made up of **triglycerides**, which are what we usually call "fats").

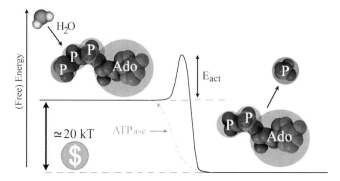

**Fig. 6.9.** Sketch of the hydrolysis of ATP in ADP (the value of the activation energy is just indicative). The label 'Ado' means "adenosine."

How is ATP produced? This is a very hard question, requiring much more chemistry than I myself have learnt. The simplest method (not truly simple at all), exploited since the dawn of life, is based on the "glucose cycle", the sugar that is produced directly by plant photosynthesis and that animals can extract, for instance, from glycogen. But the organisms respiring oxygen have learn to do it much better, either directly as bacteria do, or having it made, in our case, by tamed bacteria like the mitochondria. The main source for ATP production is in this case a complex cycle of chemical reactions based on citric acid, the "Krebs cycle", followed by a second process called phosphorylation where ADP is basically recycled and turned back into ATP by means of enzymes named ATP synthases (also belonging to the large class of ATP-ases), which are wonderful molecular motors that we shall encounter later.

Through these former steps, about 30 ATP molecules are produced from a single glucose molecule. It is very interesting to estimate how much ATP is produced by our cells as a whole. First, we may certainly say that there is a lot of ATP around at any given time, about 50 grams for an adult individual. What is really astonishing, however, is the total amount of ATP produced and consumed in a single day by any one of us. The figures we have given for the hydrolysis process in the cell are equivalent to saying that 1 kg of ATP provides about 27–28 kilocalories (the energy unit preferred by nutritionists). Taking into account the average caloric requirement of an individual (ask the

---

[11] Like **starch**, a close relative made by plants.

nutritionist for this) and the efficiency of their conversion into energy (this is rather harder), you would find a value not much lower than your *full body weight*. In a short and illuminating paper, Michael Buono and Fred Kolkhorst from San Diego University have calculated that, to run a marathon at Olympic record pace (in about two hours), it takes about 65 kg of ATP. This figure explains why living beings need cycles that produce ATP non-stop, instead of transforming food into ATP all in one go and spending it only when needed: running a marathon with a 65 kg bag on the shoulders is not such a great idea! The release of energy by ATP involves other enzymes and does not consist so much of direct hydrolysis, which would only release heat, but rather the transfer of "energetic" phosphate groups (this will be easier to see with a specific example below).

Considering that enzymes contribute to the synthesis and transformation of all molecules of biological interest, we might expect many diseases to be due to the lack, excess, or general poor regulation of some of these proteins. But in fact, there are few diseases of strictly enzymatic origin, and they are usually rare. This is because most enzymes have remained unchanged for hundreds of millions, if not billions, of years, and are the same in simple bacteria as in the most advanced (at least in my view) species on this planet, the dolphins. Unluckily, the diseases that are of enzymatic origin are sometimes very serious. Examples are Tay–Sachs disease, Hunter syndrome, Menkes disease, or Krabbe disease, where the lack of some enzyme because of genetic defects can cause mental disabilities or growth retardation. Also of this nature is phenylketonuria, one of the most common genetic diseases in babies, which affects about one infant in ten thousand, and if left untreated causes serious damage to the central nervous system, mental disabilities, and often early death. Yet even so serious an illness comes only from the lack of a rather simple enzyme that digests the amino acid phenylalanine.

To tell the whole truth, there are also less serious diseases, grouped under the somewhat misused name (since it means everything and nothing at the same time) of "food intolerance". For instance, mammals in general are "designed" to drink milk only in the very first stage of their life, and therefore **lactase**, the enzyme that breaks down lactose, the main sugar in milk, is no longer produced in grown individuals. Some human populations have, however, retained the ability to make lactase, thus allowing us to go on consuming milk and dairy products lifelong. Nevertheless, if we stop taking in these products, this power drops, causing an intolerance to lactose that manifests itself in adults with a frequency ranging from about 1% in Danes to more than 95% in Chinese people, being essentially complete for Native Americans. Once grown up, anyway, many Europeans prefer beer or wine to milk. Thus, Caucasian populations have noticeably boosted the production of the enzyme **alcohol dehydrogenase**, breaking down ethanol in wine and spirits[12]. In

---

[12] The same enzyme operating on another alcohol, methanol (or "wood spirit", because it can be obtained by burning wood), produces formaldehyde, a powerful

contrast, Asians often show an intolerance to spirits due to the lack of **aldehyde dehydrogenase**, a *different* enzyme involved in the later stages of alcohol metabolism. If you do have serious problems of food intolerance, the culprits are not so much your enzymes, but rather the DNA genes coding for them. So you should blame mom and dad, after all.

## 6.4 Truck drivers and intelligence

Enzymes are therefore very busy in the manufacturing trade. However, not all proteins play such an inventive role; many of them confine themselves to *transporting* small yet important molecules. Often these are hydrophobic compounds needing a pickup to carry them around, but it is not necessarily so. Hydrophilic molecules too may profit from transport proteins that get them to their destination more rapidly and selectively than spontaneous diffusion in body fluids would do. Thus, these "intercellular truck drivers" have just as high a social status as the chemical engineers we have just met and, like "macroscopic" truck drivers, have considerable bargaining power. When a transport protein goes on strike, this means trouble.

Oxygen is clearly among the most important stuff to be carried around, a task performed by **hemoglobin**, which makes up more than one-third of the content of red blood cells, water included. As you can see in Fig. 6.10, hemoglobin is made of no fewer than four chains, each one with a *heme group* (those four little red "spiders") containing an iron atom which can bind and carry oxygen. But hemoglobin's working principle is an ingenious one, because these four chains do not operate independently, but *cooperatively*. Let me try to make this clearer. Have you ever said, "I am so thirsty that the more I drink the thirstier I feel"? Well, hemoglobin works exactly that way. When a heme group binds to oxygen, this modifies the hemoglobin conformation very slightly – we might say it "opens up" a bit, in such a way that binding oxygen becomes easier for *another* heme. As a result, in that large "drinks bar" of the lungs, where oxygen is plentiful, hemoglobin quickly downs pints of this precious gas[13], whereas, in Prohibition times (for instance, during unassisted diving), it agrees to behave like a moderate drinker.

Hence, it is not a waste to have a quaternary structure with as many as four heme groups, for it is thanks to this that we have a process that self-reinforces when the oxygen pressure in the surroundings increases. In fact **myoglobin**, the poor relation of hemoglobin made of a single chain with a single heme group, which ensures oxygen contribution to muscles even when

---

poison leading to blindness or even to death. Because of this, the addition of methanol to wine, to evade taxes on ethanol, is not just a fraud, but a criminal act.

[13] Sometimes more than a cubic centimeter per gram of hemoglobin, which is more than *seventyfold* the spontaneous solubility of oxygen in blood.

you are out of breath, is totally insensitive to the amount of oxygen at its disposal[14].

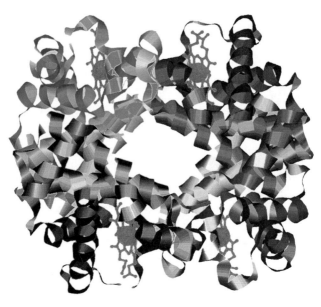

**Fig. 6.10.** Structure of human hemoglobin, with the four heme groups in red. (*Source: PDB, structure 1GZX*)

Besides the well-known example of hemoglobin, there is obviously a large number of transport proteins outside the cell. Among them, a crucial role is played by those mailmen that allow cells to communicate. In a multicellular organism cells cannot just mind their own business, but must unceasingly talk to each other[15]. In fact, any cell is constantly bombarded by messages coming from other cells. Actually, were it not so, the cell would literally commit suicide by producing proteins leading to *apoptosis*, a kind of programmed cell self-murder. Thus, willing or not, cells have to be very social fellows. But *how* do they communicate? There are two general communication strategies, corresponding in a way to telephone and ordinary mail for us. In higher organisms,

---

[14] Hemoglobin and myoglobin are nonetheless very similar, and not just because they give the red color to blood and muscles.

[15] Sometimes, it is also useful to communicate with *someone else's* cells, a role played by **pheromones**, substances stimulating in other individuals of the same species the production of those hormones we shall shortly talk about. Many vertebrates, but even more insects, are great producers of pheromones, used as signals of alarm and sexual receptiveness, or as territory and path markers. In spite of many tales, and Kevin Kline's rituals before amatory encounters in *A Fish Called Wanda*, no pheromone has so far proved to change human behavior significantly (they may help in synchronizing women's menstrual cycles, but no certain proof exists).

phone calls take place through the nervous system, used to transfer very fast messages, which are as rapidly wiped out to leave room for new stimuli. Yet there are other messages, suggestions, and commands that have to circulate for hours, if not for days. In this case, the strategy is to release in the blood, by means of the endocrine glands, chemical messengers in the form of small molecules, peptides (little groups of amino acids) or little proteins like insulin. These messengers are the **hormones**. Among them, a particularly important role is played by some **steroids**, complex organic molecules with manifold functions derived from cholesterol[16]. Steroids are, however, hydrophobic substances, thus requiring transport proteins that bind and carry them in the blood. Some vitamins, vital substances that our body is not able to produce, are hydrophobic too, and therefore they also need specific transport proteins[17].

Transport proteins are generally very specific: they carry a single substance or a few similar ones, exactly as in the macroscopic world there are trucks that transport only cars or deep-frozen food. There is, however, a protein made by the liver, **albumin**, that baffles us because it binds a lot of stuff, but rather poorly. For instance, it happily combines with most hormones, fatty acids, and drugs, but also with birilubin, the main product of bile, and with specific substances needed to fight rare but serious diseases like porphyria. In fact, it binds to almost everything, including the plastic micro-well plates used to test the action of an enzyme, so that it is used as an "insulating layer" in many biochemical methods. Albumin is so poorly characterized that, when I asked a protein-guru friend what its real purpose was, she told me that, given its hunger for binding, it might be nothing but a kind of "dustbin" to get rid of everything that is best not left lying around in the blood. Yet what is really puzzling is that this multi-purpose garbage collector is so abundant, making up more than half the protein content of the blood and about 3–5% of its total mass, a truly abnormal concentration for a protein. Our perplexity grows if we consider that albumin is not even necessary. Because of genetic defects, there are people who do not produce albumin, and still live without any serious problems[18]. So what is albumin for? We shall later see that osmosis, that

---

[16] Well known examples are cortisol (or **hydrocortisone**), the hormone that controls the stress state in the organism by increasing the pressure and sugar content of the blood, and, by reducing the immuno-response, also has a noticeable therapeutic effect as an anti-inflammatory; **estrogens**, **progesterone**, and **testosterone**, which originate sexual differences and support reproduction; or anabolic steroids like **nandrolone**, which are essential for the growth of muscles and bones, but whose side effects we had better omit here.

[17] Among them are vitamin A (better known as retinol), needed to make rhodopsin, the pigment allowing us to see; vitamin D, crucial to calcify bones, whose lack causes rickets and osteoporosis; vitamin E, an important anti-oxidant; and vitamin K, essential for blood clotting.

[18] The *metabolic*, instead of genetic, lack of albumin, on the other hand, may cause serious consequences such as edema, where liquid collects in subcutaneous tissues, making them swell. Years ago, when I used to deal with proteins, I showed a none

important effect we discussed in Chapter 2, is a very delicate matter for cells, which have to control it actively. The huge albumin concentration makes it the main contributor to the osmotic pressure of blood due to proteins, namely, to what physicians usually call "oncotic" pressure, to distinguish it from the other contribution due to salts and small molecules (and to confuse us with their cryptic technical jargon). For a physicist, so much effort to produce nothing but a trivial "osmotic load" is rather perplexing. Just to make a guess, maybe long ago transport proteins were not so specialized as today, and living creatures had to be content with simpler but more "flexible" carriers like albumin. The latter could then just be a kind of outdated model that Nature (which, like my granddad, never throws anything away unnecessarily) has recycled for other purposes. Anyway, being a biology semiliterate, I am probably talking nonsense. Do forget it, if you wish.

Besides those in charge of transport, there is another wide class of proteins that binds to other molecules, but that, instead of carrying, rather aim *to be carried* along. These microscopic hitchhikers, the **immunoglobulins** (Igs), better known as *antibodies*, are at the same time the informers of the immune system and the "Trojan Horses" it exploits to pierce the enemy defense. The immune system is a wonderful machine allowing us to figure out if something alien to our body, whether viruses, bacteria, fungi, or simply cells from another individual, even genetically very close, can harm us, and then to eliminate it effectively (almost always). So impressive are its strategies that it would take not just a book but an encyclopedia to discuss them all. Here, I only wish to mention one such strategy, based on the Igs.

Everything begins in the bone marrow, where **B cells** ("bone cells") are produced at a frenzied pace (this very instant, millions of millions of them are circulating in your body). These are lymphocytes (white blood cells), skilled in recognizing **antigens**, those molecules "marking" a specific enemy. Once it reaches maturity, a B cell can produce just *a single* Ig, and therefore is able to identify one and only one antigen. But it is mostly *the way* it matures that is very peculiar. In principle, our body might be a kind of fashion designer, making tailored suits to clothe and mark already detected enemies, but in reality it is not like that. The strategy is rather that of *prêt-à-porter*, or, considering the large-scale production, of retail trade. The bone marrow constantly produces vast quantities of immature B cells, which gradually turn from carefree youngsters to mature adults, who "marry" for life one and only one antigen (usually, but not always, a protein expressed by the alien agent). These cells are then transferred by the lymphatic system (that complex alternative circulation system, less familiar than the blood and still partly mysterious) to the spleen or to the lymph nodes, where they are accurately preselected, first eliminating (by apoptosis) those lymphocytes that recognize the proteins *we*

---

too spindly student, who was looking for a thesis topic, a rather impressive picture of a woman affected by albumin deficiency. Maybe she thought it a veiled allusion and took it badly, for I have never seen her since.

express (a rather good move), and then by allowing just a small fraction of them to ripen. The rest takes place by "natural selection". When a given antigen is actually around and binds to its bride, the B cells producing that specific immunoglobulin start multiplying at the expense of their idler companions. Many, but not all, of their progeny take care of the production and distribution of the precious antibody. The remainder of their offspring turn into "memory" B cells that, even when the infection is over, remember the face of the enemy, cutting out all the former complex preliminaries.

Immunoglobulins are generally weakly bound to the plasma membrane of B cells by other proteins (that is, the Igs are "peripheral" membrane proteins of the kind we shall shortly discuss), which is useful for training the lymphocyte during the "ripening" stage, but when the lymphocyte is activated, they are literally "shot" out into the surrounding medium. The secret of the effectiveness and of the wonderful variety of the Igs lies in their special structure. As you can see in Fig. 6.11, four chains, two of them known as "light" and the others "heavy", make a kind of "Y" which bears on its arms two binding sites. Different immunoglobulins have part of this structure in common but, on both the light and heavy chains, there is a region varying from Ig to Ig. There are thousands of versions of these variable parts and, since we can choose four of them, the number of distinct combinations is enormous (if there were exactly 1000 variants of each regions, we would in principle have a *trillion* possible combinations). But why a Y-shape and, above all, *two* binding sites and not just one? Because this structure allows an Ig to link two antigens, in fact to link many of them together, because most antigens have more than one site where an Ig can bind, building up a kind of network where the antigens get stuck like in a gel[19]. In a nutshell, immunoglobulins manage to coagulate and even precipitate certain antigens, something that makes it much easier for macrophages, the soldier cells specializing in swallowing up invaders, to get rid of them. Obviously, this strategy works only if the antigens are not too bulky. If the enemy is a bacterium, the Igs must confine themselves to binding to sites called receptors they recognize on the cell wall, where they provide a valid signal for macrophages. Bacteria can, however, be eliminated even without resorting to macrophages. In the blood there are protein complexes produced by the liver that specialize in dissolving bacterial membranes. These complexes, collectively dubbed "complements", recognize the body of the Y, instead of its arms, which dangles out of the bacterial cell wall where the Igs are bound. Those that I have sketched out very roughly in this section are just a few of the complex defense strategies of the immune system, which, as I said, is almost a perfect machine.

---

[19] It is easy to see that this requires each antigen to have at least *three* binding sites.

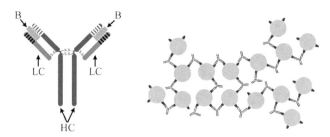

**Fig. 6.11.** Left: structure of an immunoglobulin. Heavy and light chains are respectively indicated by HC and LC, whereas B are the binding sites. Right: antigen aggregation induced by immunoglobulins.

## 6.5 Freemen of Flatland

There is a great novel by Edwin Abbott Abbott, *Flatland*, that tells us of a fictitious two-dimensional world whose inhabitants, actually geometrical figures with a number of sides proportional to their social status (females, needless to say for a Victorian writer, are reduced to simple segments), move on a plane, which is for them the one true Universe. I shall not reveal its full plot (because I hope that, if you have not read it yet, you will do so sooner or later), but I wish to use it as a starting point to tell you about those cell residents that, for all practical purposes, have to live on worlds of this kind. These are the **membrane proteins**, which make up a noticeable fraction not only of the molecules in the plasma membrane, but of those bound to the internal organelles too.

We have already talked about the basic structure of the cell membranes, highlighting the structural role of the biological lipids. However, as we said, the plasma membrane is much more than an inert bag, both because it has to control what gets in and out of the cell, and because it is the interface by which the cell communicates and interacts with the outside world. Moreover, we shall see that in the membrane there should be hooks and pulls for the complex mechanical control system of the cell. For the first of these functions, the input and output fluxes can be regulated with a simple *passive* control, opening or closing gates that allow molecules to pass through the membrane by osmosis. More often, however, this control has to be *active*. In simple words, controlling the gate opening is not enough, for some substances need to be pulled in or pushed out by force, working *against* their natural tendency to diffuse in the opposite direction. In many ways, this is still a transport process, but one where the carrier... does not move. The interaction with the surroundings is moreover multi-faceted. It does not only consist of receiving and transmitting signals of different types (chemical, electric, luminous, mechanical), but also, for instance, of allowing the cell to cling to a substrate or to recognize other cells. These roles, and many more, are played by membrane proteins. More specifically, there are membrane proteins known as *integral*, permanently lying in the membrane, and those we call *peripheral*, like the immunoglobulins, which can attach and detach from the membrane. We shall mainly deal with the former.

What special features must a protein have to obtain the right of citizenship in a membrane? The inside of the lipid double-layer, filled by the phospholipid tails, is strongly hydrophobic, so a protein such as lysozyme that, having a lot of charged and polar groups on the surface, is mostly hydrophilic and therefore highly soluble in water, would not fit in the hydrophobic double-layer. We might then guess that a membrane protein basically exposes hydrophobic residues on its surface because it has to stay within the hydrophobic inside of the double-layer. Yet a protein fully buried within the hydrophobic double-layer core can neither transport nor communicate anything. To do this, it must stick out both into the cytoplasm and outside the cell. How can membrane proteins satisfy these two conflicting requirements? To understand it, just look at Fig. 6.12, which shows **aquaporin**, a simple membrane protein. As you can see, a wide central region, made of strongly hydrophobic amino acids, can be easily spotted, whereas charged and polar residues are mostly confined in the protein's "head" and "feet". The former is just the region embedded in the membrane, while the polar tips are in contact with the cytoplasm and the surroundings. This is the solution adopted by all trans-membrane proteins, those that have to stretch both outside and inside the cytoplasm, whereas those proteins that use the membrane only as a support just need a sufficiently large hydrophobic region to anchor to the lipid double-layer.

The name "aquaporin" should give you a hint of the function of this protein. Aquaporins are the "pipes" of the cell's hydraulic system, which have the important role of regulating water flow through the membrane. The question of water exchange between the cell and the surrounding has been long debated. Since the double-layer is hydrophobic, we may expect the plasma membrane to be waterproof. Yet experiments made with *artificial* vesicles show a moderate permeability to water. The mechanism is complex and still only partly understood, but it seems to depend on water exploiting the disorder created by defects on the chains to slip surreptitiously among them. For a long time, biologists have believed that the same holds true for the plasma membrane, but this does not work. The membrane is much more complex than a simple vesicle, and above all contains enough cholesterol to hinder water transport severely. Moreover, the plasma membrane is certainly not always permeable, because fish eggs and frogspawn laid in fresh waters do not swell exaggeratedly because of osmosis. It was using frog's eggs that Peter Agre, in 1992, made a beautiful experiment. While regular frog's eggs can be dipped in distilled water with no problem, when they are forced to incorporate in their membrane aquaporins extracted from red blood cells, they burst just like those red blood cells would in the same conditions. The presence of aquaporin is therefore crucial to allow the transit of water through the mem-

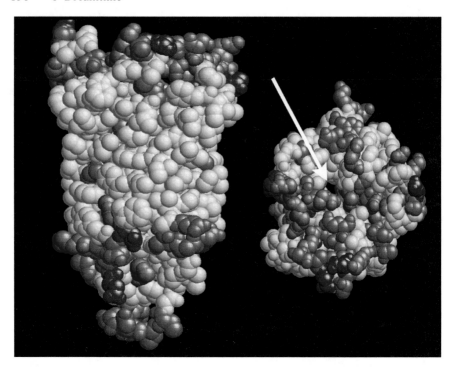

**Fig. 6.12.** Structure of human aquaporin 1, with the same amino acid color code as in Fig. 6.7. The lateral view (left) reveals the central "hydrophobic belt" that is immersed in the membrane, whereas the top view (right) highlights the central hole, indicated by the arrow. (*Source: Protein Data Bank (PDB), structure 1H6I*)

brane[20]. Looking at the protein from the top, you can make out a little hole, indicated by an arrow, which is just a channel running through the whole protein. These channels, which can be closed when needed by slightly modifying the protein structure, are so narrow that water molecules have to line up to pass through them. Besides the one I showed you, there are many kinds of aquaporins, mainly differing in the channel diameter that regulates the flow through these microscopic pipes. Even if this control is basically passive, it is nonetheless highly selective, because the chemical environment of the channel interior does not allow molecules other than water to get through.

Aquaporins are just a simple example of membrane proteins, which are generally much more complex structures, in particular if they have to detect and process signals. Among the most spectacular of the latter are the *photosynthetic systems* used by plants, algae, and many bacteria to extract energy directly from sunlight, which have a primary importance because the whole food chain depends on these organisms. Photosynthesis basically consists in

---

[20] For this discovery, Agre shared with Roderick MacKinnon the 2003 Nobel prize for chemistry.

using the energy absorbed by a pigment to produce ATP, but this operation, to work efficiently (and our man-made solar cells pale in comparison), requires a little army of proteins, organized into extremely skilled sub-units. The main pigment used in photosynthesis is chlorophyll, or rather *the chlorophylls*, since they are of two principal kinds (in bacteria there are more) with the property of absorbing light of different colors. Taken together, these pigments absorb almost all wavelengths, except those corresponding to yellow-green. This greenish shade is reflected, yielding the typical color of plants and vegetables. Besides chlorophylls, plants use other pigments such as the carotenoids (whose color should be evident from their name) that dominate in autumn, when the amount of chlorophyll in the leaves decreases.

As we have already mentioned, plants do not perform photosynthesis on their own, but make use of chloroplasts, bacteria that they tamed long ago, much as we did with mitochondria. Chloroplasts contain vast membranes in which the protein complexes responsible for each single stage of photosynthesis are embedded. These are of different kinds and complexity[21], but always contain chlorophyll-binding proteins that work like an **antenna** with the main task of capturing sunlight. These rod-shaped antennas are generally organized in a way similar to those radio towers for mobile phones scattered around roofs and open fields, or, if you prefer, the Stonehenge standing stones, with a circular arrangement surrounding another complex, the **reaction center**, where the main reactions take place. All together, the photosynthetic systems is made of *tens* of polypeptide chains.

The aim of all these reaction is to use solar energy to produce elementary electric charges, that is, electrons, which are then carried by other protein complexes, the **cytochromes**, to the internal side of the membrane. What matters more for the production of ATP is that, because charge has to balance, an excess of positive charge accumulates *outside* the cell in the form of $H^+$, used to operate the **ATP synthase**. The working principle of the latter, a protein that uses the internal membranes of chloroplasts and mitochondria basically as supports, is in itself astonishing, because it is a miniature "ion turbine", in some respects similar to the "plasma engines" for space probes that we have started to design only recently. This mechanism, sketched in Fig. 6.13, opens up a channel allowing the excess protons to pass by osmosis, and uses the osmotic flow to make a "rotor" turn. The aim of the latter is not only to provide the energy used by the part of the protein outside the membrane, the true "reactor", to turn ADP in ATP, but also to act as a "metronome" (or like clockwork, for there is even a kind of escapement), beating the time for ADP to attach, turn into ATP, and be released, by three identical assembly lines. Proton pumps of this kind are used in many other

---

[21] So far, we only know in detail about bacteria photosynthetic systems. Even for these, however, we are still unable to reproduce the working mechanism in the lab. Unfortunately, artificial photosynthesis, which would be an extremely effective way to exploit sun power, is still a dream.

situations too, for instance to turn flagella, the little tails that bacteria use to move.

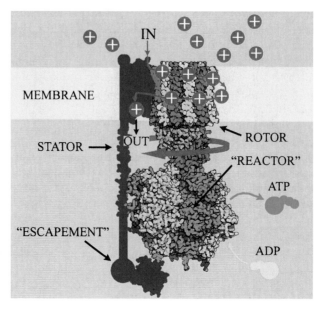

**Fig. 6.13.** Structure and operation of ATP synthase (reworked from *Wikimedia Commons*).

How do all these proteins fit in the membrane? First, their surroundings have to be quite stable, with the lipid chains suitably organized to host them. This is hardly compatible with the lipids moving freely about the double-layer. Hence, the region of the membrane close to a protein is enriched with lipids that increase its rigidity, such as sphingomyelin and, mostly, cholesterol, whose concentration is about double that in the rest of the membrane. These regions are like "rafts" floating over the liquid double-layer. Besides these flat rafts, there are also little boats with a deeper draft, called *caveolae*, which are small rigid furrows in the membrane favoring endocytosis, the absorption of substances from outside, and are quite common in adipocytes, the cells specialized in storing fats.

Till lately, the picture of the plasma membrane we had was just this, a two-dimensional ocean whereon a mosaic of little islands float, but recently we have realized that the scene is richer. In fact, the membrane looks more like a whole planet, divided into continents and oceans, large islands and internal seas. So far, we know very little about the geology of this world and about its origin[22], but it is a sensible guess that this has a lot to do with the way cells

---

[22] Some distinguished biophysicists believe that this may even have to do with the kind of spontaneous segregation generating the "Ouzo effect".

recognize each other. After all, if the other cell is a macrophage, licensed to kill intruders, it is advisable that our own cells should be easily recognized!

Because the surface of membrane proteins is mostly hydrophobic, they are almost insoluble in water, and thus extremely hard to crystallize. However, making crystals is a crucial step to understand protein structure. How, then, can membrane proteins be solubilized? The solution is once again to resort to surfactants. The relationship between surfactants and proteins is rather turbulent. Ionic surfactants such as SDS, for instance, "denature" proteins irreversibly, basically turning them into polyelectrolytes with no structure. This is not necessarily bad. Indeed, a basic technique in protein biochemistry is SDS-PAGE chromatography, allowing the molecular weight of a polypeptide to be found just by "pasting" SDS to it, and then making it drift in a polymeric gel by means of an electric field[23]. Some surfactants, in particular non-ionics like $C_m E_n$, are nonetheless rather inoffensive, and allow us to extract proteins from membranes by sticking to the hydrophobic middle region with their tails and creating a kind of "life belt" that makes them soluble. However, recipes of this kind do not always work, and because of this the number of membrane proteins whose structure is known is much smaller than for water-soluble proteins.

## 6.6 Yard workers

Unlike bacteria, eukaryote cells would be loose bags, were it not for specific structures that define their shape and allow it to be altered. Moreover, as we already said, there is a constant traffic of proteins, vesicles, and organelles within the cell which, to move around, require lines of communication just as a country needs roads and railways. Finally, in multicellular organisms, tissues are not just made of cells, a considerable part of their volume being filled by what is called the **extracellular matrix**. Thus, eukaryotes need materials that can perform all these tasks, from the trivial matter of filling the space, to much more elaborate processes involving structural changes or controlled transfer of stuff across the cell scaffold. All these functions are performed by two classes of bio-macromolecules, the *structural proteins* and the *motor proteins*.

Let us start from the extracellular matrix, which is simpler in many regards than motor proteins or those producing structural changes. Nonetheless, depending on its specific function, this matrix can take on surprisingly different structures and shapes, ranging from the transparent tissue of cornea to the very elastic one of cartilages or even, when it is suitably "calcified", the stuff that makes up bones and teeth. Usually, it is made of two main components,

---

[23] This technique is extremely effective, since the number of SDS molecules, and thus the amount of negative charge that binds to the polypeptide, is strictly proportional to the number of amino acids of the protein.

certain rather complex polyelectrolytes called *mucopolysaccharides* (or simply GAG, from the more correct term *glycosaminoglycans*) and some basic proteins universally used by animals as building materials. The polymeric part is a gel of GAG that, being charged, swells as water is drawn in by the huge osmotic pressure of the counterions. Because of this, the gel mostly provides resistance to *compression* stresses, for instance allowing the knee joints to sustain pressures of the order of *hundreds* of atmospheres. One of the most common GAGs is hyaluronic acid, currently well known to the general public because of its applications in anti-aging treatments, which is widely used as a filling material for connective tissues and is therefore a kind of styrofoam for living organisms.

But let us take a look at the proteins of the extracellular matrix, and mainly at three leading personalities among the structural proteins: **collagen**, **keratin**, and **elastin**. There are different kinds of collagen, but they have a common structure made of three chains, each with a helical structure, twisted together to form a thin rope with a diameter of only 1.5 nm, but very, very rigid (see Fig. 6.14A). These basic structures are not, however, simple alpha helices. When they are synthesized, collagen chains mostly contain two amino acids, glycine and proline. Glycine is the smallest amino acid, the only one that manages to fit into the narrow space between the chains, allowing collagen to form very tight helices. In addition, collagen is one of those proteins that are revised after having been made, by attaching some $OH^-$ groups to a fraction of the prolines (but also to other amino acids such as lysine). The extra groups make the amino acids polar and permit proline to make those hydrogen bonds that it is normally reluctant to form[24]. Collagen remains nonetheless a poorly soluble protein, and this leads to additional aggregation processes, where a large number of the former triple helices join to make a fibril, a sort of super-strong cable with a diameter as large as $0.2 - 0.3$ microns. But that is not all, since fibrils in turn combine to build the full collagen *fibers*, self-organizing so as to maximize the *tension* stress they can stand. By the way, by treating the collagen extracted from animal skin and bones with acids and hot water, to make it water-soluble, and then filtering and purifying it, we obtain a substance you are surely familiar with: gelatin, a basic ingredient of many delicious dishes.

Often, however, the extracellular matrix must not only be strong, but also very elastic. Just think of the skin (at least when you are young), the lung tissue (before you ruin it with cigarettes), the artery walls (if you have not hardened them with precipitated cholesterol), or even the bladder (on which age also leaves its mark, generating well-known pressing needs). Both characteristics are provided by (*nomen omen*) elastin, a protein sharing with collagen the feature of being made of a few kinds of amino acids, but which, unlike collagen, has no particular structure. Among all the proteins we have

---

[24] A reduced ability to do this, owing to lack of vitamin C, is at the roots of scurvy, a serious disease familiar to sailors in bygone times.

met, elastin is the most similar to a simple polymer. In fact, it truly resembles the "ideal rubber" dreamt of by Goodyear, because lysine, an amino acid it is very rich in, can form crosslinks between different chains, leading to the formation of a polymer network similar to volcanized rubber.

Collagen and elastin (together with some assistants) can form strong and elastic tissues, but living beings, in particular for protecting themselves from their surroundings, or conversely for attacking it, often need really *hard* stuff. As John Belushi used to say, "when the going gets tough, the tough get going", and keratin really is a tough character, as hard as nails. And not only nails, but also horns, hair, feathers, even tortoise shells. In a way keratin resembles collagen, for it also exploits the small size of glycine to make very tight helices. In addition, however, it is very rich in cysteine which, as we know, can bind different chains with sulfur bridges, giving keratin a notable stiffness. To make long tough fibers, the worms of *Bombyx mori* follow a different strategy, producing **fibroin**, a fibrous protein also containing a large amount of cysteine. Here, however, the protein does not form helices, but very rigid beta sheets which, bound together by **sericin**, another protein acting like a glue, make what we call silk.

Silkworm silk has truly exceptional mechanical properties, but spiders can do better. The silken threads of cobwebs, made of proteins that are close relatives of fibroin, have a breaking load comparable to that of the best steels, but are sixfold lighter, so that a thread encircling the globe would weigh just half a kilogram! Compared with a polymer fiber like Kevlar, the spider thread is threefold looser but, since it is very elastic, it is up to fivefold tougher. What is the secret of cobweb thread, making it superior even to regular silk? In part it is the protein composition, but there seems to be another rather curious reason, more related to fluid physics than to chemistry. A spider secretes its thread much faster than a silkworm and apparently this lines up the fibers, which act like a liquid crystal within the lubricating fluid where they are dispersed[25].

Why not keep spiders instead of silkworms, then? The fact is that spiders are very jealous of their own territory, to such an extent that they forget about flies and slaughter each other. It is then extremely hard to raise a spider farm for, to quote a famous movie, in the end "there can be only one" (by the way, this *Highlander* is always female). But let me suggest that my fellow countrymen should try it: this could be the future of the "Made in Italy" brand!

## 6.7 Body builders

Let us come to the real framework of the cell, the *cytoskeleton*. Actually, as a skeleton, it looks more like something borrowed from a horror movie

---

[25] Some experiments indicate that the strength of common silk increases when silk-worms are forced to secrete it faster.

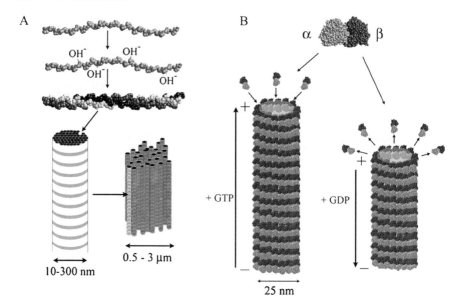

**Fig. 6.14.** (A) Formation of collagen triple helix, fibrils, and, eventually, fibers. (B) Tubulin dimer and growth/dissolution of microtubules.

than the lifeless dinosaur bones in natural history museums, for it is a very active framework. The cytoskeleton also contains structures, the so called intermediate filaments (sometimes, as in skin cells, made of keratin), but the true architects of the internal structure of the eukaryote cell are **actin** and **tubulin**, two proteins adopting a very different strategy. Instead of organizing into long helices or rigid beta sheets, they are small globular proteins, which can self-assemble spontaneously to form filaments or even very long cables. It is just this alternative strategy that allows these scaffoldings to be assembled or dismantled *very rapidly*.

We begin with actin, a very successful protein and clearly one of the treasures most cherished by eukaryotes, for it has barely changed during the whole of the evolution leading from algae to mammals. In itself, it is a small globular protein, but it has a "pocket" where an ATP molecule can bind. When this occurs, the hydrolysis of ATP "activates" actin, which begins to polymerize just like in a simple polymerization process. The only difference is that, in this case, the monomers are so big that the chain they make should really be called a "superpolymer". Two of these chains then intertwine, forming a kind of double helix with a diameter of about 7 nm and a pitch of slightly less than 40 nm, which is the actual actin filament. The filament growth, however, does not take place symmetrically from both ends, but only from the terminal where the last monomer has the ATP pocket turned towards the *inside* of the chain. Because of this, the filament is said to have a "polarity", with a "positive" end that grows and a "negative" one that does not. Clearly, this growth cannot be unrestrained, for this would seriously jeopardize the health of the cell[26]. Tuning of actin growth is actually a complex process, which also requires small proteins acting like "stoppers" on the positive end.

Actin is present throughout the whole cytoplasm, where it provides the fine framework acting as a mechanical support and a transport system for the organelles. Yet it is mainly concentrated in a thin region close to the plasma membrane which, together with other structural proteins allowing it to anchor to the membrane, forms the so-called *cell cortex*. One of the main jobs of actin is to modify the membrane shape, favoring endocytosis, exocytosis, and cell adhesion. But it is also involved in more complex tasks, such as pushing out small "feet", the pseudopodia, used by many single-celled beings to move, or forming microvilli, allowing the intestine to increase the surface area that absorbs nutrients. Moreover, the contraction of muscular fibers is a complex mechanism of tug-of-war in which another protein, myosin, pulls an actin filament (it would be nice to tell you the full details, but time is a hard master).

If actin provides the sliding panels of the cell, tubulin makes its bearing walls. In many aspects, tubulin resembles actin, for it too is a small globular protein, and polymerizes to form long filaments that are "polarized" as well, thus growing mostly from a single end. But the structure of these filaments, called **microtubules**, is very different. They are rigid hollow tubes with a diameter of about 25 nm that can be up to *tens of microns* long (see Fig. 6.14B).

They are asymmetric (polar) because tubulin is in fact a dimer; it is made of two sub-units called alpha and beta tubulin bound together, which then join to other dimers and again wind up into a helical structure that constitutes the tube wall. Because of the way this structure folds, only one kind of tubulin is exposed at each of the two ends, and the polymerization process can take place only from the "positive pole", made of beta tubulin. Like actin filaments, microtubules are very dynamic structures, since they grow by a couple of microns per second, and are dismantled up to ten times faster, so that their average lifetime is just a few minutes. In fact, a microtubule cannot keep still: either it grows, or it collapses. Where does this mad urge come from? Tubulin polymerizes similarly to actin, with the main difference that here, for once, the required energy is not provided by the hydrolysis of the same old ATP, but by *guanosine triphosphate* (GTP) a close relative where in place of adenosine there is another organic molecule, *guanosine*, which is also closely related to DNA. For microtubule synthesis, then, but also for another very important

---

[26] Phalloidins, toxins whose curious name comes from *Amanita phalloides*, the most poisonous mushroom, have precisely the effect of inducing unchecked growth of actin filaments. If you ever happen to ingest even a bit of this unhealthy substance (I hope you never will), a temporary remedy could be to swallow large quantities of raw red meat, which obviously contains plenty of actin that could "divert" the toxin from your own cells. But it would be better to rush straight to ER (and, in any case, phalloidin would not be the *only* toxin you had ingested).

job, the synthesis of proteins in ribosomes, a special currency is used, much as some "confidential" transactions are made in Swiss francs instead of the usual dollars. What happens is that, whenever GTP is bound to the positive pole, the microtubule grows, but when GDP (guanosine *di*phosphate, the analog of ADP in GTP hydrolysis) binds to that pole instead, the structure very rapidly starts to dismantle itself. In a nutshell, microtubules are always in a state of "dynamic instability", a very special situation that has attracted the attention of quite a few pure physicists[27].

Every moment, large numbers of microtubules spread out from a small organelle called the **centrosome** which usually stays at the cell center (assisted in finding it by the microtubules themselves), anchoring to the centrosome with the negative pole. Beside being the bearing walls of the cytoskeleton, microtubules also make it possible to build complex structures such as *flagella*, used by single-celled algae for moving, or *cilia*, which, for instance, allow our windpipes to sweep away filth from the respiratory tract, and egg cells to move from the ovaries to the womb. For that matter, microtubules play an essential role in cell reproduction. During this process, the centrosome duplicates, and the twin centrosomes migrate to opposite sides of the nucleus. From each of these sides, the microtubules hook one member of each pair of identical chromosomes, which have formed in the meantime by DNA duplication, pulling them away from their twin and gathering them together to form the identical nuclei of the daughter cells.

I already mentioned that microtubules, besides having a structural role, are also highways with heavy traffic. The highway tourists are proteins that, unlike myosin which just tugs the actin filaments, can literally walk on the microtubules, or indeed run marathons, tirelessly carrying with them objects that can be as large as vesicles. Oddly, there are two kinds of these indefatigable runners: dyneins, which always move to the negative end of microtubules (that is, towards the centrosome, which is close to the nucleus), and kinesins, which nearly always head for the cell's outskirts. How do these proteins move? In fact, the two chains sketched in Fig. 6.15, which are actually the little feet of a kinesin, are not simple appendixes, but proper ATP-driven motors. To these little feet are attached two long chains, twisted together, and on top of these goes the stuff to be carried. As long as a kinesin is free in solution, ADP is bound to both feet, so that, to attach to a microtubule, one of the two ends must release an ADP molecule. Now look at Fig. 6.16 to see what happens when the kinesin docks with a microtubule. Because of the way the kinesin is made, the second leg cannot hook the cable, and remains suspended. This is where ATP comes in. It binds to the attached end causing a sort of con-

---

[27] Among them Stan Leibler, one of the most successful examples of transition from basic physics to biology. I owe to a chat with Stan one of my favorite definitions of life, according to which life itself is nothing but "soft matter plus ATP" (curiously, although he made seminal contributions to understanding the dynamic instability of microtubules, Stan had forgotten about GTP).

traction and rotation of the protein. As a result, the free leg moves *forward*, getting closer to the microtubule and catching it. At the same time, ATP has done its job and can hydrolyze (the energy it releases is just what is needed to deform and rotate the protein), leaving ADP bound to the first end that docked. Thus, the ADP has to detach, and the whole cycle starts again. We must remember that all this takes place within the restless Brownian world, where nothing happens for certain. In fact, kinesin does not advance with a Prussian march, but hesitates and may even take some steps back, in tipsy fashion. But if it is slightly drunk, at least it has a GPS, for the path it follows is in fact well directed toward home. In a nutshell, kinesin exploits ATP to "rectify" Brownian motion.

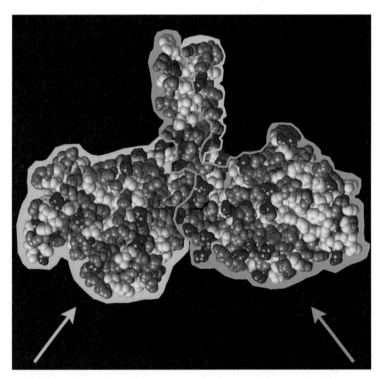

**Fig. 6.15.** "Feet-shaped" biological assembly of the terminal part of the kinesin dimer (from *Rattus norvegicus*). D and T stand for ADP and ATP respectively, whereas P is a phosphate group. (*Source: Protein Data Bank (PDB), structure 3KIN*)

It is clearly impossible to observe the details of kinesin motion directly in the traffic-congested interior of a cell. The investigations that have allowed us to derive the model we have sketched (and we have grasped only the main stages, so the real story might be rather different) are instead based on experiments in the lab, where a kinesin walks along an isolated microtubule. This has been feasible thanks to new experimental methods allowing researchers to manipulate particles or strands by trapping them under a microscope with a laser beam, and therefore to study single molecular micromachines outside their biological environment. These "laser tweezers" are the secret of many successful investigations recently performed by biophysicists.

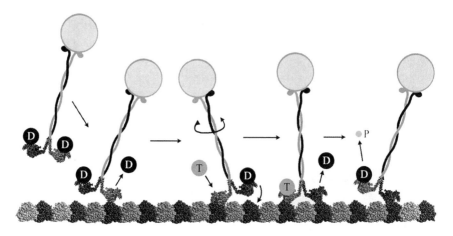

**Fig. 6.16.** Kinesin strolling along a microtubule (with a vesicle on tow).

## 6.8 A (protein-rich) lunch break

To pause after our former binge, which is not finished yet, let us have (rather paradoxically) a lunch break, in which we briefly look at other proteins, maybe not so "magic", but certainly tastier. Because they are at the same time polymers and also in some ways amphiphiles, proteins sometimes steal the job of surfactants, managing to make or stabilize structures that might appear the prerogative of the latter. To tell the truth, it is a rather unfair competition. Protein has so much inventiveness as to look like a Ferrari against the humble compact cars that are surfactants. But we will gloss over this iniquity, for without it we could not relish many familiar dainties.

We shall mostly concentrate on milk, which has escorted us all along this journey. The main milk protein, making up to 80% of its protein content, is **casein**, a very special polypeptide because it contains a huge amount of proline. As we mentioned, this amino acid disturbs the formation of alpha helices and beta sheets, so casein is mostly made of filaments; that is, it is quite similar to a common polymer, in which helices and sheet structures are absent. One of the advantages of being already "unstructured" is that it cannot denature further, which makes casein quite insensitive to temperature changes. A second important feature of casein is that it has several phosphate groups bound to it, which are linked to the protein in the Golgi apparatus *after* it has been synthesized (it is one of those examples of "post-transcription" modifications taking place in the cell's mailing office).

How does casein organize in solution? The early observations had already shown that it makes globules, with an average size of about 100 nanometers, which were given the rather unfortunate name of "micelles". I say unfortunate because they cannot be micelles in the proper sense, for the latter never reach a size larger than *a few* nanometers. These globules are actually rather poor imitations, but they do have something in common with the micelle structure. There is more than one kind of casein, and one of them in particular, $\kappa$-casein, which has a much larger content of hydrophilic amino acids, is mostly found toward the outside of the globules, whereas alpha and beta caseins, which are much richer in phosphate groups, stay inside. The internal structure of casein "micelles" has been long debated. At first, they were thought to be formed by the aggregation of many smaller micelles, for structures looking like this are actually observed in the Golgi apparatus of mammary gland cells. Today, however, these substructures are believed to be calcium phosphate grains precipitated within the globules. Casein "micelles", then, should really be described as being *polymer microgels* where small colloidal particles of this calcium salt, whose formation is induced by the presence of phosphate groups, are trapped.

Attaching groups that can bind calcium to casein is actually a very far-sighted move on the part of the mammary glands. What, indeed, is casein useful for? While it is very stable to temperature changes, it cannot withstand acids at all. Very similarly to what happens in a simple colloidal aggregation, acids cause the rapid coagulation of the globules, something that obviously takes place in your baby's stomach (and also in yours, when you drink a cappuccino) where the conditions are very acidic, thanks also to a series of enzymes making up what we usually call "rennet". Milk is then turned into a reserve of amino acids needed for the infant's growth (digesting a steak, or even blended food, is not so easy). But there is more. At the same time, the baby gets a large amount of the calcium needed to build bones, a substance that, not being very soluble in water, would be much harder to deliver in other ways. Thus, besides being a (real) "fast food", casein is also a wonderful transport protein. Curious apes that we are, we have learned to make good use of rennet taken from animals (either calves, sheep, or goats) for precipitating casein to make cheese. As a soft matter physicist, then, I must consider casein globules as poor substitutes for more elegant structures like micelles or emulsions. Yet, as a food lover, when I think of the incredible variety of cheeses they produce, ranging from Parmesan to Camembert, from mozzarella to Gruyere, I have to admit they win the taste competition every time.

Besides casein, milk contains many other proteins. Among them, we should mention **lactalbumin**, a protein which forms the starting point for the synthesis of lactose and which could moreover have a specific protective role, because in the presence of lactalbumin, tumor cells seem to die off (they undergo apoptosis, in other words). Conversely, there is another protein, **beta-lactoglobulin**, which is quite abundant in cow's milk[28], but whose biological function we just don't understand. However, we know quite well what *we* can use it for. When heated, beta-lactoglobulin is denatured to form a solid polymer gel, which is a basic ingredient of some Italian delicacies such as "tortelloni di magro", "pastiera napoletana", "cassata" and "cannoli" (if you have never tried any of them, you have really missed out!). For this gel is better known as. . . ricotta.

Remember then that milk is first of all an emulsion, containing fat (mostly triglycerides) globules, with a size ranging between some tenths of a micron and about ten microns. These emulsions are far from simple, for the layer stabilizing them, the so-called "native membrane" made in mammary gland cells, contains phospholipids and also proteins. Milk proteins stabilize many dairy products, one of the most delicious surely being ice cream (in fact a foam, for it is mostly made of air), but they are not the only stuff used for emulsifying food. We all know mayonnaise, and probably many of you know that a fundamental ingredient is egg yolk, containing many other proteins that can stabilize emulsions (besides lecithin which, being an amphiphilic molecule, is a natural emulsifier), although our understanding of the specific role of each of them is still incomplete. Egg *white*, as you know, is often whisked until stiff. This action drives proteins such as **ovalbumin**, which forms up to 60% of the albumen protein content, into intimate contact with air, and partly denatures them. The hydrophobic regions exposed in this way go to the interface of air microbubbles, stabilizing them and favoring the formation of tasty mousses and patés. Ovalbumin also denatures on heating, whereupon, like beta-lactoglobulin, it forms an elastic gel: there we have it, a hard-boiled egg. Finally, we have already seen how other food preparations make use of gelatin, namely, of a fibril protein like collagen. The list of protein applications in food could go on and on, but I had better not make your mouth water – although water, brine, and above all *ions* will still be discussed in what follows.

## 6.9 Artificial respiration

In Chapter 2 we have seen how the concepts of osmosis and osmotic pressure are crucial to understand the behavior of colloidal systems. If osmosis is in some sense the breath of colloids, for cells it is more like artificial respiration, because everything a cell can do stems in fact from a constant struggle to *control* osmosis. We have already encountered an episode of this battle when seeing how ATP synthase exploits a concentration difference of $H^+$ ions, generated for instance by photosynthesis. But this is only a pale example of the effort performed by cells to keep the internal ion concentration different from

---

[28] Not in human milk. It is the intolerance to lactoglobulin that makes it hard to bottle-feed babies with cow's milk.

the outside. The hydrophobic inside of the lipid double layer acts as a barrier to the passage of ions, whereas it is permeable to gases such as oxygen and carbon dioxide (we have already talked about water). The ion fluxes then need to be controlled either by passive channels or, in most cases, by active transport proteins.

To grasp how "anomalous" this situation is, consider some figures. Most ions are much more concentrated outside than inside the cell. For instance, the sodium ion $Na^+$ is between 10 and 30 times more concentrated, magnesium $Mg^{++}$ up to four times, and the external concentration of $Cl^-$ may be twenty times the internal one. Calcium, which is present in the cytoplasm, but only bound to proteins, is an extreme case, for the *free* calcium is less than ten thousandths of the external concentration, just a few micrograms per liter. Yet there is a specific ion, potassium $K^+$, which is a notable exception, for it is up to 30 times more concentrated *inside* the cell than outside. How can the cell create and maintain these differences, and, above all, *why* does it do it? Proteins, once again, are what enable the cell to do this, by acting like proper ion pumps, the most important of them surely being the sodium–potassium pump. Obviously, it also uses ATP to work, throwing out three $Na^+$ ions, but bringing in only *two* potassium ions for each hydrolyzed ATP molecule. These pumps have to work non-stop, otherwise there is serious trouble. So far, indeed, we have neglected the fact that proteins, which are in most cases negatively charged and constitute up to 50% of the cytoplasm, cannot pass freely through the membrane, and that the proteins keep their positive counterions close to them. These *counterions* raise the osmotic pressure of the cytoplasm. If the $Na^+/K^+$ pumps were to stop working, water would immediately enter the cell, swelling it and making it burst. Keeping the cell alive is therefore the primary role of these pumps, but in addition many membrane proteins work just by exploiting this $Na^+$ and $K^+$ imbalance, as ATP synthase does with $H^+$.

This strategy has a further, most ingenious use. Because more $Na^+$ ions are driven out than $K^+$ are carried in, a slight charge imbalance actually builds up between the inside and the outside. The charge difference is tiny, yet the cell becomes slightly negative, so that between the cytoplasm and the outside there is a slight "voltage", or better, an *electric potential difference*. It is typically only about 70 millivolts, less than 5% of the voltage provided by a common battery, but because the membrane is very thin, this corresponds to huge electric fields (*millions* of volts per meter), so high that, if the membrane were an air layer of the same thickness, lightning and thunderbolts would spark across it (luckily the membrane is a very good insulator). This "resting potential", which keeps out anions like $Cl^-$, shunned by the negatively charged cell, exists to a greater or lesser extent for all cells, but there is one kind of cell, the neuron, that particularly profits from it. As we mentioned, neurons are there to transmit nerve signals, and to do this they use a long thread-like structure called the *axon*, covered and reinforced once again by microtubules. Unluckily, we do not have enough time to discuss in detail the transmission of a nerve signal along the axon, but we can mention that it works by varying the electric potential difference across the membrane. At the starting point, the signal is generated by opening, with precise timing, specific channels that allow the $Na^+$ and $K^+$ ions to pass freely through the membrane. Thus, the membrane voltage decreases, and this generates an ion current stimulating the opening of other channels downstream. In the meantime, the channels at the starting point close, and the resting potential is restored. Thanks to this "chain" mechanism, an electric signal propagates from the neuron body, the *soma*, to the tip of the axon. Here it causes the release of vesicles containing *neurotransmitters* which are collected by other shorter filaments, the *dendrites*, of another nearby neuron, which becomes activated in its turn. Signal transmission takes place at a speed that may exceed 100 meters per second, a trifle compared with transmission in electric wires, but not bad at all for such a complicated chemical mechanism! Think about it (that is, put your neurons to work): without sodium–potassium pumps there would be no nervous system, and your top ambition might be to be a baobab (or a rose, if you prefer). ¿From all we have said, it is not difficult to imagine that the sodium–potassium pumps burn up quite a lot of the ATP we produce. But if I tell you how much, you may still be surprised: a third, no joking, *a third* of the total! All in all, we are mainly a system of ion pumps.

## 6.10 The Chieftain and his Shaman

The moment has come, in these remaining pages, to take a look at those macromolecules that form, at the same time, the historical memory and the headquarters of the cell. These are the nucleic acids, namely, DNA, the village elder of the cell, and RNA, its faithful servant, whose magic gives voice to the commands of its lord and master. It will, deliberately, be only a rapid glance, first because I do not wish to steal the job from the real specialists, the biologists (nor would I be able to), but also for "ethical" reasons, because we have already heard too much of DNA, for so long the darling of the mass media. It is time for it to step aside, maybe leaving room for the poor RNA, which is probably not so telegenic, but surely as important.

### 6.10.1 The secret of simplicity

We must say something anyway about this Chieftain, whose primary role is to contain all (or almost all) the information defining a specific living being, and to transfer it to RNA to put it to work, with no other apparent goal than the rather egocentric one of reproducing itself. I guess you all know that DNA is a "double helix" polymer. Now, we have already met a lot of helices, not only single, but also double and even triple. What is so special, then, about DNA? First, it is much simpler than a protein, for instead of twenty different bricks, it is content with only four, called *nucleotides*, which can moreover be divided into two couples with an almost identical structure. However, the secret of DNA's success lies just in its simplicity and in that "almost". Nucleotides are made of three basic elements, two of which, a **phosphate** group (negatively charged, so that DNA is a polyelectrolyte) and a sugar (**deoxyribose**), are common to all of them. It is the third component that makes the difference. This is always a "nitrogenous base" (or **nucleobase**), an organic molecule containing carbon and nitrogen rings bound together. Nucleobases in DNA are of two different kinds, pyrimidines and purines. The former have a single ring made of six atoms, whereas purines have an additional five-atom ring bound to the first, and are therefore bulkier. The rings and the chemical groups bound to them (which vary from base to base) lie on a plane, so nucleobases are definitely "flat" compounds. DNA uses two different kinds of pyrimidines, **cytosine** and **thymine**, and purines, **adenine** and **guanine**, which slightly differ in the groups attached to the rings. From now on, as is commonly done, we shall simply indicate them by their initials C, T, A, G.

Just because it has such a simple chemical composition compared with a protein, DNA, isolated in 1869 and chemically analyzed in 1919, was long ignored by biochemists, even though it was the most abundant stuff in chromosomes, those little rods that were already known to be the basis of genetic heredity. How could such a humble molecule contain the huge amount of information required to make a living being? The scene changed completely in 1943, with a seminal experiment performed by Oswald Avery and his co-workers Colin MacLeod and Maclyn McCarty. They showed that on extracting DNA from pneumonia bacteria and transferring it to a related but harmless strain, the latter suddenly became just as virulent. Ten years later, thanks to this and other important experiments, it was already clear that DNA was the carrier of genetic information, but how it did so was baffling. It was at that time that James Watson, an American biologist, and Francis Crick, a British physicist[29], after analyzing the X-ray diffraction pictures obtained by Rosalind Franklin (who, at that time, was working at King's College in London, and later moved to Bernal's lab), presented in the scientific journal *Nature* the DNA model that today we see everywhere, from T-shirts to company logos, from ornaments to sculptures. In comparison with earlier proposals, the genius of the Watson and Crick model was that the bases were *inside* the structure, bound to the sugars that, alternating with the phosphates, formed the basic framework of two chains twisted into a double helix.

It is reassuring that the bases are inside, for they are hydrophobic, hence they are screened in this way from water by the external sugar/phosphate skeleton. Moreover, this accounts for the high stability of the double helix, both because they form many hydrogen bonds in the common plane where the bases lie, which bind the two chains strongly together, and because rings

---

[29] Crick might perhaps have remained a physicist dealing with rather boring things, had it not been for a bomb that, during the Battle of Britain, destroyed the lab where he was working. Never was a bomb more fruitful.

on adjacent planes attract each other with *base-stacking* forces, further rein-
forcing the structure. Moreover, this structure explains a curious observation
made some years earlier. Chemical analysis had shown that DNA, however it
was prepared, always contained as much A as T, and as much C as G. That
is, purines and pyrimidines always seemed to come two by two. To explain
Franklin's X-ray results quantitatively, Crick and Watson had to assume that
the inside of the helix was too tight to host two purines face to face, and
too wide for two pyrimidines in the same configuration to bind by hydrogen
bonds. Necessarily, facing bases had to be of different kinds; moreover, there
had to be *only* A–T and G–C base pairs. This yields an extremely regular and
elegant model, where the two chains twist around each other with a pitch of
exactly 10.5 base pairs.

### 6.10.2 Message in a bottle

Above all, Watson and Crick's brilliant insight was about to unravel one of
the deepest secrets of life, by suggesting that the base sequence along DNA
constitutes the alphabet in which the genetic code is written. At that time, the
idea that a special molecule could contain all genetic information was already
in the air. Erwin Schrödinger, one of the founding fathers of atomic physics,
had already suggested that such complex information could be written into
an "aperiodic" structure, where a few base units repeated, not so tediously
as in a regular crystal, but rather according to a precise "code". Just one
year after Watson and Crick's exceptional result, which made this mysterious
"aperiodic crystal" a reality, George Gamow, another great theoretical physicist[30], pointed out that, to obtain the twenty amino acids using combinations
of A, C, T, and G, each of them had to be codified by at least *three* bases,
an hypothesis soon confirmed by experiments made by Sydney Brenner with
Crick himself. With two letters, each one chosen from four possible ones, we
can form just $4 \times 4 = 16$ different words, and this is not enough. However,
three nucleotides seem too much, for they yield $4 \times 4 \times 4 = 64$ possibilities,
whereas we need only 20 amino acids. Gamow and Crick tried to develop con-
sistent models to get rid of this apparent waste of information[31], but in vain.
A clever series of experiments in which the genetic code was "cracked" step
by step, pioneered by Marshall Nirenberg, showed that *all* possible triplets

---

[30] Better known for having given a sound basis to the "Big Bang" model of the
origin of the Universe, showing that from this model one can accurately obtain
the current abundance of the main atomic elements.

[31] Crick's model, on the basis of *purely logical* considerations, led him to conclude
that, to read a sequence of nucleotides from any starting point without making
mistakes, only twenty nucleotide triplets (as many as the amino acids) should
"code", whereas the others had to be just nonsense. However, to paraphrase
Shakespeare, there were more things in heaven and Earth than were dreamt of in
his model. Even if it clashes with reality, Crick's model has been defined as "the
most ingenious of all wrong models ever proposed".

(except three of them, which acted like stop signals for the transcription, a process we'll discuss in a later section, "The queen bee and her workers") are coding, so that most amino acids are coded by more than one group of three amino acids, or, as we shall say, by more than one *codon*.

You may find the codons associated with each single amino acid in the last column of Table 6.1. While reading it, however, bear in mind that codons are written there in the "language of RNA" (the molecule that, as we shall see, really provides for protein synthesis), where thymine is replaced by uracil (U), which does not change the issue, for this base also couples quite well with adenine. To translate a codon from RNA to DNA language, then, it is enough to change all U's into T's. Note that there are amino acids like methionine or tryptophane that are associated with a single codon, whereas others such as leucine can be coded by as many as six different triplets. The genetic code is, as it is customary to say, "redundant". How come there is such a waste? Why don't living systems cut back on amino acids by using just 16 of them, which can be encoded with a pair of bases per codon? After all, some amino acids have a very similar purpose. We have no idea, but one of the possible reasons has to do with nature's foresight. In case of transcription errors, for instance in reading the third base, the wrongly copied amino acid will not be too different from the correct one. So, for example, if the amino acid to be transcribed is hydrophobic, the wrong one often is too. The genetic code is then a bit of a spendthrift, but is also tolerant. Note that, because the bases are inside, the code is nonetheless well protected against unwelcome entries. As we shall see, to read it requires the double helix to be *opened*.

### 6.10.3 Double helices and strategies to pull them apart

Let us come back to the structure of DNA, which is shown in Fig. 6.17. We have seen that the double-helix conformation makes it a very rigid polymer, and DNA has a persistence length of about 50 nm, roughly corresponding to 150 base pairs (to be contrasted with the persistence length of a *single DNA strand*, which is about tenfold shorter). We also said that DNA is a negatively charged polyelectrolyte. Looking at the double helix structure, there is apparently a negative charge every 0.17 nm, but remember that Manning condensation (see Chapter 3) fixes the minimum distance between two charges (in water) at about 0.7 nm, so that the effective DNA charge is four times lower[32].

As the temperature rises, the hydrogen bonds between bases keeping the two chains together become weaker, until a specific *melting temperature* is reached, where the two chains split off. This melting temperature is obviously

---

[32] We also mentioned that charge effects increase the persistence length of a polymer, but this does not happen for the DNA double helix, which is already intrinsically much more rigid than would happen through charge effects (things are, however, different for a single strand).

higher the longer the DNA, but also depends on its composition, because a C–G pair, making three hydrogen bonds rather than two, enhances DNA stability more than an A–T pair[33]. This mechanism is at the root of the PCR (*polymerase chain reaction*) technique, where the double helix is split by heating it up to $94 - 96°C$ and then duplicated using an enzyme we shall shortly encounter. Not only the well-known DNA profiling, used in paternity testing and legal evidence, but countless other techniques in genetic engineering are due to the invention of PCR.

Another important feature of DNA structure is that the regular succession of sugar and phosphate groups allows a chain direction to be defined, for instance by assuming that a chain is directed from a starting sugar (denoted, for chemical reasons, as 3') to a final phosphate, denoted as 5'. With this convention, the two chains in the double helix have opposite directions, an important difference because, as we shall see, the enzymes that duplicate DNA or make RNA always move along a specific direction.

Watson and Crick's structure is well known to the wider audience, but you may be surprised to hear that this is not the *only* kind of double helix DNA can form. One of the alternative structures is shown to the right of Fig. 6.17. As you can see, we still have two winding strands, but the resulting double helix (dubbed A-DNA, whereas the usual one is rather oddly called B-DNA) is slightly stubbier, the base planes are not perpendicular to the main axis, and moreover there is a hole in the center (something that the hydrophobic bases surely do not like). In fact, DNA takes on this conformation only when it is extracted from cells and strongly dehydrated, while "in vivo" it is almost always B-DNA.

### 6.10.4 The great contortionist

The most extraordinary feature of DNA, however, is its huge length. Even in a simple bacterium like *Escherichia coli* (a tenant of our bowels, where it helps us to digest food), the DNA is made up of almost five million base pairs, amounting to a total length of about 1.5 mm (millions of times the length of its host). In eukaryotes this figure is very much larger[34]. For instance, we mentioned that the total length of DNA in a single one of our cells, consisting of approximatively six billion base pairs, is about two meters. This means that, if we lined up the DNA of all our cells, we could make a thread going tens of times back and forth from the Earth to the Sun, while the DNA of the entire human population would reach halfway to the Andromeda galaxy. So DNA has a serious packing problem, making those we met in Chapter 5

---

[33]  It is not surprising, then, that the DNA of thermophilic bacteria, which live at very high temperatures, is particularly rich in C–G base pairs.

[34]  Although DNA length has little to do with how much a species is "evolved" (whatever this statement means). For example, the DNA of a bean is about ten times longer than that of a human.

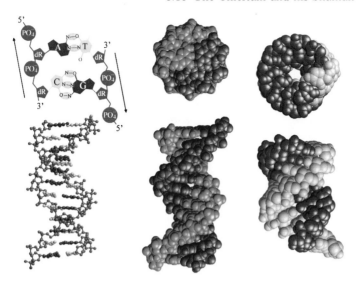

**Fig. 6.17.** Chemical (left) and space-filling (center) structure of the double helix. The A form of DNA is shown to the right. Note the "hole" in the center, absent in B-DNA.

pale into insignificance, and this problem is even more marked because it is a very rigid polymer. Just consider, for instance, the case of *Escherichia coli* where, as in all bacteria, the DNA is a single double helix closing end-to-end into a circle. Even if it were an ideal chain, its gyration radius would be $R = \sqrt{L_p L} \simeq 10\mu$m, which is much larger than the size of the bacterium itself!

How can DNA become so compact? Let us start from eukaryotes, which seem to show the most serious problem. As we said earlier, DNA is confined in the nucleus to form chromatin, of which it makes up a large fraction. However, unlike in prokaryotes, there is not a single double helix, but many of them, generally an even number except in some anomalous cases. For instance, we have 46 of them, a present half from mom and half from dad, which are actually 23 almost identical couples (unless you are a male, in which case two of them are quite unalike). About half of chromatin is not, however, DNA, but is made of a class of proteins, the **histones**, that self-organize in the form of small "spools" made of eight chains. Each nucleus contains tens of millions of histones, which are therefore much more abundant than any other protein that has to do with DNA. A specific feature of histones is that they are positively charged, so it is not too surprising that these spools have a strong affinity for DNA, which shows itself in a very distinctive manner. The double helix spirals around each spool to form a kind of "superhelix", making slightly less than two full turns, for a total of about 150 base pairs bound to the spool[35]. Thus, chromatin becomes a necklace of small beads about ten nanometers across, each bead being a spool with DNA wound around it, connected to its neighbors by short stretches of free DNA. In this way, the DNA length shrinks to about a third of the original, but this is not enough. Hence, in their turn, these necklaces twist around themselves, making a fiber with a diameter of about 30 nm, which is the chromatin we actually see under a microscope. Eventually, DNA has shortened to less than ten centimeters of chromatin. This is fine for the "normal" stages of DNA life, but not when it has to duplicate. To do this, and only in this situation, chromatin has to organize itself further to make pairs of small rods, the famous chromosomes, which have at their center a kind of "button" where microtubules hook to pull the couples apart. Hence, starting from two meters of DNA, we are left with 23 chromosome pairs amounting to a total of 120 $\mu$m: definitely an effective way to pack our genetic luggage!

For prokaryotes, which have a thousand times less DNA, the matter might seem easier, but it is not so, for bacteria do not even know what histones are[36]. Of course, they can resort to a number of stratagems, for instance turning their single circular double helix in on itself to make a "superhelix", looking a bit like a phone cable after you have used it long enough (how many times have you had to unwind it?), but this is not enough. Nevertheless DNA does self-pack quite a lot, forming an irregular blob called the *nucleoid* inside the cell, not enclosed by any membrane. Nucleoids have seriously embarrassed biologists for a long time, not only because no one knew how they formed, but also for their fleeting nature. Indeed, when they are taken out of the cytoplasm they usually collapse, as if nothing keeps them together. Since a few years ago, however, an explanation having much more to do with soft matter than with the complicated problems of cell biology has become popular, for this packing could be driven just by those "depletion forces" we met in Chapter 5. The cytoplasm is filled with little proteins (the "macromolecular crowding," remember?), which, in trying to maximize the room at their disposal, could play the same role of "depletion agents" that polymers or micelles play for simple colloidal systems. Evidence in support of this is that even a piece of alien DNA, when inserted in the bacterial cytoplasm, undergoes the same packing process. In other words, rather paradoxically, bacteria put their stuff in order by *exploiting* entropy.

---

[35] The electrostatic forces binding DNA to histones are very strong, for this figure is about equal to a full persistence length, the distance over which DNA would like to remain *straight*. Nonetheless, since this is an effect of "complexation" of opposite charges, to unwind the spools it is enough to put them into a strongly saline solution, which lowers the electrostatic attraction between chain and spool.

[36] Curiously, certain very simple *Archea*, those living in the most extreme conditions, *do* have histones. Maybe this is another clue about our ancestral roots.

### 6.10.5 The queen bee and her workers

Let us now come to the main functions of DNA, namely coding for proteins and copying itself. We will starting from the latter, which in many senses is much easier to explain. To be honest, like a queen bee, DNA does not do anything by itself. There are other characters, proteins once again, that do the real job, acting like worker bees in a hive. In Fig. 6.18 I have tried to sketch the complex series of operations required for the replication of DNA. First an enzyme, helicase, opens the road by cutting the hydrogen bonds between the two chains, thus generating a fork made, downstream of this advancing enzymatic tailor, of two separate strands which other proteins prevent from rejoining[37]. At this stage there enters the main figure, **DNA polymerase**, which starts to use the strands like "molds" to make, nucleotide by nucleotide, a new filament using *complementary* bases (for instance, when it reads ATGA... it builds TACT...), thus creating a copy of the *other* strand[38]. However, polymerase can move *only along a single direction*, to be precise from 3' to 5' (thus producing a complementary strand that *starts* from 5'), and therefore things go differently for the two strands. On one of them, called the *leading strand*, polymerase can easily follow the advancing helicase, synthesizing the copy in "real time". The polymerase on the other filament, the *lagging strand*, must move the other way round, starting from a point where helicase has already passed and moving backwards. As it moves backwards, however, helicase goes on, so a non-copied piece of DNA is left over. The poor polymerase has no choice but to come back to the new origin and finish its work. Hence, the lagging filament is copied piece by piece, and the sections are then joined by other proteins[39].

Thus, the replication process seem to be rather painful, in particular for the backward-moving polymerase. Nevertheless, in eukaryotes it is very accurate, for polymerase typically makes a single mistake every *ten million* nucleotides (and even so, other proteins later act as "proofreaders"). Duplication is, however, not rapid, at least in eukaryotes, for it takes place at a speed of about 50 nucleotides per second. Considering how many base pairs are there, it would take a lifetime to copy a full chromosome, were it not that replication begins

---

[37] While cutting, helicase would need to? unwind the DNA ahead, were it not for another protein preceding it, **topoisomerase**, which controls the twist of the double helix.

[38] In artificial DNA duplication by PCR, our polymerase does not work because it denatures at high temperature. It is therefore substituted by the corresponding enzyme of thermophilic bacteria. The fact that this enzyme also works on such a different type of DNA shows that DNA polymerase has reached such a high degree of perfection as to remain unaltered after the first stages of evolution.

[39] Moreover, the polymerase acting on the lagging strand needs a "prompt", provided in the form of a little piece of RNA to attach the first nucleotide to, which is called a primer and must also eventually be eliminated.

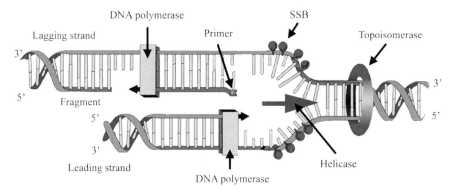

**Fig. 6.18.** Simplified scheme of DNA duplication.

simultaneously at many points, indicated by specific nucleotide sequences that act like starting flags[40].

That DNA polymerase always moves along a specified direction is far from immaterial. Already in 1971, the Russian theoretician Alexei Olovnikov proved rigorously that this has a disquieting result. In short, the DNA of a chromosome can never be fully copied, for a little piece is always lost along the way. Nature is, however, good at anticipating and, above all, taking measures, so it has provided chromosomes with "end caps" called **telomeres**, made of meaningless pieces of DNA that can be safely worn out, replication after replication, like brake pads. Yet they do wear out eventually, and this sets a limit to the number of times (typically around 50) a cell can duplicate before, so to speak, damaging the brake disks – that is, losing vital information. Telomere shortening is therefore the most intimate sign of our aging, one that no facelifts can hide. However, some molecular biologists believe that such plastic surgery could actually be done using genetic techniques: new telomeres, new life! In fact, some animals whose telomeres do not shorten, or even grow, live much longer. It is a pity that the longer a cell lives, the higher is the risk of its turning cancerous: it's your choice. Anyway, in 2009, Elizabeth Blackburn, Carol Greider and Jack Szostak were awarded the Nobel prize for medicine for their pioneering studies on telomeres, which we shall surely hear about again.

### 6.10.6 From Gladstone Gander to Donald Duck

The real protagonist of the other basic task, that of "translating" the information contained in DNA to make proteins, is not our old chieftain DNA (which, once again, just lounges about giving orders) but rather its faithful

---

[40] In prokaryotes, replication is up to 1000 times faster, at the cost of many more mistakes, but bacteria can stand these.

servant RNA. At first sight, RNA does not look too different from its super-star of a master. Apart from the replacement of thymine with uracil, their only other difference is that the sugar is not a deoxyribose but a ribose, which just has an extra OH group[41]. Notwithstanding these differences, RNA can still form double helices, at least in principle, but these are similar overall to the A form of DNA in Fig. 6.17, and this, as we said, is not a particularly good arrangement. In fact, across the whole realm of life, this never happens, except in a few viruses. While the complementary DNA strands are made more or less at the same time, and immediately twist together to form the double helix, RNA is always synthesized (by DNA, as we shall see) as a *single* strand, and in this form leaves its creator. The difference between RNA and DNA in terms of structure is huge. Not only is RNA much more flexible but it is also, so to speak, "self-adhesive". Unlike in DNA, the bases are exposed and, since the chain is flexible enough to fold back on itself, a base or even several consecutive bases can easily find their mates *far away* along the chain, binding to them and giving rise to a range of structural "motifs", often used by RNA for specific purposes[42]. Thus, RNA completely lacks the structural elegance of DNA, looking in comparison like a bleary-eyed vagrant in front of a bank manager or, if you prefer, like Donald Duck in front of his lucky cousin Gladstone Gander. Being a messy fellow makes RNA far less suited to scrupulously preserving genetic inheritance and to replicating with accuracy[43], but at the same time this richness of structure makes it vaguely resemble proteins, providing it with much more freedom of expression than its lucky and self-important relative.

We spoke about "the" RNA, but actually we should say *RNAs*, because to cast his spells the Shaman has to exist in at least three forms. We must distinguish the messenger (**m-RNA**) from the ribosomal (**r-RNA**) and the transport (**t-RNA**) RNA. Let us start with the first form, a messenger with a short and restless life, which has the job of transmitting to the world the orders of the Great Chief, just like Hermes used to do for Zeus. The synthesis of m-RNA is made by DNA in a way not too different from replication. As before, helicase, like Edward Scissorhands, opens the way by snipping the double helix. A specific sequence of nucleotides, called the **promotor**, then

---

[41] An adenine bound to ribose makes *adenosine*, like guanine forms *guanosine*. Since phosphates are there too, RNA contains all the ingredients for ATP (which once again confirms the great skill of living systems in recycling).

[42] Actually, some bonds between non-complementary base pairs may also form, for uracil is quite happy to bind to guanine too.

[43] Moreover it cannot, for we have no equivalent of DNA polymerase capable of copying RNA into RNA (the RNA polymerase we shall talk about makes it, but starting from DNA). Those viruses "working on RNA" must do it by themselves, in the simplest cases using their genome directly, as if it were mRNA, and exploiting our ribosomes to make a RNA polymerase that uses *RNA* instead of DNA as a template.

shows another enzyme, **RNA-polymerase**, where to start to transcribe[44]. However, the transcription takes place just on the strand where the promotor lies, which acts like a mold for making a *single* RNA chain (once again, however, since RNA-polymerase links complementary nucleotides, it is the *other* strand that is copied.). RNA-polymerase then proceeds more or less like its tailor colleague, always moving from 3' to 5', except that it attaches a U, instead of a T, whenever it encounters an A, till it finds a "stop" sequence. The whole nucleotide sequence read from "start " to "stop" is called a *gene*, and very often contains all the instruction needed to code for a protein, although in eukaryotes, as we shall see, it generally contains much more than this[45]. While the transcription takes place, the RNA that has already formed gets free, and the DNA upstream of the region where reading is still taking place recovers its composure, twisting back into a double helix. At this point, the fate of RNA becomes very different depending on whether the cell is a prokaryote or eukaryote.

### 6.10.7 The factory of dreams

We start from the case of prokaryotes, which is much easier. Our story leads us to meet r-RNA, which resides in ribosomes, the real factories of proteins. Prokaryote ribosomes are big complexes made of r-RNA and a lot of proteins, with a size of about 20 nm and a molecular weight amounting to *two hundred* times that of simple proteins like lysozyme[46]. Each one of them is made of two sub-units that work in close contact, giving the ribosome a distinctive shape that reminds me very much of the recording head of an old tape recorder. For the translation of the message contained in m-RNA, the ribosomal RNA asks for the help of the family's kid brother, transport RNA, or indeed of a whole family of little brothers, for there are many different, albeit similar, models of t-RNA. Each t-RNA is a short chain of less than 100 nucleotides, curled up to form a kind of hairdryer shape that connects a specific amino acid, bound to the hairdryer's nozzle, to its **anticodon** (the complementary codon), down at the end of the handle. For example, there are two distinct t-RNAs carrying cysteine, which has for codons UGU and UGC, bearing on their terminals the anticodons ACA and ACG[47].

---

[44] Eukaryotes require an additional protein complex, for our RNA-polymerase is not able to recognize the promotor by itself.

[45] Actually, there are many genes that do not code for proteins, but just for the RNA required, for instance, to regulate the activity of the genome itself.

[46] Those of eukaryotes are slightly larger and heavier. In addition, they can be divided into ribosomes that are free in the cytoplasm, which mainly make those proteins to be released within the cell, and ribosomes bound to the rough endoplasmic reticulum, which conversely deals with the synthesis and release of complex proteins often sent to their final destination by exocytosis.

[47] We might expect 61 different t-RNAs to be required (as many as the different codons, "stop" signals excluded), but actually there are fewer of them (about

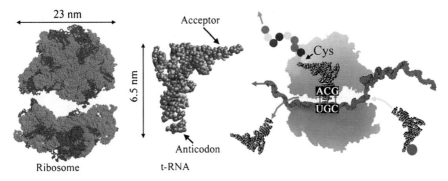

**Fig. 6.19.** The protein factory. In the ribosome structure r-RNA is shown in green, whereas in t-RNA purines are shown in pink, pyrimidines in yellow, and the sugar/phosphate backbone in light blue. The picture to the right is a schematic view of protein building, with the m-RNA "tape" in red. (*Source: PDB, structures 1FFK, 1FKA, and 1EHZ*)

How does protein synthesis take place in ribosomes? A very simplified model is shown in Fig. 6.19. Just like a tape, m-RNA gradually slides through the slit in the recording head. When the head recognizes a specific codon, it summons the t-RNA bearing the corresponding anticodon that, carried by skilled proteins, fits into a suitable socket. Now the ribosome binds the new amino acid to the peptide chain, which is growing and folding simultaneously. Finally, the t-RNA is expelled and substituted by one fit for the next codon. The whole process, which clearly requires many chemical reactions, takes place at a speed that in bacteria may reach 20 codons per second: quite amazing! Actually, what I have described is still not much more than a sketch, for many aspects of ribosomal transcription are still obscure. Indeed, even if the general structure of ribosomes has been known since the 1970s, X-ray diffraction techniques with enough resolution to allow this complex mechanism to be validated were developed only at the beginning of the new century[48]. Most likely, there will be a long way to go before we can fully scrutinize the operation of these wonderful molecular machines. One of the most interesting features, however, which helps us to understand a crucial difference between RNA and DNA, is that the ribosome active region is almost completely made of r-RNA. Thus, the r-RNA seems to be the main actor in the transcription, with the proteins acting only as backing elements, suggesting that, unlike

---

40), both because uracil flirts with guanine too, and because **inosine**, a fifth base often included in t-RNA, is not fussy about its mate, for it couples with A, C, and U.

[48] For these results, Venkatraman Ramakrishnan, Thomas Steitz, and Ada Yonath were awarded the Nobel prize for chemistry in 2009.

DNA, RNA is not just a passive subject of protein action. We shall see that there are many other clues supporting this picture.

### 6.10.8 Inner secrets of the Chieftain

Things are very different for the messenger RNA of eukaryotes because, before being translated, it has to undergo a complex "ripening" process. If prokaryote m-RNA is like a mozzarella ready to be tasted, that of eukaryotes is like Parmesan cheese, requiring time and loving care before being eaten. Two important steps are the addition of a sequence of adenines to the 3' terminal, which apparently is useful for preventing m-RNA from being digested by our enzymes during the long travel from the nucleus to the ribosome, and of an "end cap" to the 5' terminal to make the docking to the ribosome easier. But the truly crucial action, called *splicing*, takes place in advance, and consists in the removal from m-RNA of most what has been laboriously transcribed, because these nucleotide sequences are not required at all for the synthesis of the protein coded by the gene. In eukaryote genes, indeed, the coding regions, called **exons** (not to be confused with axons!), are separated by long sequences that, apparently, really are good for nothing: the **introns**. Their abundance is very variable among different species: there are fungi and lichens for which they are practically absent, whereas in the human species they form up to 26% of the entire genome. Nevertheless, they seem to be a unique feature of eukaryotes. Before m-RNA is ready to operate, it has therefore to be dismantled, throwing away introns and carefully reassembling exons.

But what are introns there for? That they do have a secret role is proved by several experiments showing that, on removing the introns from a gene, the coded protein does not fold correctly. There are moreover good reasons to believe they have at least two other tasks. The first is rather easy to grasp if we again think of proteins as prefabs. A gene made of many "building units" dispersed within simple "packaging material" can be reassembled in diverse ways, maybe choosing just some of these units, thus generating many variations on a theme, namely different but related proteins. In addition, we have the advantage of getting the whole parcel of goods in one go, then being free to take our time and try out new solutions. Finally, introns seem to play a role in that complex schedule called *regulation of gene expression*. Yes, because we have seen how the transcription of a gene works, but we did not ask ourselves when and why it occurs. Just think of a symphony orchestra without a conductor: do you think we would get out of it anything worth listening to? Genes are splendid soloists, and there are a lot of them. The trouble is that they cannot find an "external" conductor, for the only ones authorized to conduct are themselves. They have therefore to come to an agreement. The problem of how, when, and why genes express is one of the greatest open questions of modern biology. Understanding the genetic code is like learning to read notes; grasping something of the way it operates provides us with some notion of harmony and rhythm; but in fact our DNA incessantly plays a kind of music that is far more complex than a Beethoven symphony. Unfortunately, at the moment, we are far deafer than that peerless composer.

Well, the question of introns is actually the minor problem. Now that we have mapped the whole human genome, or almost all of it, we know we have just about 23,000 genes, and that, altogether, the part made of *non*-coding sequences is as much as 98.5% of the total. Introns apart, more than two-thirds of our genome seems to be just "junk DNA". To put it differently, most of the Great Chieftain is largely useless, or at least mysterious: quite a serious waste problem! Besides, junk DNA accounts for inexplicable and rather embarrassing differences in the size of eukaryote DNA. There are single-celled amoebae with more than 200 times the amount of our DNA, whereas the globefish, which has more or less as many genes as we have, has a tenth of the DNA.

What is the purpose of all this junk, sometimes called the "dark matter" of the genome? Here too, some clues question its alleged uselessness. Several of these sequences have been jealously preserved in the course of evolution, and are identical in very distant species. Lots of hypotheses on the origin and role of junk DNA have been made, but none is fully convincing. Part of it is most likely a kind of "parasite", derived from very remote viral infections. Viruses, as you probably know, are not living organisms in every respect, for they are unable to self-replicate, needing a host cell for that purpose[49]. They actually originated *after* the cells, maybe evolving from those DNA pieces that bacteria use to exchange for mixing their genes (their only and not so exciting way to have sex), and are just made of a protein shell containing DNA, or sometimes RNA (yes, viruses can work also on RNA). A fair amount of junk (at least 10%, but probably much more) is made of RNA copied ("transposed") in the form of DNA in our genome. These "transposons", besides, do nothing but replicate, managing to fit their own copies into other regions of the genome, thus increasing the amount of waste (possibly toxic waste, for they are suspected to cause serious autoimmune diseases like multiple sclerosis). Transposons integrate faster than they are got rid of by organisms, and therefore, instead of reducing, tend to increase in the course of evolution.

Anyway, it is generally believed that part of the junk DNA acts like a "spacer" between the genes, allowing the enzymes that intervene in replication and RNA transposition to work more freely, without hindering each other. Another "serious" role could be to take part in gene regulation in some still obscure way, for instance during embryo development, which remains mysterious to a large extent. Finally, other non-coding regions may serve as decoy

---

[49] Although, to say this with full confidence, we need to know what *really* defines a "living being". Is it just the faculty of self-reproducing? When in the (not too distant) future there are robots capable of making other identical robots, will you say they are "alive"? Conversely, when, talking on the phone sometime later, you cannot tell a computer (*unable* to beget any progeny) from a human being, will you regard it (him? her?) just as a "machine"?

targets against the attacks of enemies like chemical or radioactive agents, a bit like the decoys used by submarines. Many sequences could, however, just be leftovers of genes that are no longer useful (for instance, in our case, those in charge of making a nice tail grow).

How is it that DNA does not get rid of all this bulky garbage? I like to think that DNA is a bit like my grandpa, who never threw away anything, for everything sooner or later might come in handy. The "dark matter" could then be a sort of supply of sequences that are *currently* useless. Someday, however, like a rabbit pulled out of a top hat, we might get out of them novel genes giving us a competitive advantage. Were it so, instead of pointless garbage this would be the true raw material of evolution. From the biologist's point of view, in any case, junk DNA displays another interesting and useful feature. Indeed, while the genome is carefully conserved in evolution (for changes may cost dearly), mutations of non-coding sequences are widely tolerated. Therefore, the dark matter of relatively close species such as mouse and man, whose genome is 80% similar, is quite different, and this allows reconstruction of the gradual splitting of their evolutionary paths.

### 6.10.9 Time dust or stardust?

Trying to understand all the mysteries of DNA seems nonetheless a desperate venture, like an ancient Greek philosopher trying to make out how the internal combustion engine works by examining the sophisticated motor of a Ferrari – with the non-trivial difference that, while about a century has passed from the first motor vehicles to the car made in Maranello, it took at least *three billion* years of evolution for DNA to reach its current structure. Things could be slightly easier if we look instead at RNA, for we have already seen that this molecule is still much more active and adaptable than DNA. Instead of hiding its gene outfit jealously away, moreover, RNA makes a fine show of it, suggesting that it is willing to use its power, although in a slightly fuzzier manner. Today all living beings, except RNA viruses[50], use DNA to work and replicate. Yet, was there a remote past when life was the realm of RNA? In the past three decades, the idea of a "RNA world" has arisen in the imagination of biologists, mainly because of the discovery of **ribozymes**, peculiar RNA molecules that can catalyze chemical reactions exactly like enzymes do. This could shed light on the old chicken-and-egg question, now turned into the problem of nucleic acids versus proteins. Which came first, if only nucleic acids code for proteins, but absolutely require the latter to replicate? Later it was also observed that ribozymes, although rather rare, play important roles in the cell, for instance to synthesize the proteins of ribosomes.

But can a ribozyme make itself, or at least a piece of RNA? All attempts made in a lab, even when successful, are so far just Pyrrhic victories. Everything works fine provided that we use a rather large ribozyme made of, say,

---

[50] And not even all of them. To replicate, the AIDS "retrovirus" has first to transcribe it back to DNA.

150–200 nucleotides, which then manages to catalyze the synthesis of RNA pieces no longer than about 15 nucleotides (clearly unable to do the same in their turn, so that any "evolution" stops here). We might improve in the future, but RNA still remains a much more fragile molecule than DNA, able to carry a very limited genome and to copy it with much less fidelity. The RNA world may soon be seen as another creation myth like the "primordial soup" (a soup in which peptide bonds would never have formed, even if it were a *consommé*). In my view, this is like showing our Greek thinker a modern compact car, no doubt simpler, but just as highly evolved as a sport car. The problem is that there is no longer a Ford T around, not to mention a Benz *Motorwagen*. All we are left with are two perfect molecules, the winners and final survivors of a deadly fight between simpler self-replicating molecules that took place sometime and somewhere (the struggle for life in the Serengeti is nothing in comparison!). In fact, even in the first organisms that have left

**Fig. 6.20.** The Double-Helix nebula, lying just 300 light-years from the huge "black hole" that occupies the center of the Milky Way, discovered in 2006 by the Spitzer Space Telescope (infrared image, reproduced in false colors). (*Source: NASA/JPL-Caltech*)

fossils, nucleic acids were probably like the current forms. This tells us that they developed fast, very fast, for there had been little time since the planet's formation. Unless they came from somewhere else... Is the picture in 6.20, showing an interstellar nebula with a unmistakable shape, a message in a bottle from the center of the Galaxy, perhaps?

## 6.11 Back to the future

We have reached the end of our long (I hope not too lengthy) journey, which has led us from the simple mineral particles present in aquifers and in the air to the wonderful machines of life.

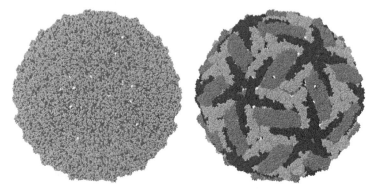

**Fig. 6.21.** Dengue fever virus. In the left panel, hydrophobic amino acids are shown in dark grey, whereas the image to the right shows in gray shades the single protein chains forming its surface. (*Source: PDB, structure 1K4R*)

Before leaving my keyboard and switching the monitor off, however, let me suggest to you a short backward trip. Take a look at the object to the left in Fig. 6.21. At first glance, it looks like a simple colloidal particle, almost a perfect sphere with no particular feature or detail. Not so, for what you see is the full reconstruction of the virus causing dengue fever, the "breakbone" haemorrhagic disease transmitted by *Aedes aegypti* mosquitos, which infects tens of thousands people in tropical countries every year. Actually, more than a sphere, it has the shape of an icosahedron, the most complex regular solid with 20 triangular faces. The darkest spots on the surface correspond to those hydrophobic regions allowing the virus to adhere to the plasma membrane and penetrate the cell. When inside, the virus exploits lysosomes, cell organelles that secrete enzymes such as lysozyme, to get rid of its complex protein shell (more evident in the picture to the right) and release its infecting RNA into the host cell. Something more than a simple sphere, isn't it?

Many viruses have a similar shape, but some of them are decidedly more complex. To a colloid scientist, for instance, the Tobacco Mosaic Virus (TMV), a mischievous RNA-virus that wages its own distinctive campaign against smoking by infecting tobacco leaves, is a wonderful example of a thin rigid rod, with a diameter of 18 nm and a length of 300 nm. In fact, it has been used as a model system in many investigations, since it forms beautiful nematic and smectic liquid crystals. We can even modify TMV, as if it were an inorganic colloidal particle, by depositing a stabilizing thin silica layer on its surface. Yet look now at Fig. 6.22. What you see is one of the many pierced disks, made of hundreds of proteins, which *spontaneously* stack on top of each other to form the TMV rod, inside which is held a spiral-wound RNA chain. Examine in detail this super-symmetric structure, the way positive charges surround and screen the negative RNA, the elaborated and functional layout of polar and hydrophobic groups: it's a masterpiece of abstract (no, actually realistic!) art.

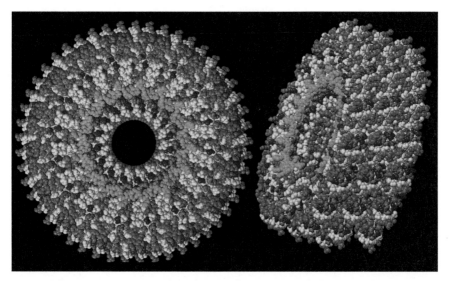

**Fig. 6.22.** Top (left) and side (right) view of the extremely rich and elegant helical structure of Tobacco Mosaic Virus. (*Source: Protein Data Bank (PDB), structure 2OM3*)

Yet, however complex this structure looks, it follows those general laws we learnt from simple colloids. For do you know how the virus gets rid of this bulky shell (known as a capsid), once inside the cell? Unlike its dengue sibling, the TMV virus does not exploit the host enzymes. As you can see, the capsid contains a lot of amino acids bearing the same charge, which can remain close to each other only because divalent calcium ions keep them together, like in the case of concrete. However, we know that calcium is almost absent in the cytoplasm. Therefore the capsid proteins start to repel each other, just enough to set a little piece of RNA free. This slips out and duplicates by exploiting the host ribosomes, and takes over. This really is an elaborate use of electrostatic forces, whose love/hate ambivalence spans the whole Middle-earth of soft matter, from cements to TMV, a land we have only just begun to explore.

In 1991, when awarded the Nobel prize "for discovering that methods developed for studying order phenomena in simple systems can be generalized to more complex forms of matter, in particular to liquid crystals and polymers", Pierre-Gilles de Gennes closed his lecture by recalling some anonymous verses written on an engraving by Jean Daullé, which reproduces a lost painting by François Boucher, *La Souffleuse de Savon*. I can do no better to close this book. And, to remember the great scientist but above all the deeply humane person de Gennes was, I shall use his original translation from the French:

*Have fun on sea and land*
*Unhappy it is to become famous*
*Riches, honors, false glitters of this world*
*All is but soap bubbles.*

**Fig. 6.23.** "Such stuff as dreams are made on". (*Pastel by Pierluigi Durat, 2009*)

# Weird words: soft matter from A to Z

*This is a collection of many of the weird words you will find in this book, including some common words used in a non-common, technical sense. It is neither exhaustive, nor very accurate, but I hope it contains most of what you need to follow the text. You may wish to browse it occasionally or read it in the end to revise some concepts. Here, too, you will find an added bonus, consisting of additional information that did not fit in the main text, which is highlighted with a ♡. Just bear in mind that, very often, to understand a word you must already have grasped some other concepts (this is the unsolvable dilemma all dictionary-makers have to face). To help you, I have indicated cross-referenced words in* CAPS. *Hence, unfortunately, rather than simply reading, you may feel like you are...playing hopscotch.*

**A**cid: An acid, in its simplest form, is a chemical species that donates a $H^+$ ION to another compound which, accepting it, behaves as a BASE. As you can see, acids and bases always come by twos. So, where is the base when we dissolve an acid, and nothing else, in $H_2O$? It is just water itself. Accepting the $H^+$ ions, a fraction of its molecules become $H_3O^+$ ions. Everyday examples of acidic stuff (and of the corresponding acid) are lemon (citric acid), vinegar (acetic acid), carbonate drinks (where the dissolved carbon dioxide becomes carbonic acid), the liquid for car batteries (sulfuric acid), and vitamin C (ascorbic acid). We have seen that DNA, RNA, and most proteins in physiological conditions are acids too, although much more complex.

**Adsorption**: In COLLOID science, adsorption is the sticking or chemical bonding of small chemical species to the surface of the dispersed particles. Because of their large surface area, colloidal particles in underground waters may be a very efficient way of transporting hydrophobic substances, such as radioactive wastes.

**Aerogel**: A colloidal GEL where the liquid solvent has been substituted by air. Aerogels are usually very tenuous structure (their density can be just a few percent larger than the air they contain), so they are the lightest solid materials ever made. Nonetheless, they can be very strong: a piece of aerosol weighing only 2 grams can support a 2.5 kg brick (for a nice picture, see *Wikipedia*). They are so light and sturdy that they have even been used to fill the inside of tennis racquets. Since they are mostly made of air, aerogels conduct heat so poorly (tens of times worse than fiberglass) that they can be used for thermal insulation of space suits. But their most distinctive feature, inherited from their colloidal origin, is that they have a *huge* internal surface. NASA's Stardust spacecraft, for instance, used large SILICA aerogel collectors to trap dust particles left behind by the comets. For the same reason, aerogels are used in detectors for sub-atomic particles. Finally, because of their enormous light-absorbing surface, aerogels made from carbon particles are probably the blackest material that can be made. ♡ Making an aerogel, however, is not so easy. One might think of just removing the solvent from a colloidal gel by evaporation, but this does not work. When most of the solvent has already evaporated, the colloidal particles are still wet and connected by thin liquid films. Because of SURFACE TENSION, these films pulls the particles toward each other, exerting very strong STRESSES on the gel network, called *capillary forces* (see SURFACE TENSION), that generate cracks in the drying solid. As a result, the gel shatters (unless drying takes place very slowly, as happens for OPALS). However, there is a clever trick to avoid these disruptive effects, which is based on a hidden but profound similarity between liquids and gases. We are used to considering liquids and gases as very different states of matter. For instance, liquid water is dense and remains in the bottle where we put it, until it evaporates, whereas vapor is tenuous, and escapes from any open container. Indeed, when we heat water, it turns from liquid to gas very dramatically, by boiling (undergoing, at about 100°C (212°F), what physicists call a *phase transition*). However, this is not always true. If we first put water under high pressure (over 218 atmospheres), then heat it to at least 374°C (705°F), and finally bring it back to atmospheric pressure keeping the the temperature over 100°C, we see the liquid transforming into vapor smoothly, without any boiling. This is because, beyond 218 atmospheres and 374°C (which is called the *critical point* of water) liquid and vapor can no longer be distinguished; they become the same thing. This does not happen just for water, but for any liquid: over a specific critical point, which depends on the liquid, there are no longer "liquids" and "gases", but only a single indistinct fluid state. The pressure and temperature of water's critical point are rather inconvenient, but there are liquids such as alcohol or ether whose critical point is more easily accessible. "Supercritical drying", which has become the standard method to obtain aerogels, involves first exchanging the original solvent for a liquid having a low critical pressure and temperature, and then in extracting the latter by going over and around the critical point exactly as we outlined for water. Since during the whole process there is no true evaporation of the solvent, but just a progressive "metamorphosis" of a liquid into vapor, no capillary stresses are generated, and the aerogel does not crack – if, of course, everything is done with care, and provided we are a bit lucky.

**Aerosol**: A COLLOID where the dispersing medium is a gas, while the suspended particles may be solid or liquid (e.g. water droplets). Because gases have a very low density, particle SEDIMENTATION in aerosols is generally much faster than in liquid suspensions (*sols*). Moreover, particle motion in gases is due more to convection by gas streams than to BROWNIAN agitation. In atmospheric aerosols, either natural (such as volcanic emissions) or man-made, particles have a wide size range, and may undergo complex aggregation and chemical reaction processes. The long-term effects of aerosol intake on health may be serious and are still poorly known, in particular for very small particles that can enter the blood circulation directly. Aerosols are also responsible for beautiful or unexpected optical phenomena such as gorgeous sunsets, rainbows, and blue moons.

**Aggregation number**: In SURFACTANT solutions, the number of molecules that join to form a MICELLE. The micellar aggregation number can range from a few tens, for simple spherical micelles, up to very large values for more complex aggregates such as LIVING POLYMERS.

**Alpha helix**: See PROTEIN (STRUCTURE).

**Amino acid**: Amino acids are the building blocks of proteins. Their basic structure is similar, and contains simultaneously a *carboxyl group* COOH, which behaves as a weak ACID, and an *amino* group $NH_3$, which is a BASE. Each single amino acid is, however, characterized by a specific *residue*, which can range from a single hydrogen atom to a complex AROMATIC group. Depending on the residue, amino acids can be charged (see CHARGES) or POLAR (and then, in both of these cases, HYDROPHILIC), or HYDROPHOBIC. Amino acids bind through PEPTIDE BONDS to form a POLYPEPTIDE, which then folds into a PROTEIN. Among them, useful ones to remember are cysteine, which CROSS-LINKS a protein by means of sulfur bridges, and proline, which strongly perturbs the formation of ALPHA HELICES and BETA SHEETS because it forms hardly any hydrogen bonds.

**Amorphous**: Any material lacking the regular structure of a CRYSTAL.

**Amphiphile**: A molecule with two well separated regions, a HYDROPHILIC head which likes water, and one or more HYDROPHOBIC tails which do not. The simplest amphiphilic molecules are SURFACTANTS, which have a single tail, but double-tailed molecules such as PHOSPHOLIPIDS are also extremely important, in particular in biological MEMBRANES.

**Antibodies**: The common name for *immunoglobulins*, those "Y-shaped" PROTEIN complexes used by the IMMUNE SYSTEM to identify and neutralize foreign objects, such as BACTERIA and VIRUSES.

**Antigen**: A molecule recognized by the IMMUNE SYSTEM. This word is a contraction of "antibody generator", because antigens trigger the production of the ANTIBODIES to which they specifically bind.

**Apoptosis**: The planned suicide of a cell in a multicellular organism. Unless they receive reassuring chemical signals telling them, "You are OK, keep quiet," most cells kill themselves sooner or later, because they are no longer useful or simply to leave room for newborn siblings. Unfortunately, tumor cells rebel against this kamikaze fate.

**Archea**: A large class of PROKARYOTE organisms that has long been mistreated and considered just a sub-class of bacteria. In fact, their DNA is more similar to our own than to bacteria, so that they might well be our real ancestors. Some Archea are the real daredevils of the microworld, being able to survive and even relish extreme temperature, salinity, or acidic conditions (hence the name *extremophiles*). Among them are the only organisms that do not exploit, directly or indirectly, the sun's power, spending their whole life well buried in the darkness of deep rocks.

**Aromatic** In organic chemistry, a molecule or a compound that contains rings made of carbon atoms bound to each other in a close "circuit", around which some ELECTRONS can, in some sense, "circulate" freely. For example, the monomers making polystyrene, polyester, and nylon, are aromatic molecules. In DNA, pyrimidines have a ring made of six carbon atoms, whereas the other kind of nucleobases, purines, have an additional five-atom ring. ♡ Nothing to do then with "aroma", at least in principle, although many fragrances such as cinnamon, vanilla, and anise are "aromatic" because they contain aromatic molecules. Actually, many aromatic compounds do smell good. But don't be misled, for, in chemistry, pleasant-smelling often does not mean healthy. For instance benzene, the first aromatic compound to be discovered, has a pleasant smell. But you are advised not to sniff it, if you value your blood, bone marrow, or prostate!

**ATP (adenosine triphosphate)**: The universal currency used by all living organisms to store, share, and transfer energy. A single ATP "coin" is made by attaching an extra phosphate group to an ADP (adenosine *di*phosphate) molecule, and is spent by a chemical reaction called *hydrolysis*, in which ATP reverts to ADP, releasing as much as 20 times the THERMAL ENERGY per molecule. The production and consumption of ATP in cells are carefully controlled by proteins respectively called *ATP synthases* and *ATP-ases*. Some delicate transactions, however, such as those needed to make MICROTUBULES, require a special, slightly different currency, called GTP (*guanosine* triphosphate).

**B**acteria: A large group of PROKARYOTE microorganisms, spread throughout every habitat on Earth, as well as in the bodies of plants and animals.

To tell the truth, plants and animals are dwarfed by bacteria, which make up most of the Earth's biomass. They can have a wide range of shapes, ranging from spheres to rods and spirals, and have typically a size of a few microns. ♡ Although bacteria are usually single-celled organisms, they often organizes into very complex "colonies" called *biofilms*. A biofilm is in fact a kind of super-organism itself, for the bacteria in it communicate and develop common strategies. Understanding why and how biofilms form is currently one of the most active subjects in biology, largely because of their role in infections.

**Base**: In its simplest form, a chemical substance that accepts a hydrogen ($H^+$) ION from an ACID. When a base compound is dissolved in water, it is the latter that behaves as an acid, for part of the water molecules become $OH^-$ IONS (as you see, a substance like water may behave either as an acid or as a base depending on what we dissolve in it – if it is lemon juice, for instance, it works as a base, whereas it behaves as an acid in the presence of lye). Well-known examples of bases are ammonia and lye (also known as caustic soda), but bases are less common than acids in everyday life. Which is good, for they are generally much more dangerous than acids. The mildest thing that can happen to you on touching a basic liquid with your fingers is getting a "soapy" feeling, because of the SAPONIFICATION of skin FATTY ACIDS. Getting it in your eyes is much, much worse, even if it is a weak base.

**Beta sheet**: See PROTEIN (STRUCTURE).

**Bjerrum length**: The distance at which two ELECTRONS or two PROTONS repel with an energy equal to the THERMAL ENERGY. Obviously, it is also the distance at which two elementary CHARGES of *different* sign (an electron and a proton) *attract* each other with the same strength. In water, the Bjerrum length is about 0.7 NANOMETERS.

**Bond**: Guessing what a bond is in chemistry looks trivial: bonds are what what bind atoms in molecules. ♡ Here, however, I'd like you to mull over this concept a little. In this book, we discussed at length *interactions*, namely attractive or repulsive forces between molecules and particles. Yet, after all, a bond is nothing but an attraction between two atoms. Why don't we just call it an "interaction"? It's a matter of strength. In physics, something is strong or weak compared with something else, and the gauge of interaction strength is the THERMAL ENERGY. If the interaction energy between two atoms or particles is much smaller than $kT$, then they can walk out easily, because they have enough kinetic energy to do it. But if the interaction strength is much larger than $kT$, they are hopelessly bound together. For instance, at room temperature, the strong *covalent bonds* formed by atoms that "share" some of their electrons are more than a hundredfold larger than $kT$: even on the surface of the Sun the thermal energy would be six times smaller than the covalent bond between two carbons in a HYDROCARBON chain! Conversely, the energy associated with dispersion forces between two atoms (not two big colloidal particles) is of the order of $kT$, so they are not real "bonds". Finally, the HYDROGEN BONDS between water molecules, or AMINO ACIDS in PROTEINS are larger than $kT$, but only a few times, so they are actually borderline between interactions and strong, durable bonds.

**Brownian motion** The very irregular motion (actually, a totally RANDOM WALK) performed by the COLLOIDAL PARTICLES due to the effect of collisions with the solvent molecules. Because the little "kicks" given by the solvent are so random in direction, the particle does not move very far away. Actually, the region it "explores" is only proportional to the square root of the time, so it grows more and more slowly as time passes. This property is typical of all those phenomena called DIFFUSION processes.

$C$atalyst: Many chemical reactions that should occur because they are energetically favored do not take place, or proceed very slowly, because they require an *activation energy*, in other words a "kick" to overcome a hill placed between the starting and end points of the reaction. A catalyst is like a bulldozer that levels the hill, or, if you prefer, digs a tunnel through it. The way it does this is by lending some energy to the reagent chemicals, which it then takes back from the products. The great point is that the catalyst is not consumed during the reaction, but fully recycled. ENZYMES are very efficient catalysts, which specialize in operating on a single, or at most a few, chemical reactions.

**Cell (or plasma) membrane**: The cell (or plasma) membrane is a flexible "bag" that contains the CYTOPLASM, thus defining the boundary of a cell by separating it from the external environment. Its framework is a LIPID DOUBLE-LAYER, mostly made of PHOSPHOLIPIDS, entrapping a lot of different membrane proteins (see PROTEINS (KINDS OF)) that perform all the communication and trade tasks of the cell. Being very flexible, the cell membrane can deform so as to allow for ENDOCYTOSIS and EXOCYTOSIS. Moreover, it provides anchorage for the CYTOSKELETON, which provide shape to the cell, and allows the cell to attach to the EXTRACELLULAR MATRIX or to other cells to form tissues. BACTERIA and ARCHEA, which have a less flexible shape, since their cytoskeleton is very limited, have a rigid *cell wall* too, which provides a mechanical support for the cell. ♡ Plant cells have a cell wall too, mostly made of *cellulose*, a POLYMER of glucose, and *lignin*, a very complex AROMATIC polymer, which allows them to organize into fibrous wood tissues.

**Chaperonin**: See PROTEIN (STRUCTURE).

**Charges**: The sources of the electric forces between bodies. There are two "genders" of electric charge, positive and negative. Charges of different kind attract each other, whereas like charges repel. An *elementary charge* is the smallest charge that can be found in nature. Electrons and protons, the basic charged constituents of an atom, bear respectively a single negative and positive charge (the proton, however, is about 2,000 times heavier). ♡ This is a very small charge: billions of electrons flow through a common lamp filament every *billionth* of second. Even a globally neutral substance such as water contains as many as $6 \times 10^{16}$ (in other words, 6 followed by 16 zeroes) free charges per liter, in the form of $H^+$ and $OH^-$ IONS (adding an ACID or a BASE enormously increases the number of the former or of the latter). Some weird objects studied by theoretical physicists, the QUARKS, do have a smaller charge, but they are constrained to live together in "lumps" with an integer number of elementary charges.

**Chlorophylls**: Pigments used by plants and algae to capture sunlight and transform it into expendable energy (see ATP) by the complex process called PHOTOSYNTHESIS. They absorb most light WAVELENGTHS, but they are not particularly efficient in the green (which is then partly reflected, making leaves look greenish). To improve the efficiency of photosynthesis, they are usually helped by other pigments, the *carotenoids*, which are able to catch green light and are particularly abundant in fall (when leaves look yellow-reddish).

**Chloroplast**: Chloroplasts are organelles of plants and algae cells that in the far past were probably free bacteria, but were tamed by their new masters and exploited to perform PHOTOSYNTHESIS. Like MITOCHONDRIA, they are full of MEMBRANES, used as docks by the protein complexes that perform the incredibly complex reactions required in photosynthesis.

**Cholesterol**: A basic component of our CELL MEMBRANES, which helps to control their flexibility, fluidity, and permeability. However, it is not an AMPHIPHILE, but a large fully HYDROPHOBIC molecule, which is fully embedded in the LIPID DOUBLE-LAYER. Cholesterol is also an important component for making steroid HORMONES and many vitamins. ♡ Being hydrophobic, cholesterol is very weakly soluble in blood and, to be carried around, requires special PROTEINS called *lipoproteins* that form globular aggregates incorporating cholesterol. When physicians talk about "cholesterol" and "hypercholesterolemia", they actually refer to these proteins, and in particular to the aggregates with a size of 20–25 nm formed by a subclass of small and light lipoproteins, technically called LDLs and commonly known as "bad cholesterol".

**Chromatin**: The entangled mess of DNA found in EUKARYOTE cell nuclei. Besides DNA, chromatin contains special protein complexes, the HISTONES, which are little spools the DNA winds around to pack better. When the cell is replicating, chromatin is organized into CHROMOSOMES by complex mechanisms that we understand only poorly at present.

**Chromosome**: Chromosomes are the little rods into which the DNA of EUKARYOTES organizes just before cell replication. They are organized in nearly identical couples, inherited from each one of the two parent cells, with the exception of those chromosomes bearing genes that mostly determine the sexual features, which have a different shape. During cell replication, these couples are pulled apart by MICROTUBULES that hook to a little "button" just in the middle of each chromosome.

**Coagulation**: The process by which a COLLOID aggregates, because it is no longer stable. For colloids made of charged particles, this loss of stability may derive from the addition of salts, whose IONS reduce the repulsive force between the COUNTERION clouds that surround each particle. The colloidal aggregates that form in a colloidal aggregation process usually have a FRACTAL structure.

**Collagen**: See EXTRACELLULAR MATRIX (ECM).

**Colloid (and colloidal particles)**: A dispersion of "particles" in a fluid, improperly called the "solvent", that can be either a liquid or a gas (in which case we speak of an AEROSOL). A colloidal particle, in the most generally accepted use of this word, can be made of any kind of material, whether a solid, a liquid immiscible with the solvent (when the colloid is usually called an EMULSION), or even a gas (in the case of FOAMS). It can have any shape, and be rigid or flexible, like a POLYMER coil or a VESICLE. It may even exist only within the solvent, as in the spontaneous aggregates of SURFACTANTS. The only distinctive feature of colloidal particles is that they must be quite large in terms of volume compared with the solvent molecules. Since the volume of a particle grows as the cube of its size, this means that even a micelle, with a size of a couple of nanometers, can be regarded as a colloid. Giving an upper limit for the size of a colloidal particle is mostly a matter of convenience. In practice, on the Earth, we just ask that they are not so big that their settling under their own weight does not overwhelm their BROWNIAN MOTION (this is what mainly distinguishes a colloid from a GRANULAR MATERIAL). An important distinction is between colloids that are intrinsically stable, such as polymer or surfactant solutions, which are also called *lyophilic* colloids, and suspensions of particles that need to be stabilized against COAGULATION, like most dispersions of solid particles, which are technically called *lyophobic* colloids (and are "suspensions" in the proper sense).

**Colloidal crystal**: An ordered structure of colloidal particles, regularly placed on a "lattice", for instance on the corners of a cube. Colloidal crystals are then similar to the simple crystals formed by atoms and molecules, but with the major difference that the typical distance between two nearby particles can be of the order of MICRONS instead of fraction of a NANOMETER, as in the atomic case. Because of this, colloidal crystals usually show gorgeous colors due to the DIFFRACTION of visible light. OPALS are beautiful natural examples of colloidal crystals, made of tightly packed silica glass spheres. Usually, colloidal crystals form at high particle concentration (for hard spheres, for instance, at a VOLUME FRACTION of about 50%), but in the absence of salt suspensions of particles that are electrically charged may crystallize at a particle concentration lower than 1%. These curious structures are physically "solids" – because they are ordered – but can be poured as a liquid.

**Copolymer**: A POLYMER made by two or more kinds of MONOMERS, which can be placed at random (for instance ABAAABABBABABABAAB..., where A and B are two different monomers) or in regular blocks (such as AAAABBB-BAAAABBBB...). Copolymers can have many more interesting features and applications than the simple polymers made of just one kind of monomer (*homopolymers*). Typical examples are the epoxy resins, which harden because of the copolymerization of two different monomers, and the acrylic fibers used for example to make synthetic garments. Moreover, the addition of a co-monomer can make industrial processing much easier, for instance by lowering the VISCOSITY of the polymer melt. Nature is much more skilled than we are in inventing copolymers with very specific functions, for proteins are in fact copolymers made of 20 different kinds of monomers, the AMINO ACIDS.

**Counterions**: The small IONS released in solution by charged COLLOIDAL PARTICLES, and also by ionic MICELLES and POLYELECTROLYTES. Because they are attracted by the particle, which bears a CHARGE of opposite sign, the counterions gather around it, forming a kind of "cloud" that has a size fixed by the DEBYE–HÜCKEL LENGTH. The number of counterions in the cloud is exactly equal to the total number of CHARGES on the particle. Usually, the clouds around charged particles prevent them from approaching too closely, and therefore stabilize the colloid against coagulation. However, when counterions bear more than a single charge (that is, when they are *multivalent*), as in the case of cements, they can "bridge" two particles, which is equivalent to say that they induce an effective *attractive force* between them.

**Creaming**: The opposite of SEDIMENTATION, where COLLOIDAL PARTICLES, instead of settling down, rise to the top because they are lighter than the surrounding solvent. Creaming is usual in oil-in-water EMULSIONS, since oils are generally less dense than water.

**Critical micellar concentration (cmc)**: The minimum SURFACTANT concentration at which MICELLES form. Below the cmc, part of the surfactant goes to the surface separating water and air, forming a layer that decreases the SURFACE TENSION of water. This reduction of tension stops at the cmc, and the formation of micelles, which are much bigger than the solvent molecules, is witnessed by a noticeable increase in the SCATTERING of light from the solution. The cmc is typically much lower for non-ionic than for ionic surfactants.

**Cross-link**: A chemical bond between two MONOMERS belonging to different POLYMER chains. Cross-links therefore act as "knots" that do not allow the chains to slide freely as cooked spaghetti, and lead to the formation of solid polymer networks. When the fraction of cross-linked monomers is not too high (typically a few percent), the network behaves as an ELASTOMER (a rubber), whereas for higher cross-link concentration we obtain a rigid resin, such as melamine. In the case of natural rubber, cross-links are provided by sulfur bridges.

**Crystal**: An ensemble of atoms or molecules arranged in an orderly way in space, much as tiles are arranged regularly on a floor. Besides atoms and molecules, colloidal particles can spontaneously order in COLLOIDAL CRYSTALS too. Crystals are the only "true" solids for physicists, for the presence of this "long-range" order is the *only* aspect that really distinguishes them from liquids (other features, such as the fact that solids retain their shape while liquids do not, become much more blurred in the case of soft materials). Crystals may have many different structures, depending on the shape of their *unit cell*, the basic motif that repeats regularly in space. The simplest crystal structure is the simple cubic, where the atoms are placed on the edges of joined cubes. Two slightly more complex structures are the body centered (BCC) and face centered (FCC) cubic, containing additional atoms respectively in the center of the cube or of its faces. But there are many more crystal structures, where the unit cell can be for instance a parallelepiped or an inclined prism. ♡ In fact, there are exactly 14 kinds of atomic *crystal lattices*. When the unit cell contains more than one kind of atom, as in the case of molecular crystals, one can obtain as many as 230 kinds of structures differing in their "symmetry" (the type and number of different rotations and translations that bring back the crystal to its original shape).

**Cytoplasm**: The content of a living cell, enclosed by the CELL MEMBRANE, which contains all the organelles responsible for the specific cell functions. A peculiar feature is that ions, such as sodium ($Na^+$), chlorine ($Cl^-$), magnesium ($Mg^{++}$), and especially calcium ($Ca^{++}$), are much less concentrated inside the cytoplasm than outside the cell. The cell therefore has to work hard to keep the OSMOTIC PRESSURE imbalance between the inside and the outside, using very efficient PROTEIN "pumps".

**Cytoskeleton**: The "scaffolding" that provides the cell with its shape and internal structure and enables it to change them. It is mainly present in EU-KARYOTE cells, where it is composed of ACTIN and intermediate filaments and MICROTUBULES, although recent evidences show that some bacteria too have a simple form of CYTOSKELETON.

**D**ebye–Hückel length: The typical size of the COUNTERION "cloud" that forms in solution around a charged COLLOIDAL PARTICLE. The Debye–Hückel length gets smaller and smaller on the addition of salt. More precisely, it is inversely proportional to the square root of the number of IONS in solution.

**Denaturation**: The process by which a PROTEIN loses its structure (meaning its overall spatial arrangement, or tertiary structure, but often also at the level of secondary structure, owing to the disruption of ALPHA-HELICES and BETA-SHEETS), or the double helix of DNA splits into separate strands. Denaturation may occur because of temperature changes or the addition of strong ACIDS and bases, or certain ORGANIC solvents. A classic example of denaturation occurs when an egg is boiled: the egg white, from transparent, becomes white because its major component, ovalbumin, denatures on heating.

**Depletion forces**: Attractive forces between COLLOIDAL PARTICLES induced for instance by the addition of POLYMERS. They originate from the attempt by the polymer coils to increase their entropy by gathering the colloidal particles together, so increasing the volume available to the coils and gaining freedom of motion. Depletion forces are likely to play an important role in many biological organization process, ranging from DNA packing in NUCLEOIDS to the formation of ACTIN bundles.

**Diffraction**: The SCATTERING of light, X-rays, or any kind of radiation from ordered structures such as CRYSTALS. Diffraction occurs along specific directions, giving rise to the so-called *Bragg peaks*. An analysis of the position and intensity of the Bragg peaks allows the structure of a crystal to be reconstructed, together with the positions of all atoms in its elementary cell. Diffraction (of X-rays) is therefore a fundamental tool for PROTEIN crystallography.

**Diffusion**: The physical process by which a substance "spreads around" in the surrounding medium. For instance, it is the way a drop of ink injected into water grows in size (and fades) with time. In fact, it is the macroscopic manifestation of the RANDOM WALK that the molecules of the substance perform because of collisions with those of the medium. A peculiar feature of diffusion processes is that the size $\ell$ of the region that the diffusing stuff explores grows only as the square root of time, so that the growth gets slower and slower as time goes by. One can write $\ell \simeq \sqrt{Dt}$, where $D$, called the diffusion coefficient, is characteristic of the specific substance that diffuses. ♡ Mass, however, is not the only physical quantity that "diffuses": heat, for instance, does the same, and its diffusion coefficient is called "thermal diffusivity". The latter is much higher for (most) metals than for non-metallic materials – this is why your fingers get burnt more easily if you touch a piece of hot iron than a similarly heated marble surface.

**Dispersion (van der Waals) forces**: Basic attractive force acting between all kind of materials that make them stick together. Dispersion forces can be summed up in the sentence "Like prefers like", meaning by that that this "microscopic glue" is stronger between objects made of the same materials (glass with glass, iron with iron, and so on). In the macroscopic world, van der Waals forces are mostly responsible for the existence of friction, which in its turn is essential, in the absence of cement, to prevent buildings falling. In Middle-earth, dispersion forces rapidly lead to COLLOID aggregation if the particles are not suitably stabilized or do not like the solvent very much, as POLYMERS or SURFACTANT aggregates do, because of solvation effects.

**DNA**: See NUCLEIC ACIDS.

**Ductility and malleability**: Ductility is the mechanical property of a material telling us how much tensile STRAIN it can sustain, for instance how much it can be stretched into a wire without breaking. Similarly, malleability refers to the ability of a material to deform under compressive stress, for example to be shaped into a thin sheet by hammering. Among the different kind of materials, metals are by far the best performing in terms of deformability, because their internal defects can rearrange so as to avoid concentrated stress. However, while some metals like gold are both ductile and malleable, others, such as lead, are only malleable.

**E**lastic modulus: For small deformations, all materials behave elastically, meaning that the STRAIN of the material is proportional to the applied STRESS. The ratio of the applied stress to the obtained strain is called elastic modulus. Depending on the kind of stress we apply, however, there are different elastic moduli. For instance, the *Young modulus* is the ratio stress/strain when we stretch or compress a material along an axis. Similarly, the *shear modulus* describes the resistance of an object to deform when subjected to a shear force. Finally, the *bulk modulus* tells us how much the volume of an object changes when it is equally pressed in all directions.

**Elastin**: See EXTRACELLULAR MATRIX (ECM).

**Elastomer**: A polymer material that behaves as a rubber, showing an extremely high elasticity. In contrast to a spring, an elastomer owes its distinctive properties to ENTROPY. An elastomer is actually very similar to a polymer melt, but the chains are connected by CROSS-LINKS.

**Electric field**: An alternative way of describing electric forces between CHARGES. Instead of saying that two particles attract or repel with a given force, one can say that the first particle generates a "field" that modifies the space around it, and that the second particle feels a force proportional to the strength of the field in the place where it lies. While for electrostatic forces the electric field is a convenient concept, it becomes a key quantity when the charges move, or when one deals with more complex "electromagnetic" phenomena such as light. ♡ The strength of the electric field is easily related to the everyday concept of voltage. Consider for instance a standard US plug. When it is connected via a socket to the line power, its two pins are at a voltage difference $V = 115$ volts. Then, the electric field $E$ is simply $V$ divided by the space between the pins, which for a US plug is slightly less than half an inch (1.2 cm), so $E \simeq 100$ V/cm.

**Electron**: See CHARGES.

**Electrophoresis**: The motion of charged COLLOIDAL PARTICLES or POLYELECTROLYTES induced by an ELECTRIC FIELD. For instance, if we place a colloidal suspension between two metal plates (called *electrodes*), one bearing a positive and the other a negative CHARGE, the particles move towards the plate with the opposite charge (if they are positive, to the negative electrode, and vice versa). Electrophoresis is used to evaluate the charge of colloidal particles, but its most useful applications are in biochemistry, where many versions of this experimental technique have been developed. For instance, by making a mixture of PROTEINS migrate in a polymer GEL kept in an electric field, it is possible to separate the single components on the basis of their size or charge. A very important technique in gel electrophoresis is SDS-PAGE, which allows precise values of the MOLECULAR WEIGHT of a protein to be obtained by adding to the solution the anionic SURFACTANT SDS. The number of SDS molecules that attach to a protein is directly related to its molecular weight. ♡ The reason why colloidal particles move in an electric field is not so trivial as it may seem at first glance. Indeed, it is true that a negative particle is attracted by the positive electrode, but remember that a charged particle is surrounded by a cloud of oppositely charged counterions, so that, seen from the electrode, the ensemble (particle + cloud) looks neutral. The particle movement is in fact the result of a complex "sliding" effect of the particle with respect to the counterion cloud. Very roughly, the counterions surrounding the particle are pulled in the other direction by the electric field, and leave the cloud. But the "cloud" is actually a very dynamic object, so that they are immediately replaced by new ions that lie around the new position occupied by the (displaced) particle. One could say that the particle constantly builds up a brand-new counterion cloud while it drifts. The drift speed ("electrophoretic velocity") that the particle attains is proportional to the particle charge and to the applied electric field, inversely proportional to the solvent VISCOSITY, and faster the more charged the particle is.

**Electrostatic forces**: The forces that exert between electric CHARGES, at least when they do not move (if the charges move relative to each other, there are additional forces, because a moving charge creates a magnetic field – these are called electro*magnetic* forces). In vacuum, two charges attract or repel with a force which is inversely proportional to the distance between them, so if the distance between the two charges halves, the force between them gets fourfold larger. This is exactly like gravity, the force that keeps the Universe together. But electric forces are much, much stronger than gravity: at any distance apart, the gravitational attraction between two electrons is negligible compared with their electric repulsion. So why don't we see huge electric effects in our daily life? Simply because, although any substance contains an enormous number of charges, positive and negative charges usually balance, so that objects are globally "neutral" (this is called *charge neutrality*). Electrostatic forces between charged colloidal particles are a bit more tricky. The particles become charged by releasing into the solvent a lot of particles of opposite sign, which are called COUNTERIONS, but these gather around the particle to form a little cloud that "screens" the electrostatic forces between two colloidal particles. As a result, these forces decrease much faster with distance than in vacuum and, in fact, become negligible once the two particles are more than a few DEBYE–HÜCKEL LENGTHS apart.

**Ellipsoid**: Take an ellipse, one of those squeezed circles describing the orbit of a planet, and rotate it along one of its two axes (in a circle, these are both the same length, given by the diameter). The solid shape you obtain (a kind of egg, but more regular) is an ellipsoid, or to be precise an *ellipsoid of revolution* or *spheroid*. If you spin it around the longest axis, you get a *prolate* ellipsoid, which is a stretched sphere, whereas if you choose the shorter axis you obtain an *oblate* ellipsoid, a sphere flattened at its poles like the Earth.

**Emulsion**: A COLLOID made of small droplets of oil in water ("oil-in-water", or O/W emulsions) or water in oil ("water-in-oil", or W/O emulsions), with a typical size ranging between a few and a few tens of microns, usually covered by a SURFACTANT that reduces the SURFACE TENSION between the two liquids and provide a barrier against droplet coalescence, i.e. the merging of two droplets when they meet. An emulsion, however, is not stable forever, either because the droplets, albeit slowly, eventually coalesce, or because of a more complex effect, known as OSTWALD RIPENING, by which the largest drops grow at the expenses of the smallest, which gradually disappear. Moreover, they do not form spontaneously, and usually require vigorous stirring to be produced (for a very special kind of emulsion that forms spontaneously and lasts virtually forever, see MICROEMULSIONS). Emulsions have widespread applications in cosmetics (beauty and suntan creams, ointments, lotions), farming (insecticides and pesticides), paints, and drug formulation. Many foods, such as milk or mayonnaise, are emulsions too; in this case, the stabilizing role of the surfactant is played by special PROTEINS. Very viscous and hard-to-break emulsions, stabilized by rigid colloidal particles that stick to the droplet interface, are often formed by crude oil or as an effect of oil spills into the sea.

**Endocytosis and exocytosis**: Endocytosis and exocytosis, respectively, are the processes by which cells take in or expel big molecules such as PROTEINS, or even entire organism such as BACTERIA. Both processes are often mediated by lipid VESICLES. In endocytosis, these vesicles originate from the CELL MEMBRANE, engulf the stuff that must be taken it, and detach from it. Most of them typically have a size of about 100 nm, and are decorated and reinforced by the protein *clathrin*. There are, however, smaller vesicles that form spontaneously without the aid of clathrin from tiny, flask-shaped pits in the membrane called *caveolae*. Much larger particles, like small bacteria, cell debris, and even cells that have killed themselves by APOPTOSIS, can be taken in by *phagocytosis*, but only specialized cells such as macrophages are able to do this. In simple exocytosis processes, vesicles (often produced by the GOLGI APPARATUS) dock to and fuse with the cell membrane, releasing their content in the extracellular medium. A very special form of exocytosis is performed by neurons: when a nervous electric signal reaches the tip of their axons, they shed from this tip small vesicles containing chemicals known as *neurotrans-*

*mitters*, which are collected by the dendrites of other nearby neurons (these dendrites are filaments extending from the body of a neuron, similar to, but much shorter than, the axon).

**Endoplasmic reticulum (ER)**: A complex interconnected network of LIPID MEMBRANES occupying a large portion of the cytoplasm of EUKARYOTE cells. There are two kinds of endoplasmic reticulum, smooth and rough. The latter, in particular, provides the docking for a large number of RIBOSOMES which synthesize PROTEINS.

**Entropy**: A fundamental, but rather difficult concept in physics. As we have extensively discussed in Chapter 3, entropy is strictly related to the "freedom of motion" a physical system has. For instance, the entropy of a gas is higher the more room the gas molecules have to move about, and the wider the range of speeds they can attain. But the room available for a system (either as physical space or "velocity" space) is not the only measure of entropy. For instance, the fantastic elasticity of rubber is related to the number of different shapes a polymer chain can take for a given end-to-end distance $\ell$ between its terminals. The smaller the value of $\ell$, the more the chain shape can fluctuate, and the polymer entropy is just related to the number of these different "configurations" that the chain can assume. In general, entropy is associated with the total number of ways a physical system has to move about or rearrange. ♡ Entropy is not therefore a "measure of disorder", at least directly. Nonetheless, what we would call a "disordered" system often (although not necessarily; remember the COLLOIDAL CRYSTALS) has a large entropy. Let us see why with a simple example. Suppose that in your home there are several bookcases. It is surely easier to find a book you are looking for, if each member of your family (your spouse, your children, and yourself) has a personal bookcase, and does not dare to invade other people's bookshelves. You will say this is a neat or "orderly" way of keeping books. But now look this situation from the viewpoint of your books. If they are stored in a specific bookcase, they have surely less "freedom of motion" (although they do not move by themselves) than when they are put in at random. In other words, they have less entropy and are more ordered. This example, although trivial, tells us something about another deep aspect of entropy: entropy is related to *information*. If your beloved book can be anywhere, you need much more information to find it than when you already know it must be in your own bookcase. The higher the entropy of the books, the more you have to go around asking your spouse and children, "Where the dickens have you put my book?" (although, of course, it was *you* who moved it). Entropy is therefore a measure of the *total amount of information we need to describe a physical system fully*. This great intuition, due to Claude Shannon from Bell Labs, is at the basis of all modern communication systems.

**Enzyme**: A protein acting as a CATALYST. Like all catalysts, enzymes work by lowering the activation energy for a chemical reaction, thus dramatically increasing its speed, but they are far more efficient and selective than simple catalysts, for they are specialized for a single reaction, or at most a few reactions. The molecules at the beginning of the process are called substrates, and the enzyme converts them into different molecules, called the products. Typically, the reaction takes place at a selective place in the protein structure, called *active site*. Enzymes are also used in some household products, such as in washing powders to break down protein or fat stains on clothes.

**Eukaryote**: An organism whose cells have a nucleus that keeps DNA well separated from the rest of the CYTOPLASM. Compared with their PROKARYOTE counterpart, eukaryote cells are huge – typically, thousands of times bulkier, in terms of volume – and much more complex, for they contain a lot of specialized internal organelles, sometimes even larger than a prokaryote cell, kept in place or moved around by an intricate and active scaffolding, the CYTOSKELETON.

**Extracellular matrix (ECM)**: The external structure that, in multicellular organisms, provides support and anchorage for the cells, separates different tissues, and forms the medium through which the intercellular communication via chemical signals takes place. The main proteins proteins of animal ECM are *collagen* (the main component of aspic or gelatin), which forms very strong fibers that can stand very strong tension stress, *elastin*, which is akin to an "ideal rubber", and keratin, a very tough protein of which nails, horns, and hair are made. In the ECM there are also non-protein components, such as complex POLYELECTROLYTES called GAGs, of which hyaluronic acid is the best known because of its applications in anti-aging treatments.

**Extremophile**: A living organism that thrives in what we would call extreme or even unbearable environmental conditions. Extremophiles are usually PROKARYOTES, and very often ARCHEA (but sometimes also BACTERIA). ♡ There are extremophiles that love to settle in very acidic or basic waters (the *acidophiles* and *alkalophiles*), or which like it hot (the *thermophiles*, which in hydrothermal vents can withstand temperatures as high as 120°C), or salty (the halophiles, which live like pickled herrings), or sweet (the *osmophiles*, which would happily grow in maple syrup). There are even Archea that live in microscopic spaces within rocks, and whose sole sources of energy or carbon are the minerals surrounding them. According to the latest discoveries, these strange creatures that can live without sunlight seem to populate rocks down to great depths, so much so that they could possibly outnumber bacteria to form the greatest biomass on the Earth.

**F**atty acids: A class of extremely important molecules, involved in a large number biological processes, derived from fats contained in animal fats or plant oils. Fatty acids are composed of moderately long HYDROCARBON chains (up to 20–30 carbon atoms) terminating with a COOH group, which makes them weak ACIDS. ♡ Our body is able to make almost all the fatty acids we need, except a couple of them, known as "essential" fatty acids, which are, however, crucial for producing HORMONES that regulate blood pressure and clotting, the immune response, and the inflammation response to injury infection. Luckily, we can obtain them from vegetable oils or from the longer "omega-3" fatty acids contained in fish.

**Fcc structure**: See CRYSTAL.

**Foam**: Any soft material made of a large number of gas bubbles trapped in a liquid or in a solid to form a "cellular" structure. The most common example of foams is soap froth, where thin water films covered on both sides by layers of SURFACTANT molecules surround the air bubbles which, after most of the water has drained, take on a polyhedral shape. The draining process by which a liquid foam progressively changes its structure is governed by the Plateau laws for SURFACE TENSION. Solid foams are often formed by polymers such as polystyrene ("styrofoam") or polyurethane foam, but foams made of metals such as aluminum have great practical interest too, because they are both light and resistant. Moreover, because they are mostly made of air, which conducts heat very poorly, they areexcellent thermal insulators. ♡ A daily example of solid froth is leavened bread, where the bubbles in the dough are produced by yeast, whereas delicious liquid froths can be obtained by bubbling air through milk, as anyone who has ever drunk a cappuccino knows.

**Fractal**: In geometry, an object with a weird shape, so that it is actually something in between a line and a surface, or between a surface and a solid. To understand this, we have to define what we mean by "spatial dimension" of an object. For smooth curves and surfaces, this is easy: we need, respectively, just one or two numbers ("coordinates") to fix a point on a straight line or on a plane. So, a straight line is one-dimensional, and a plane two-dimensional. But what happens if the line or the surface are so rough that, even if you look at them through more and more powerful lenses, you still find a finer and finer structure, a finer and finer scenery, no matter how much you magnify them? Giving precise "coordinates'" may not be easy, and we need to resort to a different way to state what their spatial dimension is. Hausdorff proposed to draw a sphere around a given point, and measure how much of the curve (or of the surface) lay within the sphere. For a smooth curve, the enclosed part grows in proportion to the sphere size; for a smooth surface, it grows as the size squared. In other words, they are respectively proportional to $R^1$ and $R^2$, where $R$ is the sphere radius. So, if we define the "dimension" $D$ of an object as the power at which $R$ has to be risen, we find $D = 1$ for a regular curve, and $D = 2$ for a smooth surface, which agrees with the usual definition. But the point is that, using this definition, those weird lines and surfaces we referred to may have a dimension which is *larger* than 1 or, respectively, 2, and not necessarily a whole number. Fractals are mathematical abstractions, but many physical objects, such as snow flakes, mountain ranges, coastlines, or even our lungs, have roughly a fractal shape. We have also seen that the clusters that form in COLLOIDAL AGGREGATION are approximately fractal, unless you look at them so closely that you see the single particles.

**Free energy**: Take the amount of energy a physical system "contains" (its *internal energy*), and subtract from it the product of (absolute) TEMPERATURE and ENTROPY. The difference is the free energy. The adjective "free" is added because the free energy $F$ is the maximum work a machine can do when it is supplied with a given total amount of energy $E$ (and is always less than that). $F$ is thus in some sense the maximum *useful* energy. A system exchanging heat with the environment reaches equilibrium when $F$ is as low as it can be in those conditions.

# G AG: See EXTRACELLULAR MATRIX (ECM).

**Gel**: An AMORPHOUS solid that, like a GLASS, lacks a long-range order, but, in contrast to a dense glass, is a made of a tenuous matrix of interconnected COLLOIDAL PARTICLES or POLYMER chains, immersed in a fluid (which is usually a liquid solvent, but can also be air in the case of AEROGELS). Although most colloidal glasses (which are always very concentrated SUSPENSIONS) do not flow like liquids because the particles find hard to move through such a crowded environment, gels are mechanically solids because the particles stick to each other. Thus, *attractive* forces between the particles are always responsible for the formation of gels. So the particle clusters that grow during COLLOIDAL AGGREGATION eventually form a soft gel, unless the latter is rapidly crushed by cluster SEDIMENTATION. Similarly, DEPLETION FORCES induced by polymers or surfactants often lead to the formation of colloidal gels. Semidilute polymer and POLYELECTROLYTE solutions also form gels, where the chains are connected by CROSS-LINKS (a kind of rubber, with the solvent inside) or simply stick because they do not much like the solvent. Polymer gels are used in many applications. For instance, contact lenses are made of silicone gels, whereas gels of sodium polyacrylate, a polyelectrolyte also known as "waterlock", are used to make diapers, since they can absorb a great deal of water to reduce the osmotic pressure of the COUNTERIONS. But there are high-tech applications too. Polyacrylamide gels are extensively used in biology as supports for gel ELECTROPHORESIS. There are even "smart gels" that can sense changes in temperature or chemical composition of the environment, or that are responsive to specific antigens and can therefore be used as bio-sensors. ♡ Let me tell you about a kind of gel that, conversely, is a very annoying problem in oil production. Crude oil contains waxes (*paraffins*), which are prone to form little CRYSTALS at low temperature. Unfortunately, these crystallites often stick to each other because of dispersion forces, and readily form very rigid gels even when the wax percentage in oil is low. Thus, wax deposits can form on the walls of pipelines where the temperature of the external environment is low, a situation commonly encountered in offshore and Arctic production. Needless to say, when these deposits clog a pipe, it's a disaster, in particular if the pipeline lies deep in the sea. Oil companies invest considerable effort into evaluating the risk of deposit formation and defining effective countermeasures.

**Genetic code**: The set of rules by which information encoded in DNA is first translated into messenger RNA (mRNA), and then from mRNA into chains of AMINO ACIDS, which then fold to form proteins. The code is based on associating to a sequence of three NUCLEOTIDES, called a *codon*, an amino acid. This code is *redundant*, in the sense that the same amino acid can be coded by more than one set of codons, but *unambiguous*, because each codon is associated to a single amino acid. However, just a small fraction of the nucleotides contained in DNA is "coding". A large part of it is used for regulating the functions of the genes, which is genetic information too, or is "JUNK" DNA, which is apparently (but probably only apparently) useless.

**Genome**: The whole of the information needed for all vital tasks, which is encoded in DNA (or, for some viruses, in RNA) and is transmitted by living organisms to their progeny.

**Glass**: A material that behaves mechanically as a solid, but lacks the regular arrangement of a CRYSTAL. In fact, if we look at the microscopic structure of a glass (how the molecules or particles it is made of are arranged), it closely resembles a liquid. Thus, strictly speaking, a glass is not a "real" solid, much as the Parmesan cheese sold in the United States is often a poor imitation of what is produced around my father's native town. One of the main differences between a real solid and a glass is that, when heated, the former melts abruptly at a precise TEMPERATURE, whereas the latter *progressively softens* across a range of temperatures. Often this range is narrow, so we can approximately speak of a *glass transition temperature* $T_g$, but this is not true for all glasses. A practical way of defining $T_g$ is to look at the fluid VISCOSITY, which shoots up on cooling a liquid toward the glass transition, and taking $T_g$ as the temperature below which the viscosity gets so high that it would take ages for the liquid to flow. Simple glasses, like those used for window panes, are made of small molecules, but POLYMER melts can turn into glasses too. COLLOIDAL SUSPENSIONS also form glasses, but in this case what controls their formation is not temperature, but particle concentration. Looking at colloids makes easier to grasp why glasses form. When the particle concentration is progressively increased, it becomes harder and harder for a particle to move around, because of the hindrance due to the others. Beyond a given concentration (which, like the glass transition temperature for simple glasses, is not precisely defined), our particle gets trapped in a "cage" formed by its neighbors, which is very hard to escape, unless by some complex cooperative motion of the entire particle assembly. When this trapping experience is shared by most of the particles, the colloid motion is "arrested". There is, however, another possible scenario: the particle concentration may not be so high as to enable the formation of closed cages, but the particles can stick to each other, because of some attractive force, and may therefore... remain trapped in an *open* cage! Structures of this kind, which are more "open" than the glasses made by HARD SPHERES, are called *attractive glasses*. If the particle sticking is very pronounced, attractive glasses can form even at very low particle concentration. But, at this point, we really don't know how to distinguish these from GELS.

**Glossary**: What you are now reading (which therefore has a lot to do with soft matter).

**Golgi apparatus**: A very complex network of folded and stacked LIPID MEMBRANES found in EUKARYOTE cells, named after the Italian physician Camillo Golgi who first identified this internal structure thanks to novel cell-staining methods. The Golgi apparatus is, at the same time, a kind of stockroom for "tanks" where PROTEINS and other macromolecules are stored, called *cisternae*, and the "post office" or "mailroom" of the cell, where the same proteins are marked with the recipient's address, assembled in vesicles, and dispatched to their final destination.

**Granular materials**: Any substance, like rice, sand, or washing powder, made of a large collection of separate grains. Granular materials are often regarded as COLLOIDAL SUSPENSIONS that have "lost their suspending agent", but this is not rigorously true, since a "solvent" is actually there: air, exactly as in AEROSOLS. The real difference is that, compared with aerosol particles, these grains are big and *heavy*. Hence, their low THERMAL ENERGY is totally overwhelmed by their weight, so that the structures they form do not much resemble those they would take on if the grains were able to arrange themselves: burdened by gravity, the grains "jam" into bridges or arch-like configurations leading to very peculiar mechanical properties. For instance, their PRESSURE is not uniform as in a liquid, and they press more on the side walls than on the bottom of the container. Let me give you some advice (which I hope you won't need). If you ever happen to be sentenced to death by being buried under a dune, asked to be placed directly below the top: in terms of the pressure you feel, it is much better than half-way to the side (all other inconveniences, unfortunately, are the same). Granular materials also flow (indeed, they are called "granular fluids" too), but very differently from a stream of water. Their complex flow behavior leads to fascinating effects resembling snowslides and traffic jams, or to segregation phenomena like those observed in a bag of mixed nuts when it is shaken. Granular fluids hold many other surprises but, as in the movies, I am afraid you will need to wait for my next book ("Soft Matter strikes back") to discover them.

**Gyration radius**: A measure of the size of the coils that POLYMER chains form in dilute solutions, which is in some sense the "mean distance" of all the monomers from the coil center. For "ideal" chains (see POLYMER SOLUTIONS), the gyration radius $R_g$ grows as the square root of the number of monomers, whereas for real chains in a good solvent $R_g$ is proportional to $N^{3/5}$, so that the coils are bigger (more "expanded") than those formed by ideal chains. A nice property of the gyration radius is that it can be accurately measured using light SCATTERING techniques, even if the the polymer coils are too small to be seen by any microscope.

**H**ard **Spheres**: A very simple ideal physical system, made of moving, spherical particles that do not interact, except when they bump into each other. They are "hard" because, when they collide, they just rebound without changing shape, like billiard balls. Investigating a system of hard spheres may seem a dull hobby for theoreticians, but it is quite the opposite, for there is a lot of physics to be learned from hard spheres. Indeed, when the spheres fill a large enough part of a container, they can spontaneously organize into ordered CRYSTAL structures, or show "jamming" effects that turn the assembly of spheres into a GLASS. Moreover, they are not so "ideal" as they may look, for real colloidal particles can be made very similar to tiny perfect billiard balls. Using some tricks, the same hard-sphere particles can be made "sticky", which is like saying that they feel a strong attraction, but only when they touch each other. Sticky hard spheres show even more intriguing effects, such as the formation of colloidal GELS.

**Hardness**: A measure of how much a resistant solid material resists changing its shape when it is subjected to a force (or STRESS). Be careful: hard does not mean the same as *tough*. Materials like ceramics may be very hard, but they are generally fragile; they break easily because they cannot yield (they are not plastic). Agreeing to yield, when it is sensible to do so, can be a virtue for inanimate stuff too.

**Histones**: Protein complexes shaped like small "spools" around which the DNA double helix likes to spiral, making almost two full turns, because they bear an opposite electric CHARGE. Each nucleus contains tens of millions of histones, which provide a very convenient way for eukaryote cells to pack DNA.

**Hormone**: A chemical substance, but sometimes also a small protein (a short amino acid chain called a *peptide*) released in the blood by the endocrine glands or by some organs like the pancreas, the body that sends out messages affecting cells in other parts of the organism. As messengers, hormones are slower but much longer-lasting than nervous signals. In other words, if the nervous system is the telephone network of the body, hormones provide for ordinary mail. Generally, they are needed only in very small amounts, so that not only the deficiency but also the excess of a specific hormone can have serious consequences. Target cells that must respond to a hormone are provided with specific "mailboxes", which are MEMBRANE PROTEINS called receptors. When a hormone binds to the receptor, it activates a complex mechanism by which the cell reads the mail content, and follow the orders or suggestions it contains. ♡ As hormones, in the text we have mostly mentioned steroids, those complex ORGANIC molecules made from CHOLESTEROL and secreted from the gonads or the adrenal glands, which have a huge impact on sexual behavior, and not only on that. But there are other basic kinds of hormones such as *adrenaline, noradrenaline,* and *dopamine,* which are released by the adrenal glands in response to physical or mental stress. Their main effects are to increase heart rate, blood pressure, and sugar level in the blood for preparing the body for "fight or flight". Many stimulant drugs imitate these fast-acting, thrilling substances. Much quieter, slowly released (but long-lasting) hormones are those that stimulate growth and cell reproduction, which are peptides secreted by the pituitary gland at the base of the skull. The major consequences of *growth hormone* deficiency in children are short stature and delayed sexual maturity. Another fundamental protein-based hormone is *insulin,* made by the pancreas. When the sugar (glucose) level in the blood rises too much, insulin causes cells in the liver and muscles to take it up and store it as glycogen, a large and branched polymer similar to starch. When insulin is absent, glucose is not taken up by body cells and the body begins to use fat as an energy source, slowly "burning" adipose tissues. A well-known consequence of insulin deficiency is diabetes (mellitus).

**Hydrocarbon**: A chemical compound entirely made of hydrogen and carbon atoms. There are many kinds of hydrocarbons. Some of them, of which the simplest is benzene (an important precursor in the production of PLASTICS, ELASTOMERS, and dyes), contain carbon rings, and are therefore AROMATIC compounds. Non-aromatic (*aliphatic*) hydrocarbons often have the shape of linear or branched chains. Linear hydrocarbons chains often constitute the HYDROPHOBIC tails of SURFACTANTS. Aliphatic hydrocarbons may contain only single bonds between the carbon, or some carbons along the chain may be connected by double bonds. The former are known as *alkanes* or more commonly, when the chain is sufficiently long, as *paraffins,* the main components of wax. Double bonds are often responsible for the formation of cross-links between different hydrocarbon chains. Aliphatic and aromatic hydrocarbons are the main components of crude oil.

**Hydrogen bond**: When a hydrogen atom in a molecule is strongly bound to atoms such as oxygen, nitrogen, or fluorine that tend to pull ELECTRONS strongly toward themselves (these are called *electronegative* atoms), it is partly stripped of its sole electron, becoming positively charged, whereas the electronegative atom becomes slightly negative. Then, if the hydrogen sees *another* electronegative atom nearby, belonging to a different molecule or to a different part of the same molecule, it is attracted toward it, forming a hydrogen bond (or *H-bond*). These bonds are much weaker than those that keep a molecule together, but they are still stronger than the THERMAL ENERGY (2–10 $kT$). H-bonds shared by water molecules form an extensive network that gives liquid water its very peculiar properties, and also play a crucial role in determining the three-dimensional structure of PROTEINS and NUCLEIC ACIDS. ♡ POLAR molecules display a positive and a negative region in their structure, too, so they also attract each other via forces that are called "dipole interactions", but the energy associated with these forces is only 1–4 $kT$. What is so special about hydrogen that makes H-bonds so strong compared with dipole forces? They cannot just be due to charge attraction between a "stripped" hydrogen and an "overdressed" oxygen or nitrogen atom. This is a rather hard question in quantum physics (or chemistry), but I'll try to give you at least a hand-waving and rather inaccurate explanation. It is a matter of size. Adding or subtracting an electron to any kind of atom, except hydrogen, does not much change its size, which is always around a tenth of a nanometer. But hydrogen has only *one* electron, so, if you strip it away, you are left just with a single PROTON (the nucleus of the hydrogen atom), which is 150,000 times smaller than the original atom! These tiny objects can rapidly hop back and forth between the two electronegative atoms they connect; they are in fact *shared* among them, and this gives the H-bond its strength.

**Hydrophilic**: A substance, material, or chemical group that loves water. Typical hydrophilic substances are salts, which dissociate into IONS, or stuff like sugar, which forms hydrogen bonds with water. An example of hydrophilic material is (clean) glass, whereas plastics are generally very HYDROPHOBIC.

**Hydrophobic**: A substance, material, or chemical group that does not like water. To be more precise, the water molecules like each other more than they like those of the hydrophobic stuff, which they try therefore to expel. Hydrophobic substances are therefore very poorly soluble in water, whereas they often dissolve in oils.

**I**mmune system: The rich and varied army that defends a living being against invaders, such as bacteria, viruses, or even "internal rebels" such as tumor cells. In a simple organism like a clam or an ant, this army is mostly made of fixed, ready-to-use defense systems, but big animals like the verte-brates have had to develop better strategies to cope with the ability of germs and other enemies to evolve rapidly, learn to cheat the immune system, and successfully infect their host. This "adaptive" immune system is composed of highly specialized cells, which mature in a complex way that enables them to recognize "intruders" by detecting specific ANTIGENS by means of ANTI-BODIES, and to remember and attack them each time they pop out in the future. The most important cells in the adaptive immune system are called *lymphocytes*, and in particular B and T cells. Each one of us has trillions of lymphocytes (taken together, they can weigh as much as our brain), which make up a large fraction of our white blood cells, but also circulate in the *lymphatic system*, an alternative networks of vessels where a clear fluid (the *lymph*, from the Latin *lympha*, "water") flows, which is mostly reserved for their traffic.

**Interference**: The physical effect by which two waves, when they overlap, reinforce or weaken each other. Take for instance two waves with the same strength, or *intensity*: *constructive* interference can yield a fourfold stronger wave, whereas *destructive* interference results in a total annihilation of the combined wave. Interference takes place only when the two waves have the same or a very similar WAVELENGTH, which also fixes whether they interfere constructively or destructively in a given place. So, in the case of the reflection from a surface of white light, which is the sum of light waves with many colors (corresponding to many wavelength), it may happen that a given color gets reinforced, whereas another one is canceled. This is the origin of the beautiful color of soap bubbles, and of butterfly wings too. ♡ While finishing this book, I discovered that a super-clever application of interference may rapidly put an end to colloids as e-paper makers. Qualcomm, a US communication company, has developed a totally new system, based on what are called "interfero-metric modulators" (IMOD), which are micro-electromechanical systems (or "MEMS") made of two electrically conductive plates. One is a thin film stack on a glass substrate, the other is a membrane that reflects light, suspended over the substrate. Between the two, there is an air gap. If no voltage is applied, the plates are separated, and the light hitting the substrate is reflected back. When a small voltage is applied, the plates are pulled together by elec-trostatic attraction and the light is absorbed, turning the element black. Like the systems we have discussed, this is the kind of "bistable" device needed to make e-paper with negligible power consumption. Moreover, being based on interference, it readily lets colors be obtained by varying the thickness of the gap between the plates. And it is fast, much faster than "traditional" e-paper (a tradition that dates back no more than a few years, actually). Be prepared: the stuff you will be reading on in a few years will be totally different even from current brand-new e-readers !

**Intron**: A DNA region that is not coding for a protein. They are a unique feature of eukaryote organisms, although their abundance is very variable among different species. In the human species they form up to 26% of the entire genome. These non-coding part of genes have to be removed by a process called *splicing* from messenger RNA to make it operational.

**Ion**: An atom or a molecule that is positively or negatively charged (see CHARGE) because it has lost or, respectively, gained one or more electrons. A positive ion is called a *cation*, whereas a negative one is an *anion*. Many substances, and in particular salts, dissociate in ions when dissolved in water. So, for instance, NaCl (sodium chloride) becomes $Na^+ + Cl^-$.

**J**unk **DNA**: A large part of DNA (in humans more than two-thirds of the total) that is neither coding for proteins, nor a non-coding part of a gene (an INTRON). We still don't know why this huge amount of non-genic DNA is there; in fact, it is dubbed the "dark matter" of the genome, in analogy with the huge amount of mysterious stuff that seems to fill most of the Universe (there, the situation is even worse). Part of it seems to be a kind of "parasite"

DNA, coming from long-ago viral infections, and a at least 10% is made of "transposons" that do nothing but replicate and fit their own copies into other regions of the genome, but this accounts only for a moderate fraction of this wasted DNA. That we don't know what junk DNA is there for, however, does not make it meaningless. On the contrary, experience teaches scientists that what is more challenging to understand is eventually more thrilling.

**K**eratin: See EXTRACELLULAR MATRIX (ECM).

**Kinesin (and dynein)**: Protein marathon runners of the MICROTUBULES. Rather stubborn marathoners, to tell the truth, for kinesins insist on running only toward the cell membrane, whereas dyneins move the other way around, for each of them thinks of a microtubule as a curious one-way highway (the good thing is that they never bump into each other).

**Kissing number**: The number of particles that, in a given configuration, touch a selected one. For spheres, the maximum kissing number is 12, which is obtained for CRYSTAL structures such as FCC or HCP, whereas the minimal kissing number to obtain a stable structure is 6.

**Krafft temperature**: The minimum TEMPERATURE at which a surfactant forms micelles. Below the Krafft temperature, the surfactant precipitates from the solution as a CRYSTAL.

**Kuhn segment**: See PERSISTENCE LENGTH.

**L**aplace pressure: The extra PRESSURE within a air bubble in water, or a water drop in air, required to balance SURFACE TENSION, which tends to shrink the bubble or drop. So, for instance, in a rain droplet the pressure is slightly larger than in the surrounding atmosphere. This extra pressure is twice the surface tension divided by droplet (or bubble) radius. In the case of a soap bubble, where there are actually two surfaces separating a water film from air, the Laplace pressure is twice as large.

**Latex**: A suspension of COLLOIDAL PARTICLES made of a polymeric material. A well-known example is the latex of the caoutchouc tree, from which natural rubber is made.

**Lipid**: A word used by biologists to indicate HYDROPHOBIC compounds like fats (TRIGLYCERIDES), but also, less properly, AMPHIPHILIC molecules such as PHOSPHOLIPIDS.

**Lipid double-layer**: The double layer of amphiphilic molecules constituting the thin "skin" of a biological membrane or of a LIPOSOME. The HYDROPHILIC heads of the two layers are exposed to the aqueous medium, whereas the tails, facing each other, make up the HYDROPHOBIC interior of the double layer. A lipid double-layer forms a barrier to the passage of water or IONS.

**Liquid crystal (LC):** A material in-between a liquid and a solid, originating from the fact that the molecules or the particles composing it are either geometrically anisotropic (for instance, they may have a rod-like shape) or interact via forces that strongly depend on the direction (for instance, they are much stronger if the molecules have a specific relative orientation). There are different kinds of liquid crystals, ranging from *nematics*, where the molecules are aligned but positioned as randomly as in a liquid, to *smectics*, which have a layering order and are mechanically similar to wax, to *cholesteric*, which display a complex helical arrangement along ordered layers and are extensively used because of their peculiar POLARIZATION properties. ♡ Simple LCs are pure substances which change their structure on varying the temperature $T$. For instance, on lowering $T$, a liquid made of molecule with directional interactions may become a nematic LC, or even a smectic LC at still lower $T$. Because their structure (or, as we shall say, their *phase*) can be tuned with temperature, simple LCs are said to be *thermotropic*. We mentioned, however, that POLYMER or SURFACTANT solutions can form LC structures too. In this case, what regulates the formation of LCs phases is not only $T$, but also the *concentration* of the polymer chains or of the surfactant molecules. Hence, they are called *lyotropic*, because their state depends on where they "lie" in a graph where the axes are temperature and concentration (called the phase diagram). Tobacco Mosaic Virus is an example of a biological material forming lyotropic liquid crystals at rather low concentration (because the virus has a very elongated shape).

**Living polymers:** Very long, spaghetti-like MICELLES formed by certain SURFACTANTS, in particular when they bear a positive CHARGE. They look like polymers, but because they are spontaneous, weakly bound aggregates, they can break and re-form very easily. This is why they are said to be "living".

**Liposome:** In the jargon of biologists, a VESICLE made of biological lipids. Liposomes are used in the pharmaceutical industry to encapsulate drugs that must flow undisturbed in blood vessels to be eventually released at the target tissue.

**Lysosomes:** Sort of "micro-stomachs" contained in most animal cells. Lysosomes are globular organelles, with a size that can exceed one micron, containing enzymes which, working in a rather acidic environment, "digest" all the stuff (including VIRUSES and BACTERIA) that have been taken into the cell by endocytosis, and sometime even bulky waste from the cell itself.

**M**acrophages: White blood cells whose main role is to phagocytize (engulf and then digest in LYSOSOMES) pathogens such as BACTERIA or dead cell material. A macrophage can eat up tens of bacteria, before it kicks the bucket (not through indigestion, but because it digests itself).

**Macromolecule**: In general, a molecule with a very large MOLECULAR WEIGHT. Very often used as a synonym for POLYMER.

**Malleability**: See DUCTILITY.

**Membranes**: In biology, the structures that bound the cell (PLASMA MEM-BRANE), separate the NUCLEUS and the internal organelles from the rest of the CYTOPLASM, or serve as docking surfaces in the ENDOPLASMIC RETICULUM and in the GOLGI APPARATUS. The framework of a biological membrane is a LIPID DOUBLE-LAYER, where PROTEINS and other biological molecules are usually trapped or attached.

**Metabolism**: The set of chemical processes that must take place in an organism to maintain life, allow growth, respond to environment, and ensure reproduction. Metabolism can be subdivided into a first series of processes (called *catabolism*) aimed to obtain energy, for instance by PHOTOSYNTHESIS or from food through cell RESPIRATION, followed by those processes required to build up basic constituents of like PROTEINS and NUCLEIC ACIDS (*anabolism*).

**Micelles**: The spontaneous aggregates formed by a SURFACTANT in water. In a micelle, the surfactant HYDROPHILIC heads face the water, which they like, shielding the HYDROPHOBIC tails from contact with the solvent. Micelles can have different structures and sizes, mostly depending on the shape of the surfactant molecules that make them. In particular, it is useful to define a *geometric parameter* $g = v/(A\ell)$, given by the ratio between the volume $v$ of the molecule and the product of the area $A$ taken by a surfactant head times the length $\ell$ of the tail. Spherical micelles, for instance, can form only when $g$ is smaller than $1/3$, whereas for larger vales of $g$ cylindrical micelles or even VESICLES form. When $g > 1$, the surfactant is generally insoluble in water, and cannot therefore form aggregates. However, the surfactant is soluble in oil, forming what are called *inverted micelles*. Here the tails face the solvent, whereas the heads form the core of the aggregate. The fact that surfactant form micelles is crucial for detergency. Fats and oily substances, which are not soluble in water because they are hydrophobic, can dissolve into the core of a micelle, and be carried away from skin or fabrics.

**Microemulsion**: Like simple EMULSIONS, microemulsions are colloidal suspensions made of droplets of oil in water, or water in oil, stabilized by a SURFACTANT. In contrast to simple EMULSIONS, however, microemulsions do not require strong stirring and mixing, but form spontaneously and remain stable virtually forever. They are also very transparent, because the droplets are tiny, with a typical size of some tens of nanometers, so they scatter little light (see SCATTERING). The secret of making a stable oil-in-water microemulsion is to swell the surfactant MICELLES with oil, generally adding another component, such as a long-chain alcohol, that helps the micelle to increase its size without breaking. Because of their stability, microemulsions are replacing traditional emulsions not only in many cosmetic applications, but also in hi-tech areas, such as the formulation of anti-rejection drugs used after transplants, or for producing diesel fuels containing a little water, which noticeably reduces the emission of pollutants.

**Microfluidics**: A novel set of lab techniques based on controlling and manipulating fluids in very small hydraulic circuits, made of channels that are typically tens of micrometers wide. These methods enable tiny pumps, valves, and pipes to be designed that operate on liquid volumes that can be as small as a few picoliters (a millionth of a millionth of a liter). In many senses, the print-head of a common inkjet printer is already a microfluidic device, but the most promising applications of microfluidics are in biotechnology and medicine, for they allow proper biolabs the size of a microscope slide to be made, in which fast testing can be made on tiny sample amounts.

**Micron (micrometer)**: A thousandth of a millimeter or, if you prefer, $10^{-6}$ meters.

**Microtubule**: Rigid hollow tubes, with a diameter of about 25 nm and up to tens of microns long, made of polymers of the PROTEIN *tubulin*, which are the bearing walls of the CYTOSKELETON, and also make it possible to build complex structures such as the *cilia* that allow egg cells to move from the ovaries to the womb. Besides this, microtubules also provide the "highways" on which lorry-like proteins such as KINESIN move. Tubulin is actually made of two sub-units, called alpha and beta tubulin, bound together. These "dimers" then join together, winding up into a helical structure that forms the tube wall. The structure of tubulin gives a microtubule a peculiar "asymmetry", called *polarity*. In fact, only one kind of tubulin is exposed at each of the two ends of the microtubule, and the polymerization process takes place only from the end made of tubulin beta. Microtubules are very dynamic structures, with an average lifetime of just a few minutes. Actually, a microtubule cannot stay how it is: either it grows, or it collapses (this is called *dynamic instability*).

**Mitochondrion**: The "power plant" of an eukaryote cell, which uses oxygen to produce ATP, the main energy currency in living beings, from food by the process known as RESPIRATION. Mitochondria (the correct plural of this Greek word) have a size of a few MICRONS, and are thought to have been in the far past free organisms, similar to bacteria, that the primeval eucaryotes have taken in and "tamed". This is suggested also by the presence of a mito-chondrial DNA, totally alien to the DNA of the host (and coming only from the mother).

**Molecular weight**: Very roughly, the mass of a chemical substance measured in units of the mass of an atom of the simplest element, hydrogen. ♡ Chemists, who are much more precise than physicists, more accurately define it in terms of the mass of the most abundant kind ("isotope") of carbon, called "carbon-12", but for our purposes this makes little difference. What is more important is that the molecular weight, being the ratio of two masses, is a pure number, and does not need any new unit to be stated (although you may sometimes find it given in "daltons", which makes physicists turn up their nose).

**Monomer**: The basic building block, or "repeating unit" of a POLYMER. In other words, monomers are small molecules that join together to form those long chains which are called polymers. When they join, two monomers make a *dimer*, three a *trimer*, some an *oligomer*, and so on, until there are so many that we just call the resulting molecule a polymer. So, for instance, by opening the double bond that keeps the two carbon atoms together, ethylene monomers $C_2H_4$ join to form polyethylene, which has the generic chemical formula $CH_3 - (CH_2 - CH_2)_n - CH_3$, where n, the number of monomers, can be as large as tens of thousands. Among natural substances, the true champion of monomers is the sugar *glucose*, which forms many kinds of polymers, and in particular cellulose and starch, constituting about 3/4 of the weight of all vegetables.

**N**anometer: A billionth of a meter ($10^{-9}$ m) or, if you prefer, a thousandth of a micron. Looks a very small size, but is still several times larger than most small molecules.

**Newton's law**: There are actually many laws due to this giant of science, and above all the basic laws that rule classical physics (the physics before the advent of Einstein's relativity and quantum mechanics), but the one we are referring to concerns the motion of simple fluids such as water or air. Consider, for instance, water confined between two plates, which have a surface area $S$ and are a distance $d$ apart. If we try to set it in motion by pushing one of the two plates along the plane with a speed $V$, the force $F$ that we have to apply is given by $F/S = \eta V/d$, where $\eta$ is called the VISCOSITY of the fluid (the larger is $\eta$, the stronger we have to push). Many complex fluids, however, do not satisfy Newton's law, because their viscosity may change depending on how strongly or how fast we push: therefore, they are called non-Newtonian or VISCOELASTIC.

**Nucleic acids**: Commonly regarded as the most important molecules in biology (although proteins may disagree), namely RNA, or *ribonucleic acid*, and DNA, or *deoxy*ribonucleic acid. Don't be misled by the word "acid": These are huge POLYMERS (or rather, POLYELECTROLYTES), performing incredibly complex tasks that simple acids, like those present in lemon juice or vinegar, could never dream of doing. DNA, the king of the biological jungle, is responsible for "keeping the memory" of life. True, it does not make anything in particular, for it is RNA, its faithful servant, that puts DNA's will into action, by translating the genetic code into a shape (its shape) suitable to be used. DNA and RNA share a very similar chemical structure. They both have a support "skeleton" made of sugars and phosphate ($PO_4^-$) groups, which keeps together the NUCLEOBASES, whose sequence is the language in which the genetic information is written. The small difference in their chemical composition, and in particular the difference between the ribose sugar of RNA and the deoxyribose (a ribose which has an oxygen atom less), makes a huge difference in terms of their overall structure. Two filaments ("strands") of DNA can join to form a very long and rigid double-helical structure, with the nucleobases kept well shielded inside. It is this rigidity that enables DNA to protect the genetic information, much as a hard disk stores computer data. RNA, conversely, can form only a very poor helix, and in fact, in living organisms, it never does. RNA is then forced to live as a single, very flexible strand, with the further disadvantage that its nucleobases, which are dangling around, stick to each other very easily. This is too inconvenient for a reliable long-term information carrier. Flexibility, however, has its own advantages. Because of it, RNA can attain many shapes, which allow it to perform very different tasks. So messenger RNA, or m-RNA, "reads" DNA and carries the information to the ribosomes, where ribosomal RNA (*r-RNA*) builds proteins by assembling the amino acids carried there by many kinds of different transfer RNAs (*t-RNA*). Both DNA and RNA are nonetheless so complex that they must have been preceded, in the history of life, by much humbler biomolecules capable of performing only simpler tasks. The problem is that these forerunners have been swept away by evolution, leaving no trace – and leaving us with the firm suspicion that it will be extremely hard to understand how life actually began.

**Nucleobases**: See NUCLEOTIDES.

**Nucleoid**: A fuzzy region within BACTERIA cells, without a surrounding MEMBRANE, where the genetic material is packed, generally in the form of a circular piece of DNA. How the nucleoid forms, and what keeps it together, is still unknown, but it seem to originate from segregation effects similar to those brought in by DEPLETION FORCES in simple COLLOIDS.

**Nucleotides**: The "monomers" making the most important biological polymers, the NUCLEIC ACIDS DNA and RNA. A nucleotide is composed of a sugar group (the sugars *ribose* for RNA, and *deoxyribose* for DNA), a phosphate group ($PO_4^-$), which gives the nucleic acids their negative CHARGE, and a *nucleobase*, a flat ORGANIC compound containing nitrogen, which is chemically a BASE. It is the kind of nucleobase that makes the difference between nucleotides. There are two basic kinds of nucleobases, pyrimidines and purines. The former are ring-shaped molecules, where the ring is made of six carbon atoms, whereas purines are bigger, because they contain a second carbon ring. DNA contains two different purines, *adenine* (A) and *guanine* (G), and two pyrimidines, *cytosine* (C) and *thymine* (T) (in RNA, thymine is replaced by a close relative, uracil, U). A purine likes to bind side by side to another purine or to a pyrimidine through hydrogen bonds, and the same is true for a pyrimidine. However, there is only one way to fit properly into the space within the DNA double helix: a purine must bind to a pyrimidine (two pyrimidines together are too small, and two purines too bulky). The only possibilities are A–T and C–G, so these couples are called *complementary bases*.

This one-to-one correspondence between bases is at the root of the GENETIC CODE.

**Nucleus**: Of course, we mean the nucleus *of a cell* (atoms and the Earth have nuclei too, but they are totally different things), or better of an EUKARYOTE cell, for simple PROKARYOTES have no nucleus. In eukaryotes, the nucleus is the cell headquarters, since it is there that DNA is contained in the form of CHROMATIN. Moreover, it is in the nucleus that key biological functions like DNA transcription and replication take place. Since the first of these tasks is a joint venture of DNA and RNA (an advantageous deal for DNA, for it is RNA that actually does the job), the nucleus needs a storage compartment for RNA too, and in particular for that RNA which is later packed in RIBOSOMES (rRNA) to make proteins. This is the *nucleolus* which, unlike the nucleus which is separated from the rest of the cell by two membranes, is a non-membrane-bound structure.

$\text{O}$**CP** (**Ordered Close Packing**): The maximum fraction of the total volume of a large container (or *packing fraction*) that a lot of objects, all of the same shape, can reach if they are arranged in a regular way, as in a CRYSTAL. Usually, the packing fraction for OCP is larger than at *random* close packing (RCP), namely, when the objects are thrown in the container at random. This is true for spheres, where the value for OCP is exactly $\pi/\sqrt{18} \simeq 0.74$ (corresponding to the face-centered cubic crystal structure), whereas it is only about 0.64 at RCP, but also for disks, rods, and ELLIPSOIDS. However, we do not know whether this is true for all shapes. If there were objects that packed better randomly than ordered, they would always prefer to form GLASSES rather than crystals. If you can find one (and prove that it is so), you might easily get, if not a Nobel prize in physics, at least a Field Medal in mathematics (which is much less money, but no less prestige).

**Opal**: A natural COLLOIDAL CRYSTAL. Opals form by slow precipitation of SILICA, which assembles into spherical pellets of well-defined size, followed by water evaporation, taking place over geological timescales. Like in artificial colloidal crystals, the iridescent colors of the most valuable opals are due to Bragg SCATTERING. However, natural opals known as *potch* do not show colors but just a whitish appearance, because they lack the regular structure of a crystal. ♡ What we have discussed is just a general chemical process that leads to opal formation. However, the specific geological conditions leading to opal formation are still much discussed among geologists. In the most widely accepted model, silica solutions are thought to percolate down from their source rock until they are blocked by the barrier of a layer impermeable to water, for instance made of clay. There, a change from BASIC to ACIDIC conditions, and the presence of multivalent salts, leads to silica precipitation. In this model, a source of silica such as sandstone is required for opal to form. An alternative model proposes that opals already formed from mineral-bearing waters subsequently rise along ground faults toward the surface, forced by the underground pressure or by large-scale tectonic processes. Finally, some geologists suggest that BACTERIA may have played a role in the formation of opals. At the time that the Cretaceous sediments were deposited, clay layers provided an ideal habitat for microbes. Waste acids and ENZYMES excreted by them may have caused chemical changes and weathering of clay, which promoted opal deposition.

**Organelle**: In biology, a small subunit, contained in (or sometimes bound to) a EUKARYOTE cell, that has a specific function. Usually, organelles are separated from the CYTOPLASM by a LIPID DOUBLE-LAYER (this is, for instance, the case for MITOCHONDRIA or for the CHLOROPLASTS, which are supposed to have been independent organisms in a far past), but the term "organelle" is generally used also for structures like the RIBOSOMES that are not bound by membranes. ♡ Besides those we have encountered in this book, there are other organelles, performing very specialized tasks. Some of them (like the *peroxisomes*, needed to transform hydrogen peroxide, $H_2O_2$, a poisonous product of the cell metabolism, into harmless water) are present in all eukaryote cells, but many of them are found only in specific organisms or tissues. For instance, the cells of our skin contain *melanosomes*, where melanin, a strongly light-absorbing polymer that determines skin color and protects us against UV rays, is stored (be careful to distinguish this word from *melanoma*, a terrible form of skin cancer occurring when the cells producing melanin go astray – to use a euphemism). Similarly, in sperm cells, the GOLGI APPARATUS turns into a cap-like structure, the *acrosome*, which covers the front halves of the spermatozoa and contains ENZYMES allowing them to fuse with the female egg and fertilize it (failure of this process causes male infertility). Green algae and other single-celled organisms that carry out PHOTOSYNTHESIS even contain an organelle, the *eyespot apparatus*, which allows them to sense the direction and intensity of light (the simplest form of eye, if you like).

**Organic (molecule or compound)**: Any molecule or chemical compound that contains carbon, with the exception of a few very simple molecules like carbon dioxide ($CO_2$), and of pure carbon (coal, graphite, or diamond), which, by a curious custom of chemists, are regarded as *in*organic. Nothing to do with life, therefore, although all living organisms are basically made of organic molecules and water.

**Osmosis**: Take a bag made of a material that is impermeable to salt (or sugar, or PROTEINS, or COLLOIDAL PARTICLES) but lets the solvent pass through freely; fill it with a salt (or sugar, or protein, or colloidal particle) solution; seal it, and put it in a container filled with the solvent. Some solvent molecules will enter the bag from the outside reservoir, and the pressure inside the bag will increase. The extra pressure eventually reached inside the bag is just what we call osmotic pressure. If the salt solution is diluted, this excess pressure $\Pi$ is given by $\Pi = ckT$, where $c$ is the sugar concentration (in molecules per unit volume) and $kT$ the THERMAL ENERGY. This is called the *van 't Hoff law*. The fact that solvent molecules enter the bag means that the salt solution gets less salty: that is, osmosis tend to *dilute* the solution. If we want to get the opposite effect, that is, use *reverse osmosis*, for instance to get fresh water from the sea, we have to spend energy. Osmosis plays a key role in biology. In fact, cells are masters in controlling the flow of water and salt in and out of the CELL MEMBRANE.

**Ostwald ripening**: The mechanism by which an EMULSION progressively coarsens, even if the droplets it is made of hardly COALESCE (for instance, because their density is close to that of the external phase, so that there is little SEDIMENTATION or CREAMING). The reason is that the LAPLACE PRESSURE is slightly higher within the smaller droplets. Therefore, the internal fluid tends to be expelled from them and, passing through the external solvent, to enter the larger ones. As a consequence, the bigger droplets grow at the expense of the smaller ones. Ostwald ripening is particularly significant for emulsions made of droplets with a small average size and a large distribution of droplet radii.

**Overlap concentration c\***: The concentration at which, in a POLYMER solution, the coiled chains start to overlap. It is approximately given by $c^* \simeq N^{-4/5}/a^3$, where $N$ is the number of monomers, each of size $a$, in a polymer chain. The overlap concentration marks the transition from the dilute solution regime, where the solution is made of separated polymer coils, to the semi-dilute regime, where the polymer solution resembles an intricate mesh.

**P**CR **(Polymerase Chain Reaction)**: The method used in molecular biology to amplify a piece of DNA by orders of magnitude, generating thousands to millions of copies of a particular DNA sequence. ♡ The methods consists of several steps, taking place at different TEMPERATURES. In short, DNA is first heated to more than 90°C (about 194°F) to split the double helix and get the two filaments (strands) free. The solution is then cooled just below DNA's *melting temperature* (typically between 50 and 65°C, about 120 to 150°F, at which the DNA fragment unwinds and separates into single strands through the breaking of HYDROGEN BONDS between the bases. This allows specific NUCLEOTIDE sequences that act as a starting point for DNA synthesis, called *primers*, to attach to the strands. Finally, the temperature is again raised to about 80°C (176°F), enabling the ENZYME DNA polymerase to make new DNA strands complementary to the original template. Since our own DNA polymerase would undergo denaturation at these high temperatures, it is substituted by the corresponding enzyme of thermophilic bacteria. All these steps are repeated many times in cycles. It is PCR that makes possible DNA profiling, used in parental testing and for legal evidence, DNA cloning, and countless other techniques in genetic engineering. Because of this,

Kary Mullis, who developed PCR in 1983, was awarded the Nobel Prize in chemistry just ten years after, a very fast track for becoming Nobel laureate.

**Peptide bond**: A strong chemical bond formed between two amino acids when the ACID terminal COOH of one molecule reacts with the BASIC group $NH_3$ of the other one, releasing a molecule of water. This "dehydration" (also known as *condensation*) reaction, which leads to the formation of POLYPEP-TIDES and PROTEINS, takes place only in the absence of water, unless suitable ENZYMES are used and energy is supplied (this happens unceasingly in our RI-BOSOMES). In fact, peptide bonds in proteins are not stable, meaning that in the presence of water they break spontaneously, but this process is extremely slow.

**Persistence length**: POLYMERS are flexible, but only to a limited extent. Generally, a monomer cannot be bent wherever you like with respect to the preceding one. So, over a given length, which is called the persistence length (or *Kuhn segment*), the polymer looks rather straight. The persistence length of polymers can vary a lot: most polymers are very flexible, so that their persistence length is just a few monomers, but some of them (DNA being an extreme case) can be almost as rigid as a rod. In POLYELECTROLYTES, the ELECTROSTATIC FORCES increase the polymer rigidity, so that their persistence length is longer than it would be in the absence of CHARGES.

**Phospholipid**: An extremely important class of double-tailed AMPHIPHILES that are the major component of all CELL MEMBRANES, as they can form LIPID DOUBLE-LAYERS. In the cell membranes there are many kinds of phospholipids, and their different shape may allow the membrane curvature to be controlled. For instance, lecithin (or *phosphatidylcholine*) has a rather bulky HYDROPHILIC head, and is akin to a "cone" with the tip inside the double-layer, whereas cephalin (or *phosphatidylethanolamine*) has a much smaller head, and is therefore shaped like an "inverted cone", with the tip turned toward the cytoplasm or the outside of the cell. By combining these two shapes, any membrane curvature can be generated.

**Photosynthesis**: The method used by plants and algae to extract energy from sunlight. It involves the work of many coordinated PROTEIN complexes, including in particular an "antenna", which catches the light, and a "reaction center", where it is transformed in useful energy. Photosynthesis is a far better mechanism to exploit solar energy than all those we have so far devised. Actually, neither plants nor algae do anything by themselves: the photosynthetic reactions take place within the CHLOROPLASTS.

**Plastic**: A word with two distinct meanings, which should not be confused. As a *noun* (and in the common usage), a material made of an AMORPHOUS or partially crystalline POLYMER that mechanically behaves as a solid, and which can be either THERMOPLASTIC or THERMOSETTING. As an *adjective*, "plastic" is any material which undergoes a permanent change of shape (*plastic deformation*) when strained beyond a certain point. Most metals, although they have nothing to do with polymers, are "plastic" in this sense, whereas some polymer materials are fragile, and break before deforming appreciably, so they have little "plasticity" in the technical sense.

**Plateau laws**: The basic laws governing the organization of soap films, found by the great blind scientist Joseph Plateau, which are consequences of the requirement of minimizing the elastic energy associated with surface tension. According to these laws, when soaps films are supported by a frame, each edge of the frame supports a single film and three films always meet in a line making angles of exactly 120°, whereas four films must always meet in a point, making angles of slightly less of 110°, like in a tetrahedron, the pyramid with equilateral triangular faces.

**Polar**: In chemistry, a molecule or a compound where ELECTRONS are not uniformly distributed, so that there is a region which is slightly negative, and another one which is slightly positive. Unlike an ion, however, a polar substance is altogether neutral (not charged). For example, water is polar because the electrons tend to be concentrated more around the oxygen atom than around the two hydrogens. Polar substances are generally soluble in polar liquids. So, for instance, ethanol (ethyl alcohol), which is polar, dissolves quite well in water, as any drinker knows.

**Polarization**: In optics, how the direction along which the electric and magnetic fields of a light wave vibrate (these fields being what light is actually made of) is oriented with respect to the direction in which the wave is going. For instance, if we set three axes $x, y, z$ along the three directions of space (left/right, up/down, and forward/backward), and the light moves forward, namely along $z$, the vibration can take place along $x$ or $y$, or a combination of the two directions. ♡ Polarization is responsible for many optical phenomena, for instance for the fact that the sky looks darker at the zenith (your vertical) than near the horizon (LIGHT SCATTERING from the sky has rather complex polarization properties). Moreover, while direct sunlight is randomly polarized (that is, it does not vibrate along a preferential direction), the light reflected from a surface *is* polarized. For example, if the surface is horizontal, the reflected light is horizontally polarized, at least partially. This is why you use "polarizing" sunglasses (or polarizing filters on cameras) to reduce reflections. A *polarizer* is a device allowing only one direction of light polarization to pass through. So, using glass polarizers with their "polarization axis" set along the vertical, you can prevent the reflected light reaching your eyes or the camera sensor. Polarizers are fun. If you take two polarizing lenses, put them face-to-face, and turn one of them while you look through this combined device, you will see less and less light, until, when the axes of the two polarizers are crossed, they will not transmit any light. However, what is really curious is what happens when you insert a *third* polarizer in between them: even if the original polarizers are kept crossed as before, you will magically see light passing through again! Polarization is therefore a curious property, which does not apply only to light. Indeed, all "transverse" waves (all waves that vibrate in the direction perpendicular to their propagation, like seismic waves or those on a rope) can be polarized. Sound waves in air or water, conversely, have no polarization, because they consist of pressure peaks and troughs traveling along the direction of wave motion.

**Polyelectrolyte**: A POLYMER in which all or parts of the MONOMERS are charged. PROTEINS, NUCLEIC ACIDS, and many water-soluble polymers such as sodium polystyrene sulfonate, are actually polyelectrolytes. A curious feature of polyelectrolytes is that their real total CHARGE is generally lower than the value one would guess from their chemical composition. Indeed, the COUNTERIONS that the polyelectrolyte release into the solvent re-associate to the chain, partly neutralizing the charge of opposite sign, until the free charges on the chain are at least a BJERRUM LENGTH apart. This effect is called *Manning condensation*. The effective charge of DNA, for instance, is about four times lower than we would expect from the distance between the charged phosphate groups on the double helix. Polyelectrolyes are generally more rigid than what they would be if they were not charged, for the ELECTROSTATIC FORCES between the monomers increase their PERSISTENCE LENGTH.

**Polymer**: A MACROMOLECULE formed by joining many chemical units, all identical or of a few different kinds, called MONOMERS. Polymers are the basic component of PLASTICS and ELASTOMERS. As we have seen, the development of macromolecular chemistry has led to the synthesis of a large variety of different polymers with a wide range of properties. However, natural polymers are abundant too. Besides natural rubber, discussed in Chapter 3, other polymers of biological origin that we have long exploited are gelatin, obtained from animal bones, amber, which is fossilized tree resin used both in perfumes and as jewelry, and shellac, a resin secreted by the lac bug, which is dissolved in alcohol and used for instance as a wood finish. PROTEINS, NUCLEIC ACIDS, and *polysaccharides*) such as starch and glycogen, acting as sugar (glucose) reservoirs, or cellulose and chitin, respectively used as building materials for wood and insect shells, are polymers too. ♡ Most of the polymers we have discussed are linear chains, but there are also polymers, such as glycogen, which are extensively "branched". Here, however, I would like to mention a class of molecules that are also made by connecting many identical units, but hardly resemble polymers at all: the dendrimers, whose name comes from the Greek word δενδρον (dendron), meaning "tree", although they look more like globular bushes or tumbleweed. Dendrimers are synthesized starting from a central unit, which has a few (at least three) functional groups symmetrically arranged around the core where identical chemical groups, called *dendrons*, can be attached. Each of the dendrons, however, has functional groups too, to which a "second generation" of dendrons can bind, and so on and so forth. Typically, up to 9–10 of these "generations" can be added in sequence. For instance, if the core bears three functional groups, each dendron of the first generation binds with one functional group to the growing core, leaving two "berths" for the next generation, so that the molecular weight of the dendrimer approximately doubles at each successive generation. Because of the symmetric arrangement of dendrons, over the generations the structure comes to resemble a kind of snowflake, with a beautiful spherical symmetry around the core and a size that can reach 10–15 nanometers. From the practical point of view, the important point is that the surface of the dendrimers still bears a lot of active functional groups, which can be chemically modified for a specific application. Because of this they have already been used for drug delivery and as microscopic "chemical sensors" of specific molecules.

**Polymer solutions**: When dissolved in small amounts in a liquid they like (a *good solvent*), POLYMER chains, provided they are flexible (i.e., that they do not have a very large PERSISTENCE LENGTH), take on a coiled shape, like a tiny ball of wool. The coil actually resembles a RANDOM WALK path, with the important difference that, while a drunkard performing a random walk can retrace his (or her) steps, a monomer cannot be placed where another monomer already lies (it is then a *self-avoiding* random walk). Because of this, the coil is more "expanded" than for an ideal random walk: whereas the region explored in a random walk grows as the square root of the number of steps, the size of a polymer coil (its GYRATION RADIUS) increases as the number of monomers to the exponent $3/5$, which is also the FRACTAL dimension of the segmented line describing the polymer chain. On increasing the amount of dissolved polymer beyond the so-called OVERLAP CONCENTRATION, the coils begin to intertwine, and the solution gradually becomes a kind of mesh, where the single chains can no longer be distinguished. This state, where the chains form a uninterrupted network, but where regions of higher or lower polymer density can still be recognized, is called a *semidilute* solution. In simple words, the chains look like interwoven spaghetti in a pot of boiling water. The typical length scale that characterizes the network is no longer the radius of a single polymer coil, but rather the mesh size. On increasing the concentration further, the solution becomes more and more uniform (fewer fluctuations in the polymer density), until all the solvent is gone and, provided the temperature is higher than the GLASS transition temperature, we are left with a *polymer melt*, which is like spaghetti after it has been strained. A curious property of a polymer melt is that the shape of the single polymer chains is exactly a simple random walk (since it is entirely surrounded by other chains, what could a monomer "avoid"?), so that chains are said to be "ideal". Both in semidilute and in concentrate solutions, the only way to move around for the polymer chains is to slither like snakes (this is called REPTATION). All we have said holds true if the solvent is a good one. If the solvent properties change, so that the polymer does not like it anymore, the chains start to shrink in distaste, and the coils become more and more compact, until they become solid balls: these are what we call colloidal LATEX particles. These "compaction" effects are also extremely important for DNA.

**Polypeptide**: A POLYMER made of AMINO ACIDS bound by PEPTIDE BONDS. PROTEINS are made of one or more long polypeptide chains.

**Pressure**: In mechanics, the force per unit area exerted on a surface, where, more precisely, we have to consider only that part of the force that is perpendicular to the surface (so it is what is called a normal STRESS). We can also speak of the pressure of a gas, a liquid, or even a solid, referring to the force that these substance exert on the walls of the container or, for a fluid, on a submerged body. Since a force is measured in newtons, pressure is measured in newtons per square meter, or pascals (Pa) (the corresponding unit in the US common system is the psi, pounds per square inch). However, in practical use, pressures are often stated in "atmospheres", with reference to the pressure that the air exerts at the sea level and normal temperature, which is about 100,000 Pa and corresponds to bearing a weight of about one kilogram on each square centimeter of your body. ♡ Because of the weight of the overhanging fluid, the pressure in a liquid increases with depth. On a body lying at depth $h$ below the surface of a liquid of density $d$, the pressure increases by $dgh$, where $g = 9.8$ m/s$^2$ is the acceleration due to gravity, with respect to the atmospheric pressure. So, for instance, the pressure you feel underwater increases by about one atmosphere every ten meters of depth. This means that, in the deepest parts of the ocean, the pressure exceeds 1000 atmospheres. This value is well above the critical pressure of water (see AEROGEL), so that, close to the "hydrothermal vents", which are very hot springs on the ocean bottom, water can be above its critical point. Such "supercritical" water is very different from ordinary water, being able for instance to dissolve even most HYDROPHOBIC substances. For EXTREMOPHILES living close to hydrothermal vents, life may not be easy!

**Prokaryote**: A simple organism that lacks a cell NUCLEUS to store DNA. There are two "domains" (or "superkingdoms", for this is really the highest rank subdivision of living organisms) of prokaryotes: BACTERIA and ARCHEA. Most prokaryotes are single-cell organisms (although they often live in very organized colonies, see BACTERIA), but a few of them such as *myxobacteria*, a kind of bacteria that are mostly found in the soil, have multicellular stages during their life cycle, in particular when food is scarce (maybe because of their more complex behavior, their GENOME is much bigger than for most bacteria).

**Protein (structure)**: After all, proteins are POLYMERS, but they are much more sophisticated than their humble siblings such as polythene. To take their shape, they have to solve two basic problems. First, some AMINO ACIDS are HYDROPHILIC and others HYDROPHOBIC. Since these two kinds of MONOMERS are more or less randomly placed along the chain, the protein cannot just arrange itself with the water-loving groups outside, screening a fully hydrophobic core from contact with the solvent, as SURFACTANTS do. Second, all amino acids like to make HYDROGEN BONDS, much more between themselves than with water. Thus, proteins face a difficult dilemma, which can be solved only through a complex origami strategy. Two typical ways of satisfying the amino acid hunger for hydrogen bonding are to form screw-like assemblies, called *alpha helices*, or planar structures, dubbed *beta sheets*, where the amino acids are hydrogen-bonded on a flat surface. Helices and sheets have then to arrange, together with some filaments made by those residual amino acids that could not be fitted in these structures, so as to minimize the contact of the hydrophobic regions with water. This has two consequences. First, a protein is much more compact than a polymer coil, more resembling a solid COLLOIDAL PARTICLE (but with more complicated surface properties, because the same protein can be either positively or negatively charged, depending on the acidity of the solvent). Second, and more important, this challenging *folding* process leads to one stable structure, or at most a few structures. In other words, a given amino acid sequence usually leads to a unique folded protein, which is called its *tertiary structure* (alpha helices and beta sheets are elements of the *secondary* structure, whereas the *primary* structure is the amino acid sequence). This is what gives a protein the specificity needed to perform its task, but is not the end of the story, because many proteins, to work properly, have to organize into complexes formed by several chains (which is the *quaternary* structure). What is almost incredible is that all this folding process is extremely fast, far faster than it would be if the chain attempted to fold randomly. Although we know that there are "assistant" proteins, called *chaperonins*, which help in giving birth to a protein, many aspects of this fundamental biological process are still poorly known.

**Protein (basic kinds of)**: Proteins perform all basic tasks required for life, and therefore are highly specialized. They can, however, be grouped into a few large families. First, there are the ENZYMES that, acting as catalysts, control all the biochemical reactions taking place in a living being. Then there are *transport* proteins that carry through the body specific substances (such as oxygen for hemoglobin), messages (like HORMONES, which give warning of stressing or exciting conditions), or various stuff (such as albumin, which is not too fussy about what it carries). Sort of transport proteins are also many *membrane proteins*, which control what gets in and out of the cell (first of all water, whose flux is regulated by aquaporins), although they do their job without moving, because they are trapped into the cell membrane with their hydrophobic "belt". But membrane proteins are burdened by many other tasks, since they are the interface by which the cell communicates and interacts with the external world. So, large membrane protein complexes allow the CHLOROPLASTS to perform PHOTOSYNTHESIS, or MITOCHONDRIA to produce ATP. Membrane proteins also enable *motor proteins* to hook to the membrane. Examples of the latter are actin, a protein organized into filaments acting as pulls and stay wires, and tubulin, forming long and tough cables, the MICROTUBULES, used by other motor proteins such as KINESIN to travel around the cell. Some proteins, such as collagen, elastin, and keratin, play the humbler role of building materials for the EXTRACELLULAR MATRIX, and are therefore dubbed *structural proteins*, whereas a much more delicate role is played by those proteins, like DNA and RNA polymerases, which assist the nucleic acid in replication or transfer of the genetic information.

**Proton**: See CHARGES.

**Purine**: See NUCLEOTIDES.

**Pyrimidine**: See NUCLEOTIDES.

**Q**uark: A mysterious object that can never be found in isolation, but always in the company of one or two of its siblings to form atomic particles (which are not, then, so "elementary" as we once thought) such as mesons, short-lived particles found in cosmic rays (they last much less than a millionth of a second), or PROTONS, the constituents of the atomic nucleus that give it its positive CHARGE, which are much more stable (maybe not immortal, but certainly living more than a billion billion billion years). To understand this book, you could not care less about quarks (so, instead of a ♡, I should really have put a ♠ in front of this item). But I couldn't find any relevant words starting with "Q", except maybe...

**Quat**: A short name for "quaternary ammonium cation", the positively charged head group of some SURFACTANTS used in hair conditioners and bactericides.

**R**andom walk: The mathematical description of BROWNIAN MOTION, and of many other physical processes (including the way an inveterate drunkard moves about). It is useful also to investigate the shape of a polymer chain in solution, although here the description is much harder, for two monomers cannot be placed in the same place. In other words, it is the motion of a drunk who cannot retrace exactly his or her own steps (a *self-avoiding* random walk). ♡ Here I cannot resists telling you one of the most curious properties of a random walk. Suppose that you are not a drunkard, but a prisoner who manages to escape from jail. However, you do not have the slightest idea of what direction to go, so you actually *move* randomly around like a drunk (to make things worse, suppose it is also pitch dark). Provided that you choose a direction randomly at each step (for instance, by tossing a coin), one can prove that, sooner or later, you will return *exactly* to where the jail is. Given all I have told you about random walks, you may not be too astonished. The real surprise comes when you imagine being imprisoned in a "space" jail, so that, trying to escape, you can move not only back and forth, left and right, but also *up and down*. A rigorous calculation then shows that there is a probability of about 1/3 that you escape from the jail *forever* (this is a very difficult problem that was fully solved only in the fifties of the last century).

The quest for freedom can be then very different in space rather than on a flat surface!

**RCP (Random Close Packing)**: See OCP (ORDERED CLOSE PACKING).

**Refraction**: In optics, the *index of refraction* (or refractive index) $n$ of a transparent material is related to the velocity of light $v$ in that medium by $v = c/n$, where $c \simeq 300,000$ km/s is the light speed in vacuum. Hence, the higher $n$, the slower the light travels. The index of refraction of air is practically 1 (the value for vacuum), whereas in water is about 4/3 and in a solid like diamond can be as high as 2.4. When a light beam crosses the surface that separates two materials with different refractive index it is refracted, meaning it is deviated from the original direction of propagation, unless it strikes the surface perpendicularly.

**Reptation**: The snake-like motion performed by a polymer chain in a melt or in a concentrated solution to slip through the other chains.

**Respiration**: From the point of view of the cell, respiration is something very different from "breathing". Basically, it is a very efficient way to convert the chemical energy contained in nutrients into ATP by exploiting oxygen. Respiration is therefore the complex of biochemical reactions involved in what is called "aerobic" metabolism, the favorite dance of MITOCHONDRIA, the cell organelles where respiration processes occurs.

**Rheology**: The science that studies how matter flows. This is often a rich and fascinating topic for soft matter and complex fluids, which often show a behavior intermediate between liquids and solids.

**Ribosome**: Ribosomes are the cell factories where PROTEINS are assembled from AMINO ACIDS. Ribosomes, which are big protein/RNA complexes, translate the genetic information encoded in messenger RNA (m-RNA) into proteins. They do this by "reading" the sequence of nucleotides in mRNA and using it as a template for the correct sequence of amino acids of a particular protein. The amino acids are attached to transfer RNA (tRNA) molecules, which enter the ribosome and bind to the messenger RNA sequence. The attached amino acids are then joined and released as a POLYPEPTIDE that rapidly folds into a protein.

**Ribozyme**: A ribozyme (from "ribonucleic acid enzyme") is a piece of RNA that wants to play the ENZYME, because it acts as a CATALYST. At the same time, it is still a NUCLEIC ACID with hereditary information written inside. The double role played by ribozymes has driven some biologists to suggested that, in the distant past, cells may have used RNA alone, both as genetic material and as a substitute for proteins. At the present stage of our knowledge, however, this "RNA world" hypothesis looks rather bold and probably questionable.

**RNA**: See NUCLEIC ACIDS.

**S**aponification: The chemical reaction transforming animal fats (TRI-GLYCERIDES) or vegetable oils (in particular olive oil) into soap by making them react with a BASE such as lye.

**Scattering**: The process by which light, or radiation in general, is diffused around by the interaction with molecules or suspended particles. Although molecules do scatter a little, the effect is much stronger for large particles (in fact, the scattered intensity grows with the sixth power of the particle size), so that it is much more pronounced in COLLOIDAL SUSPENSIONS or in AEROSOLS (fog is nothing but scattering from airborne water droplets, and clouds are opaque because light is scattered many times before getting out of them). Light scattering depends on the color of the incident light too, and is stronger the shorter the light WAVELENGTH. The color of sky, for instance, is due to air molecules that scatter blue preferentially. The scattered radiation may therefore be colored even if the incident light is white, an effect that becomes fascinating when scattering takes the form of DIFFRACTION by COLLOIDAL CRYSTALS. But these colors originate from a process that is totally different from the *absorption* of light, which mostly accounts for the color of pigments. Nonetheless, owing to the size-dependence of scattering, the color of a dye solution may depends on how big the dispersed particles are. ♡ Physicists have discovered that, looking at how scattering *changes* with time, they can extract much more information than from the intensity of the scattered light alone. To understand why, note first that the light reaching a detector looking at the sample is the sum of many scattering contributions, due to all the particles illuminated by the incident light. To make things easier, suppose there are just two scattering particles. Since light is a wave phenomenon, their contributions will produce INTERFERENCE at the detector position. At a fixed moment, they may interfere constructively or destructively, it doesn't matter. What really matter is that, if the particles *move* relative to each other, after some time their interference pattern *changes*. For instance, if at time zero they produced fully constructive interference (a maximum of intensity), one can show that when they move apart by a distance comparable to the light WAVELENGTH, the two scattering contributions *cancel* each other. This simplified example can be rigorously generalized to the scattering by a large ensemble of particles. So, by carefully analyzing how the intensity measured by the detector changes with time, one can detect the particle motion and in particular calculate their DIFFUSION coefficient. This observation lies at the roots of *Dynamic Light Scattering*, one of the most powerful methods in soft matter science.

**Sedimentation**: The progressive settling of suspended particles due to their own weight. The settling speed grows with the particle size $R$ (more precisely, as $R^2$) and with the difference in density between the particle and the solvent. In fact, when particles and solvent have exactly the same density, sedimentation is suppressed. This is often a sought-after condition both in basic science, for sedimentation significantly affects COLLOID aggregation or the formation of colloidal GELS, and in practical applications, because EMULSION coalescence is enhanced by sedimentation (or creaming). Unfortunately, for large particles, it is a difficult condition to meet exactly. Biologists, in contrast, try to *enhance* the sedimentation of small (and almost density-matched) particles like PROTEINS, for instance by using a centrifuge, because from the settling speed they can obtain the size of these MACROMOLECULES. Settling processes are extremely important in geology. Particle sedimentation within water basins, or the deposit of materials carried by rivers and streams in their flow, leads to the formation of rocks such as limestone or sandstone that often embed natural resources like coal, fossil fuels, or mineral ores. ♡ Some living beings, like those drifting organisms capable of very little independent movement which constitute a large fraction of plankton (the basic food of fish larvae, but also of many whales), have to *fight* against sedimentation, as gravity settling drives them away from the water surface, where they get their power supply from sunlight.

**Silica**: The main component of common glass, but also the primary raw material of porcelain, many ceramics, and Portland cement. It should neither be confused with *silicon*, which is the semiconductor element on which most electronics chips are based, nor with *silicone*, which is a polymer. Chemically, it is silicon dioxide, $SiO_2$. Because it likes water, silica is often used as an additive in the production of foods, for instance to absorb water from hygroscopic products.

**Solvation**: The process of attraction and association of molecules of a solvent with molecules or ions of a solute. As ions dissolve in a solvent they spread out and become surrounded by solvent molecules, which leads to stabilization of the solute species in the solution.

**Stabilization (of colloids)**: The methods by which lyophobic colloidal particles (see COLLOID) are prevented from aggregating because of DISPERSION FORCES. If dispersed in water, particles may be stable, even if the material they are made of dislikes the solvent, because they have a surface CHARGE, which induces the formation of a COUNTERION cloud preventing the particles approaching too closely. The addition of a lot of salt, however, spoils the party, and the particles aggregate and precipitate. In oils, where this does not work (ions cannot be free in a non-polar solvent), stabilization can be obtained by attaching to the particle surface short POLYMER "hairs" or by adsorbing SURFACTANTS, which again keep two particles far enough apart. A useful trick to improve colloid stabilization is to use particles and solvent with similar refractive indices (see REFRACTION), because this reduces the strength of attractive dispersion forces.

**Strain**: The fractional amount (or the percentage) of deformation of a solid body due to a stress force, both when it consists of a stretch or a compression (normal strain), or of a tilt that makes plane solid layers slide over each other (shear strain).

**Stress**: The amount of force exerted on a body per unit area, which is also (because of the law of reciprocal actions) the resistance the material offers to being deformed. *Compressive* and *tensile* stresses, which respectively tend to compact or to stretch the body, are applied perpendicularly to the surface of the material. However, the force can also be applied tangentially to the surface, in which case we speak of a *shear* stress. In solids, the deformation (the STRAIN) of a solid is proportional to the applied stress if the latter is small enough, and we say that a solid behaves elastically. However, beyond a limiting value, which is called the *yield stress*, a solid begins to deform plastically; that is, irreversibly. Fluids also resist compression elastically, but flow freely in the presence of shear stresses. In this case, for simple liquids, the shear stress is proportional to the *strain rate*, i.e. to *how fast* the shear deformation increases with time.

**Surface tension**: The force that make the surface of a liquid elastic like a rubber balloon, so it resists being deformed from a flat shape. It is given by the force per unit length required to "tear off" the surface itself. An equivalent and probably more useful concept of surface tension comes from looking at it as the energy per unit area required to create the surface. Besides telling us explicitly that building a surface has a *cost* (due to the different state of a molecule on the surface, which is unable to satisfy its attraction toward the others as much as its companions in the bulk), it yields a better idea of the surface tension of *solids*, and also allows us to define the *interfacial tension* between two different materials as the energy per unit area of their contact surface. ♡ Many effects, including those related to how much a liquid "wets" a solid, are due to surface tension. Here, however, I'd like to discuss an important phenomenon without which tall trees could not exist, and where surface tension is probably *not* the main driving force. To live, trees have to suck water from their roots and carry it to the top. But some of you may know that, if you try to make water rise in a tube immersed in water just by sucking up, the latter never rises by more than about ten meters. Why? Try to understand it from the fact that a 10 m high column of water corresponds to the PRESSURE of one atmosphere. In fact, to get it to the top of very tall buildings, we need water pumps. How is then that trees are so tall, although they have no pressure pumps at their disposal? Many textbooks claim that this can happen because of surface tension, and in particular of an effect called *capillary action*, but things are probably not so simple. To see what capillarity is, just do a simple experiment: dip into a glass of water a thin glass tube, with an internal diameter of, say, 1 cm, and look. You will see that water rises *spontaneously* in the tube up to a modest but appreciable height of about 3 mm. Had we used a tube with a diameter of one *milli*meter, the water rise would have been tenfold larger. This effect is strongly related to what we have called LAPLACE PRESSURE. Indeed, if you look carefully, you will notice that the surface of water in the tube (the so-called "meniscus") is not flat, but rather resembles a hemisphere curved upwards. Because of that, water pressure just beneath the meniscus is slightly *lower* than the atmospheric pressure. Water in the glass tube then has to rise until the pressure at the base of the column, which is increased by the weight of the overhanging fluid, becomes level with the atmospheric pressure, which is the pressure at the top of the surrounding water in the glass (at the same level, still liquids must always have the same pressure). The vessels carrying water in tree trunks (forming the so-called *xylem*) are much thinner than the fine tubes we have considered. Yet they are still too large. In conifers, xylem vessels have a diameter of about 10 $\mu$m, which would give a capillary rise of about 3 meters: not bad, but still far too small. Capillarity, however, although not sufficient to account for water rise in trees (unless they do it through much thinner interstices), is the reason why paper towels absorb liquids, or sponges take up water in their small pores. Anyway, how can redwoods survive and flourish? A possible candidate as the real driving force is OSMOTIC PRESSURE, which might suck water up the trunk for the same reason it swells eggs soaked in pure water. An approximate calculation actually shows that the osmotic stress could make water rise up the trunk by more than *one hundred* meters, a figure comparing rather well with the height of the tallest trees. But many aspects of this interpretation are still controversial. As far as I know, the best solution is to ask the redwoods.

**Surfactant**: A simple amphiphilic molecule, or AMPHIPHILE, with just a single HYDROPHOBIC tail attached to a HYDROPHILIC head. Soaps, derived from fatty acids, are the simplest example of surfactant molecules. The word "surfactant" comes from the fact that, when dissolved even in tiny amounts, they strongly reduce the SURFACE TENSION of water, at least until their concentration reaches the critical micellar concentration. Surfactant can be anionic or cationic, if the head group bears, respectively, a negative or a positive CHARGE, or non-ionic, when the head group is uncharged but still hydrophilic, generally because it forms HYDROGEN BONDS with water. A more complex example is polymeric surfactants, where both the hydrophobic are short POLYMER chains. If you want to know more about surfactants, there is a full chapter (Chapter 4) waiting for you...

**T**acticity: The way the side groups are disposed with respect to a POLYMER chain. So polymers can be *atactic*, if the side groups are placed at random, *isotactic* if they are placed all on the same side of the chain, or *syndiotactic* if they alternate on one or the other side. Because atactic polymers generally have very poor mechanical properties, chemists have devised very clever tactics to control tacticity, in particular by exploiting specific CATALYSTS in the polymerization process.

**Telomere**: The terminal part of a CHROMOSOME that is not copied completely during replication, and therefore progressively wears out. This is not

because of negligence by the cell: complete duplication of the DNA constituting a chromosome is strictly impossible. Telomeres, which just contain totally
meaningless DNA, are then the sacrificial victims of cell replication. They are
also a sort of ID card of the cells, telling how many times they have replicated
– and, unfortunately, since they cannot breed forever, how long they have left
to live.

**Temperature**: In its simplest physical meaning, temperature is a way to state
precisely what we mean by "equilibrium". Think for instance of throwing a
hot steel ball into a water pool. Right after you do it, many things happen: the
water surrounding the ball agitates or even boils, and we know that the ball
"cools". But a little later everything becomes still, nothing seems to happen
anymore, and we say that the ball has "equilibrated" with water. To describe
this final situation, we say that the ball and the surrounding water "have the
same temperature": a rather abstract concept, don't you think? The matter
becomes much clearer if we look at its microscopic meaning. In the microworld, temperature is a measure of the kinetic energy (the energy associated
with motion) of atoms or molecules (see also THERMAL ENERGY). In physics,
it is convenient to measure temperature in absolute – or kelvin – degrees. If
you are used to degrees centigrade (or Celsius), just add approximately 273 to
the temperature in Celsius, and you get it in kelvins. If you are more familiar
with the anthropocentric (or, better, horse-centric) scale of Fahrenheit. . . well,
just get used to the Celsius scale, and then do the same.

**Thermal energy**: The average kinetic energy that an atom or a molecule has
at a given TEMPERATURE $T$, and also the kinetic energy that a COLLOIDAL
PARTICLE acquires owing to collision with the solvent molecules, which give
rise to its BROWNIAN MOTION. It is approximately equal to $kT$, where $T$ is
measured in kelvins (absolute degrees), and $k$ is called the Boltzmann constant. At room temperature the thermal energy is about $4 \times 10^{-21}$ Joules,
which is much smaller than the energy that binds the atoms in a molecule,
or even of the energy of the HYDROGEN BONDS that keep water molecules
together. ♡ It is worth expending some more words on energy in general, this
fundamental quantity both in science and in everyday life. We have seen that
mechanical energy is the sum of kinetic energy, the energy associated with motion, and potential energy, "stored" energy that forces can convert into kinetic
energy by performing work. However, physicists soon discovered that mechanical energy is not always conserved, but sometimes gets mysteriously lost, for
instance when a bouncing ball rebounds lower and lower, until it comes to
a halt. Since physicists are basically conservative people (believe me, we are:
maintaining with loving care the few concepts we trust is the only way to
cope with this complex world), they devised *ad hoc* a new form of energy, the
so-called "internal energy" of a body, to account for the missing part. Luckily,
at the end of the XIX century, this invisible stuff turned out to be nothing but
the kinetic (thermal) and potential energy of atoms and molecules, which also
gave a meaning to heat and temperature. So energy got the high status of an

eternal, immutable quantity. The new physics of the past century introduced some important changes in the concept of energy. Relativity showed that, to account properly for the behavior of matter moving almost as fast as light, *mass* has to be considered as a form of energy that a body possesses even if it does not move. More precisely, this energy is the product of the usual mass times the square of the light speed. This means that, if we were able to fully convert 0.7 grams of matter into energy, we could make a bomb as powerful as Little Boy, the atomic bomb dropped on Hiroshima (and unfortunately, we were). Atomic physics yielded an even more surprising result, telling us that energy is not rigorously conserved. In simple words, quantum mechanics tells us that you can do anything, provided you do it fast enough. So you can create energy from nothing. Unfortunately for practical uses, it lasts for a ludicrously short time: over human time scales, the total energy remains an unchanging quantity. But what's so special about energy that gives it this status? To do science and not magic, physicists are bound to believe that the laws of physics do not change in time (I mean, we believe that apples fall from the trees today obeying the same rules as the apple that Newton was looking at more than three centuries ago, and that this will be true as long as there are apples and trees). Here comes the great revelation: if this is true, then the physical quantity that we call total energy *has* to be constant. Believe me, although it sounds more like metaphysics than real physics, this is the real secret of energy.

**Thermoplastic**: A polymer is said to be thermoplastic when it softens on heating, progressively turning from a GLASS into a liquid. Most polymers of industrial interest, such as polystyrene (PE), polypropylene (PP), polyvinylchloride (PVC), and polyethylene terefthalate (PET) are thermoplastics, since thermoplastic polymers can easily be melt-processed.

**Thermosetting**: A thermosetting plastic is a polymer material that cures on heating, hardening irreversibly into a plastic or a rubber. This means that thermosetting material cannot be melted and re-shaped after it is cured, and is also harder to recycle. The curing effect of heat usually consists in generating CROSS-LINKS between the polymer chains. Examples of thermosetting materials are vulcanized rubber and melamine resins.

**Toughness**: In material science, toughness indicates how much *energy* a material can absorb before breaking. Toughness is *not* equivalent to strength (or HARDNESS). Brittle materials can be strong (they can stand large STRESS without breaking) but, since they can elongate or deform very little before breaking (they stand very limited STRAIN), they are not tough.

**Transcription**: The process of creating an "RNA copy" of DNA, ensuring that the genetic information contained in the latter is faithfully preserved. This task is primarily performed by RNA polymerase, an ENZYME that reads the NUCLEOTIDES one by one along one of the two DNA filaments and produces a sequence of complementary NUCLEOBASES (since it makes RNA, every time it finds an adenine, it attaches uracil (U) instead of thymine (T), which would occur in a DNA complement). ♡ We have seen that DNA *replication* is an extremely accurate process, where errors are extremely rare and usually rapidly corrected. Errors in transcription are much more tolerated, and actually happen all the time in cells. An error in transcription results in a defective messenger RNA, and therefore causes RIBOSOMES to make an imperfect PROTEIN. Most of the times, however, these proteins simply do not work, and cells are able to detect and dispose of them rapidly. Moreover, numerous copies of the same DNA sequence are simultaneously present in a cell, so there are enough correct versions of the DNA message around, making the wrong m-RNA copy fairly "expendable". Here is the difference with replication: a wrong DNA copy would *always* produce defective m-RNA, so that *all* of the proteins made out of it would be non-functional. As a result, errors in replication must be carefully avoided.

**Triglycerides**: The main constituents of animal fats and vegetable oils, forming a long-term energy stock for the body. They form by combining glycerol with three molecules of FATTY ACIDS.

# V alence

: Roughly, the number of bonds the atoms of a given element form. For example, hydrogen, oxygen, and carbon, are mono-, di-, and tetravalent, for they respectively bind to 1, 2, and 4 other atoms. ♡ For many elements, the valence may vary between different compounds: iron, for instance, can be trivalent in oxides such as $Fe_2O_3$, the main component of pigments like ochre or sienna, or divalent in compounds such as copperas ($FeSO_4$), used since the Middle Ages for making iron gall ink.

**Vesicle**: A globular structure made of a double layer of AMPHIPHILES that folds on itself, forming a closed spherical "bag" that separates an internal from an external environment (both water solutions, usually). The amphiphilic molecules that form vesicles are generally double-tailed, for they must have a big HYDROPHOBIC body and a relatively small HYDROPHILIC head. Vesicles are much larger than MICELLES, with a size that ranges between tens of NANOMETERS and tens of MICRONS. Large vesicles are very flexible, so their shape changes spontaneously and unceasingly. Vesicles – or LIPOSOMES, as they are usually called in the jargon of biologists – of all sizes are ubiquitous in living systems, where they are used as means of transport by the cell, which is itself bounded by a giant and very complex vesicle, the CELL MEMBRANE.

**Viscoelasticity**: The peculiar feature of many complex fluids, such as colloidal suspensions or polymer and surfactant solutions, of showing at the same time the flowing property of a liquid (namely, they yield when subjected to a STRESS) and the elasticity of a solid (they *deform* under stress, but do not flow). Viscoelastic materials are then simultaneously "viscous" and "elastic", in the sense that, like a fluid, they can be attributed a value for the VISCOSITY, but, like a solid, they also have an ELASTIC MODULUS. There are several kinds of viscoelastic materials. Paint, toothpaste, or nail varnish, are for instance *shear thinning* (or *pseudoplastic*), because their viscosity decreases the faster they flow. In other words, they are reluctant to flow when they are strained weakly, but conversely yield when forced to do it. Other materials, such as Silly Putty, solutions of corn starch, or even GRANULAR MATERIALS like sand, do exactly the opposite, getting harder the faster they flow, and are called *shear thickening* (or *dilatant*). These are examples of materials whose viscosity depends on the shear force we apply. But there are also fluids whose viscosity depends on *how long* we shear them. Many biological fluids, and in particular the synovial liquid lubricating bone joints are *thixotropic*: the more we shear them, the less viscous they get. Some others, like gypsum, which gets harder when kneaded, do the opposite, and are called *rheopectic*. These are just some examples of the complex behavior of what are called "non-Newtonian materials", which often display curious and amazing effects.

**Viscosity**: A basic quantity that quantifies how hard it is to set a fluid (either a liquid or a gas) in motion, and also the resistance that fluid puts up against a body moving through it (for instance the friction of air on a moving car). NEWTON'S LAW, which holds for simple fluids, states that the viscosity is just the ratio between the STRESS applied to a fluid and its STRAIN rate (the rate at which the fluid "yields").

**Volume fraction**: The fraction of the total volume of a colloid occupied by the colloidal particles, generally represented by the Greek letter $\Phi$ (Phi), used to indicate how much a colloid is concentrated. Another useful unit of concentration is the number of particles per "unit" volume (for instance, per cubic centimeter), which is equal to $\Phi$ divided by the volume of a single particle.

**Vulcanization**: The chemical process, accidentally discovered by Charles Goodyear, by which natural rubber is turned into a stable and performing elastic material. Basically, it consists in heating the rubber in the presence of sulfur. This induces the formation of "sulfur bridges" that CROSS-LINK different chains of polyisoprene, the POLYMER which is the basic constituent of the caoutchouc LATEX.

**W**avelength: The basic quantity that tells us how "long" a wave, of any kind, is (namely, what is the distance between two subsequent crests). In the case of *light* waves, the wavelength is strictly related to the color we sense. Visible light has a wavelength that extends from about 0.4 (blue-violet) to about 0.7 (deep red) MICRONS.

**Z**eolites: Micro-porous COLLOIDAL PARTICLES, with a size of a few NANOMETERS, obtained by precipitating solutions of aluminum and silicon salts. A unique feature of these particles is that they have extremely uniform micropores that can trap specific ions or molecules. Because of this, zeolites are extensively used as "molecular sieves". Large amounts of zeolites (up to 50% by weight) are added to washing powders for subtracting calcium and magnesium IONS and avoiding SURFACTANT precipitation.

# Index of common things (or almost so)

**A**

acrylic fibers   67
adhesive tape   64
adsorbers   119
aging   198, 216
albumen   206
Alzheimer's disease   177
anisette   134
anti-rejection drugs   135
antioxidants   130
art restoration   135
asphalt   138
atmosphere   49

**B**

baby colic   65
bacteria   169
bakelite   63
beauty creams   132
bitumen   136
blue moon   53
bone joints   92
Brazil nuts   163
broccoli   36
bubble bath   118
bullet-proof jackets   65

**C**

caoutchouc   59
cellulose   64
cement   38, 95

cheese   205
chemotherapy   128
chicory   123
Chinese cinnamon   116
cholesterol   117, 157
clay   40
coal *briquettes*   160
cobwebs   199
coffee   43, 162
concrete   38
corn starch   91
crude oil   136

**D**

dandelion   59
degreasers   118
dewdrops   103
diapers   79, 96
diesel oil   135
drilling muds   137
drunkard   20
dunes   164

**E**

e-reader   47
earthquakes   92, 163
egg   26
egg yolk   43, 206
eggplant   28
enamel   45
epoxy glue   67

R. Piazza, *Soft Matter*, DOI 10.1007/978-94-007-0585-2,
© Springer Science+Business Media B.V. 2011

**F**

fabric softeners   121
fertilizers   119
ficus   59
floor detergents   118
fog   40
food additives   122
food intolerance   186
forest fires   163

**G**

gelatin   198
glass   152
gold   44
grain silo   162
grease   120
gum arabic   45
gutta-percha   59
gypsum   92

**H**

hair conditioners   120
hourglass   163

**I**

ice creams   133, 206
immune system   190
insecticides   132

**J**

jam (traffic)   163
jojoba   120

**K**

ketchup   90
kitchenware   64

**L**

lacquer   120
LCD screens   46, 156
lecithin   117
lotions   132
lubricants   129

lungs   52
lye   117

**M**

M&M's™   158
Macassar   120
Macintosh (raincoat)   60
mad cow   178
marathon   186
mayonnaise   131, 206
melamine   64
microwave cookware   64
milk   41, 133, 204
mousse   206
muesli   162
muscles   187
mushrooms   201
mustard   133
Mylar   63

**N**

nail varnish   89
nerves   207
non-stick pans   66
nuclear waste   16
nylon   65

**O**

O-ring gaskets   66
oak galls   45
oil spills   138
olive oil   106, 130
opal   149
oranges   140
ouzo   134
ozone layer   49, 66

**P**

paint   89
paté   206
paternity testing   212
PET bottles   63
photographic films   64, 133
pigment   43
Plexiglass™   65
Polaroid™   73

pollen    17
polycarbonate    65
polythene    57
pond skaters    100
PVC    64
pyramids    38

**R**

rainbow    54
raindrops    43
red blood cells    90, 187
rennet    205
rice    162
ricotta    154, 206
rubber    59

**S**

Saint Januarius' blood    92
sand    162
sanitary napkins    79
Saturn's rings    164
seaweed    119
shaving foam    114, 118
silicone    65
silk    199
Silly Putty$^{TM}$    91
snowflakes    36
snowslides    163
soap    116
soap bubbles    108
starch    75
steroids    189
stock market    163
styrofoam    64
sunset    52

suntan creams    132

**T**

table salt    33
tap water    33
tea    43
tea-tree    120
tears    174
Teflon$^{TM}$    66
The Little Prince    166
thickeners    119
toothpaste    89, 118
tube-free tyres    85

**V**

varnish    45
vinaigrette    133
virus    224
viscose    65
vitamin C    177
volcano    49

**W**

washing powder    118, 123
wax    68, 114
white blood cells    190
wood spirit    186

**X**

xanthan gum    90

**Y**

yoghurt    37